Design Optimization Under Uncertainty

Weifei Hu

Design Optimization Under Uncertainty

Springer

Weifei Hu
State Key Laboratory of Fluid
Power and Mechatronic Systems,
School of Mechanical Engineering
Zhejiang University
Hangzhou, China

ISBN 978-3-031-49207-5 ISBN 978-3-031-49208-2 (eBook)
https://doi.org/10.1007/978-3-031-49208-2

This Springer imprint is published by the registered company Springer Nature Switzerland AG
The registered company address is: Gewerbestrasse 11, 6330 Cham, Switzerland

Preface

Design optimization under uncertainty (DOUU) studies the theories, methods, and technologies to design reliable products considering various uncertainties ubiquitous in loads, geometry, material properties, manufacturing processes, and operational environments. Traditional deterministic design optimization processes treat these uncertainties by some simplified rules, such as using safety factors and considering limited operation conditions, which do not directly account for the random nature of the design variables, the constrained performances, and the objective function. This may lead to a non-reliable design, or too conservative and expensive design. Hence, it is an increasingly important task to involve uncertainties in the optimization processes for designing complex engineering equipment, such as large-scale wind turbines, automobiles, ships, and airplanes, to name a few.

This book introduces the fundamental concepts of probability and reliability, the classical methods of uncertainty modeling, time-dependent and time-independent reliability analysis methods, model verification and validation, two main categories of DOUU methods (e.g., reliability-based design optimization and robust design optimization), the state-of-the-art approaches of physics informed methods for DOUU, and a comprehensive survey of engineering applications of DOUU. Each chapter begins with the fundamental theories and methods in a lucid, easy-to-follow treatment and then elaborates on the corresponding advanced approaches using detailed methodologies, mathematical models, numerical examples, tables, and graphs. References and exercises are presented at the end of chapters. The book is ideal for both educational and research needs for readers from undergraduate students, graduate students, and faculty to engineering designers.

Writing a book on DOUU has always been a dream during my graduate study at South Korea and USA. My advisers planted and watered the seed of writing this book. I would like to appreciate Professor KK Choi, my PhD adviser at University of Iowa, Iowa City, USA. Professor Choi not only taught me the knowledge of reliability-based design optimization, but also shaped my philosophy about academic research. Sincerely thank Professor Olesya Zhupanska and Professor James Buchholz who co-advised me at University of Iowa, and Professor Dong-Hoon Choi who initially led me into the field of design and optimization during my master study at Hanyang Uninversity, Seoul, South Korea. Grateful acknowledge goes to Professor Sara C. Pryor and Professor Rebecca J. Barthelmie for their numerous encouragements, guide, and support during my postdoctoral research at Cornell University. I would also like to thank my graduate students Weiyi Chen, Tongzhou Zhang, Feng Zhao, Jiquan Yan, Xiaoyu Deng, Jianhao Fang, Qing Jiao, Jiale Liao, and Sichuang Cheng at Zhejiang University, Hangzhou, China. Without their input, this book would not have been possible. Special thanks go to Michael Luby and Brian Halm for their support in publishing my second Springer book and Olivia Ramya Chitranjan for providing answers to my many questions during the manuscript preparation. Finally, I would like to thank all my family and friends for their love and support.

Hangzhou, China
October 23, 2023

Weifei Hu

Contents

About the Author

Weifei Hu received the BS degree in 2008 from Zhejiang University, Hangzhou, China, the MS degree in 2010 from Hanyang University, Seoul, South Korea, and the PhD degree in 2015 from University of Iowa, Iowa city, Iowa, USA, all in mechanical engineering. From February 2016 to September 2018, Dr. Hu was a postdoctoral fellow at Cornell University, Ithaca, New York, USA. He is currently holding a ZJU100 Young Professor position at the School of Mechanical Engineering, Zhejiang University. His research interests include design optimization under uncertainty, digital twin, artificial intelligence, and wind energy. His work has been published in over 100 peer-reviewed journals and conference proceedings, e.g., *Journal of Mechanical Design*, *Mechanical Systems and Signal Processing*, *Structural and Multidisciplinary Optimization*, *Journal of Intelligent Manufacturing*, *Journal of Manufacturing Systems*, *Renewable and Sustainable Energy Reviews*, *Applied Energy*, *Renewable Energy*, *Wind Energy,* and many ASME and AIAA proceedings. He is now serving as an associate editor for the journal *Wind Energy Science* and an editorial board member for the journals *Wind Energy* and *Journal of Intelligent Manufacturing and Special Equipment.*

About the Author

Chapter 1
Basic Concepts of Probability and Reliability

Nomenclature

$A, B, \ldots$	Random event
$\mathrm{Cov}(X, Y)$	Covariance of random variables X and Y
$D(x)$	Variation of random variable X
$E(X)$	Expectation of random variable X
$\boldsymbol{F_X(x)}$	Cumulative distribution function
$\boldsymbol{f_X(x)}$	Probability density function
$\mathbf{g}(\cdot)$	Performance function
J	Jacobian of transformation
$\boldsymbol{N(\mu, \sigma^2)}$	Normal distribution
$P(\cdot)$	Probability of an event to occur
$\boldsymbol{P_f}$	Probability of failure
$\boldsymbol{P_X(x)}$	Probability mass function
X	Random variable
$\boldsymbol{U(a, b)}$	Uniform distribution
$\boldsymbol{\Gamma}(\cdot)$	Complete gamma function
$\boldsymbol{\gamma}(\cdot, \cdot)$	Lower incomplete gamma function
μ	Mean of random variable X
ρ_{XY}	Correlation coefficient of random variables X and Y
$\sigma(X)$	Standard deviation of random variable X
$\Phi(x)$	Cumulative distribution function of standard normal distribution
$\varphi(x)$	Probability density function of standard normal distribution
Ω	Sample space
ω	Sample in sample space
$\emptyset$	Impossible event

W. Hu, *Design Optimization Under Uncertainty*,
https://doi.org/10.1007/978-3-031-49208-2_1

1.1 Probability Theory

1.1.1 Definition of Probability

Probability models are constructed for **experiments** that produce random outcomes when repeated under the same conditions. The collection of all possible outcomes of a random experiment is called the **sample space**, denoted as Ω. Each outcome of a random experiment is called a **sample**, denoted as ω [1].

The definitions of the sample space and the sample are tightly related to the utilization of the set, which is a collection of objects meeting certain well-defined conditions [2]. Set theory is often used for analyzing the sample space and the sample [1]. Here, any subset of the sample space, Ω, is defined as an **event**. A **certain event** is an event that contains all the samples in the sample space. In other words, since the sample space Ω contains all the samples, as defined previously, it can be used to denote a certain event. An **impossible event** is an event that contains no samples. The null set, denoted as $\emptyset$, is often used to represent an impossible event [3]. Some commonly used relationships between random events are listed as follows [1, 4]:

1. Subset of random events: If all the samples in event A also belong to event B, then random event A is said to be a **subset** of random event B, which is denoted as $A \subset B$.
2. Union of random events: An event whose samples all belong to *at least one of* the random events A and B is called the **union** of random events A and B, and is denoted as $A \cup B$. Furthermore, an event whose samples all belong to at least one of a series of random events, $A_1, A_2, \ldots, A_n$, is denoted as $A_1 \cup A_2 \cup \ldots \cup A_n$ or $\bigcup_{i=1}^{n} A_i$.s.
3. Intersection of random events: An event whose samples all belong to *both* random events A and B is called the **intersection** of random events A and B, and is indicated as $A \cap B$ or AB. Furthermore, an event whose samples all belong to all of a series of random events, $A_1, A_2, \ldots, A_n$, is denoted as $A_1 \cap A_2 \cap \ldots \cap A_n$ or $\bigcap_{i=1}^{n} A_i$.
4. Mutually exclusive events: When the members of random events A and B are pairwise disjoint, then these random events are said to be **mutually exclusive**, and denoted as $AB = \varnothing$. Moreover, they are said to be **mutually exclusive and exhaustive** if $AB = \varnothing$ and $A \cup B = \Omega$. This definition can also be expanded to situations with multiple random events, i.e., when the members of a series of random events, $A_1, A_2, \ldots, A_n$, are pairwise disjoint, then these random events are said to be **mutually exclusive**, and noted as $A_1 \cap A_2 \cap \ldots \cap A_n = \emptyset$. These events are further said to be **mutually exclusive and exhaustive** if $A_1 \cap A_2 \cap \ldots \cap A_n = \emptyset$ and $A_1 \cup A_2 \cup \ldots \cup A_n = \Omega$.

5. Complement event: A random event that consists of all samples that do not belong to random event A is called the **complement event** of A, and noted as $\overline{A}$, i.e., $\overline{A} = \Omega - A$.

Suppose the random experiment is repeated n times under the same conditions. If n_A denotes the number of times random event A occurs during these n experiments, the **frequency** $f_A = n_A/n$ reflects how often event A happens. If, with the increase of n, f_A stabilizes or converges at a constant p, then p is said to be the **probability** of the random event A, and denoted as $P(A) = p$. It is clear that the following rules of probability exist: $0 \leq P(A) \leq 1$, $P(\Omega) = 1$, and $P(\emptyset) = 0$.

Example 1.1 As a random event experiment, consider a single roll of a dice and observe the number of points that appear.

1. Its sample space and sample points.
2. Examples of certain and impossible events in this experiment in randomness.

Solution

1. Sample space $\Omega = \{1, 2, 3, 4, 5, 6\}$, sample points ω are 1, 2, 3, 4, 5, 6.
2. The event that a die roll falls in the range of 1 ~ 6 is a certain event, and a die roll of 7 is an impossible event.

1.1.2 Basic Probability Theorem

Each random event A in the sample space Ω is assigned a real number $P(A)$ that satisfies the following basic conditions:

1. Nonnegativity: For any random event $A \subset \Omega, P(A) \geq 0$.
2. Normality: $P(\Omega) = 1$.
3. Additivity: If the random events A_1, A_2, A_3,... are mutually exclusive, i.e., $A_i \cap A_j = \emptyset (i, j = 1, 2, \ldots; i \neq j)$, then $P\left(\bigcup_{i=1}^{\infty} A_i\right) = \sum_{i=1}^{\infty} P(A_i)$.

Based on the definition of probability, the following important properties can be obtained:

1. The probability of the impossible event is 0, i.e., $P(\emptyset) = 0$.
2. Finite additivity: If a finite sequence of random events A_1, A_2, A_3,..., A_n are mutually exclusive, i.e., $A_i \cap A_j = \emptyset$ $(i, j = 1, 2, \ldots, n; i \neq j)$, then $P\left(\bigcup_{i=1}^{n} A_i\right) = \sum_{i=1}^{n} P(A_i)$.
3. Suppose there are random events A and B, with A being a subset of B; then,

$$P(B - A) = P(B) - P(A), P(A) \leq P(B) \tag{1.1}$$

4. For any event A, $P(A) \leq 1$.
5. Probability of the opposing event: For any random event A, $P(\overline{A}) = 1 - P(A)$.
6. Addition: For any two random events A and B, the following relationship exists:

$$P(A \cup B) = P(A) + P(B) - P(A \cap B) \tag{1.2}$$

1.1.3 Conditional Probability

Suppose there are two random events, A and B, in a random experiment, and $P(B) \neq 0$. $P(A|B)$ is denoted as the conditional probability of the occurrence of random event A under the condition that random event B occurs, which is calculated as

$$P(A|B) = \frac{P(AB)}{P(B)} \tag{1.3}$$

According to the formula of conditional probability, three important formulas can be derived.

1. Multiplicative formula:

$$\begin{aligned} P(AB) &= P(A|B)P(B), \text{if } P(B) \neq 0; \\ P(AB) &= P(B|A)P(A), \text{if } P(A) \neq 0; \end{aligned} \tag{1.4}$$

2. Full probability formula: Assume that there is a random event B in a random experiment and the sample space Ω can be divided into A_1, A_2, A_3,..., A_n, and $P(A_i) \neq 0$, $i = 1, 2,..., n$, then

$$P(B) = \sum_{i=1}^{n} P(B|A_i)P(A_i) \tag{1.5}$$

3. Bayesian formula: Suppose there is a random event B in a random experiment and the sample space Ω can be divided into mutually exclusive events $\mathbf{A} = \{A_1, A_2, A_3, \ldots, A_n\}$, and $P(B) > 0$, $P(A_i) > 0$ for $i = 1, 2,..., n$, then

$$P(A_i|B) = \frac{P(B|\mathbf{A})P(A_i)}{\sum_{i=1}^{n} P(B|A_i)P(A_i)}, i = 1, 2, \ldots, n \tag{1.6}$$

Example 1.2 A factory is producing some product parts. Through historical statistics, it is found that when the machine works well, the produced parts are 99% qualified, while when the machine does not work well, the produced parts are 60% qualified. The probability that the machine is well working when producing a part is 95%. What is the probability that the machine is well working given that a qualified part is produced?

Solution
Suppose the random event A is "the part is qualified" and the random event B is "the machine is well working." According to the given information in the example, we can obtain $P(A|B) = 0.99$, $P(A|\overline{B}) = 0.6$, $P(B) = 0.95$, and $P(\overline{B}) = 0.05$, and we need to solve $P(B|A)$ by the Bayesian formula:

$$P(B|A) = \frac{P(A|B)P(B)}{P(A|B)P(B) + P(A|\overline{B})P(\overline{B})} = \frac{0.99 \times 0.95}{0.99 \times 0.95 + 0.6 \times 0.05} = 0.9691$$

Thus, given that the first part produced is a qualified product, the probability that the machine is well working is 0.9691.

1.1.4 Independence

For random events A and B, if $P(A) = P(A|B)$, it means that whether random event B occurs or not has no effect on the probability of occurrence of random event A. Here, the two events are said to be **independent** of each other. This leads to a definition of the independence of random events.

If two random events, A and B, satisfy $P(AB) = P(A)P(B)$, the random events A and B are said to be **mutually independent**. Independence means that the occurrence of one random event does not affect the probability of the occurrence of another random event; while mutually exclusive means that the two random events cannot occur simultaneously.

The definition of the independence of multiple random events is obtained by extending the concept of the mutual independence of two random events to multiple random events. Suppose $A_1, A_2, A_3, \ldots, A_n$ are n random events. If $P(A_1A_2\ldots A_n) = P(A_1)P(A_2)\ldots P(A_n)$ is satisfied, then $A_1, A_2, A_3, \ldots, A_n$ are mutually independent of each other.

Example 1.3 Consider an example of a batch of product parts with a qualifying rate of 0.9. What is the possibility of k product parts being qualified among 10 randomly selected parts?

Solution
Since the 10 product parts are randomly selected, whether a single part is qualified or not is totally independent from any of the other 9 parts. Denote $A_i (i = 1, 2, \ldots, 10)$ as the event of "the i-th part being qualified," then $P(A_i) = 0.9$, $P(\overline{A_i}) = 0.1$.

If there are k product parts being qualified, it indicates that $10 - k$ parts are unqualified. These events are independent of each other. Additionally, the choice of k product parts involves a number of combinations; thus, the possibility of k mechanical parts being qualified among 10 randomly selected parts is

$$\begin{aligned} P &= C_{10}^{k} \prod_{i=1}^{k} P(A_i) \prod_{j=1}^{10-k} P(\overline{A}_i) \\ &= C_{10}^{k} \times 0.9^{k} \times 0.1^{10-k} \end{aligned}$$

1.2 Random Variable

Consider a random experiment with sample space Ω. A **random variable** is defined as a *function* X that assigns to each sample point ω in the sample space, one and only one, real number $x = X(\omega)$. In this book, a capital letter, e.g., X, is used to represent random variables, while its corresponding letter in lowercase, e.g., x, stands for a **realization** of the random variable. In this section, two types of random variables are introduced, namely, discrete random variables and continuous random variables.

1.2.1 Discrete Random Variable

A random variable is a **discrete random variable** if its space is either finite or countable. In general, a discrete random variable can only represent values on discrete points, so the probability of a discrete random variable can only be calculated on these discrete points.

The randomness of a discrete random variable is described by the **probability mass function** (PMF), denoted as $P_X(x)$. Suppose X is a random variable with the range of a series of discrete values $x_k(k = 1, 2, \ldots)$; then, the probability mass function of X can be defined as

$$p_X(x_k) = P(X = x_k), k = 1, 2, \ldots \tag{1.7}$$

According to the definition of probability, the probability mass function needs to satisfy the following two conditions.

1. $p_X(x_k) \geq 0$, $k = 1, 2, \ldots$
2. $\sum_{k=1}^{\infty} p_X(x_k) = 1$.

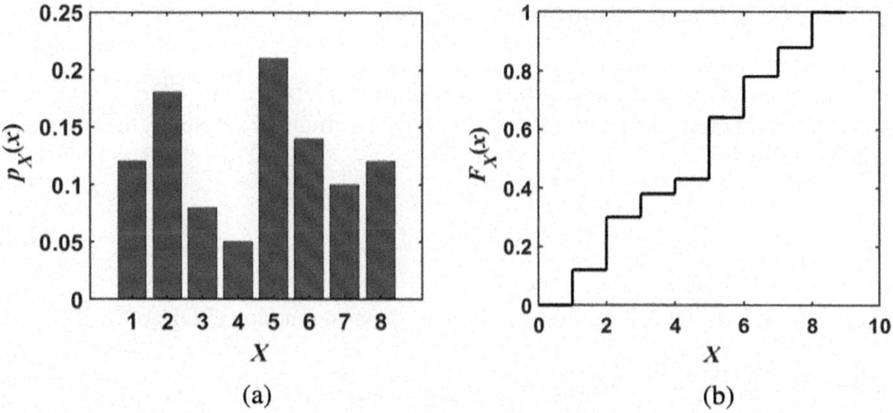

Fig. 1.1 (**a**) Probability mass function and (**b**) Cumulative distribution function of a discrete random variable

The **cumulative distribution function** (CDF) of a discrete random variable can be obtained by summing the probability mass functions, and the expression of the CDF is

$$F_X(x) = P(X \leq x) = \sum_{x_k \leq x} p_X(x_k) \tag{1.8}$$

The probability mass function of a discrete random variable is a discrete function consisting of a series of discrete values, while the cumulative distribution function of a discrete random variable is a step-shape function, as shown in Fig. 1.1.

Example 1.4 Suppose X is the number of times that a coin is tossed heads up on two consecutive tosses, what is the probability mass function of X?

Solution

Assume that A represents heads up and B represents tails up, the sample space of two consecutive coin tosses is $\Omega = \{AA, AB, BA, BB\}$, and the probability of each sample ω in the sample space is $(0.5)^2 = 0.25$. The probabilities of the random variables X being 0, 1, and 2 are calculated as

$$p_X(0) = P(X = 0) = P(\omega = BB) = 0.25$$

$$p_X(1) = P(X = 1) = P(\omega \in \{AB, BA\}) = 2 \times 0.25 = 0.5$$

$$p_X(2) = P(X = 2) = P(\omega = AA) = 0.25$$

1.2.2 Continuous Random Variable

A random variable is a **continuous random variable** if its CDF, $F_X(x)$, is a continuous function for all real $x \in R$. For most continuous variables, the following equation holds:

$$F_X(x) = \int_{-\infty}^{x} f_X(t)dt \tag{1.9}$$

The function $f_X(t)$ is called the probability density function (PDF) of X. If $f_X(x)$ is also continuous, then

$$f_X(x) = \frac{dF_X(x)}{dx} \tag{1.10}$$

The PDF $f_X(x)$ has the following properties:

1. Nonnegativity: $f_X(x) \geq 0$.
2. Normative: $\int_{-\infty}^{+\infty} f_X(x)dx = 1$.
3. For any real numbers x_1, x_2 ($x_1 < x_2$), there exists $F_X(x_2) \geq F_X(x_1)$, and

$$P(x_1 < X \leq x_2) = F_X(x_2) - F_X(x_1) = \int_{x_1}^{x_2} f_X(x)dx \tag{1.11}$$

4. Continuity: If $f_X(x)$ is continuous at $x = x_0$, then we have $F'_X(x_0) = f_X(x_0)$.

The cumulative distribution function $F_X(x)$ of a continuous random variable has the following properties:

1. For any real number a, the $P(X = a) = 0$.
2. Normality: $F_X(-\infty) = 0$,$F_X(+\infty) = 1$.
3. Right continuity: $F_X(x)$ is a continuous function such that $\lim_{\Delta x \to 0^+} F_X(x + \Delta x) = F_X(x)$.

An illustration of the PDF and CDF of a random variable X is shown in Fig. 1.2.

Example 1.5 Suppose the random variable X has a PDF

$$f(x) = \begin{cases} kx, & 0 \leq x < 3 \\ 2 - \frac{x}{2}, & 3 \leq x < 4 \\ 0, & \text{otherwise} \end{cases}$$

Try to find: (1) Constant k, (2) CDF of X, and (3) $P\{1 < X \leq 5\}$.

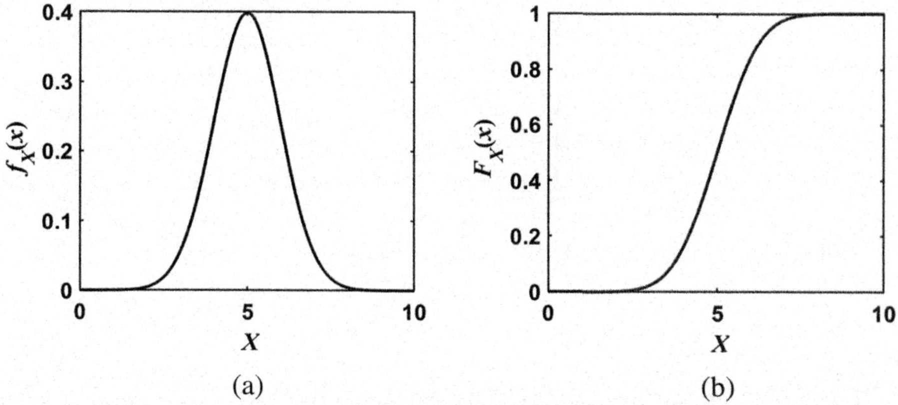

Fig. 1.2 (**a**) Probability density function and (**b**) Cumulative distribution function of a continuous random variable

Solution

1. According to the probability density function property $\int_{-\infty}^{+\infty} f_X(x)dx = 1$, it is known that

$$\int_0^3 kxdx + \int_3^4 \left(2\text{-}\frac{x}{2}\right)dx = 1$$

Solving the above equation gives $k = 1/6$; then, the probability density function of X is

$$f(x) = \begin{cases} \frac{x}{6}, & 0 \le x < 3 \\ 2 - \frac{x}{2}, & 3 \le x < 4 \\ 0, & \text{otherwise} \end{cases}$$

2. The CDF of random variable X is obtained by an integration calculation

$$F(x) = \begin{cases} 0, & x < 0 \\ \int_0^x \frac{x}{6} dx, & 0 \le x < 3 \\ \int_0^3 \frac{x}{6} dx + \int_3^x \left(2 - \frac{x}{2}\right) dx, & 3 \le x < 4 \\ 1, & x \ge 4 \end{cases}$$

$$F(x) = \begin{cases} 0, & x<0 \\ \frac{x^2}{12}, & 0 \le x < 3 \\ -\frac{x^2}{4} - 2x - 3, & 3 \le x < 4 \\ 1, & x \ge 4 \end{cases}$$

3. $P\{1 < X \le 5\} = F(5) - F(1) = 1 - \frac{1}{12} = \frac{11}{12}.$

1.2.3 Transformation of Random Variables

Suppose there is a random variable X whose distribution is already known, then there is a random variable $Y = g(X)$. Usually, the distribution of Y is of interest. The calculation of the distribution of Y is separately explained when the random variable X is discrete or continuous. When X is a discrete random variable, and g is a one-to-one mapping function, then the PMF of Y is obtained as:

$$p_Y(y) = P(Y=y) = P(g(X)=y) = P\left(X = g^{-1}(y)\right) = p_X\left(g^{-1}(y)\right) \quad (1.12)$$

Example 1.6 Consider a die that is rolled consecutively until a six-point appears for the first time. Let X be the number of rolls when the six-point appeared and its sample space is $\Omega_X = \{1, 2, 3, \ldots\}$. Define Y as the number of rolls *before* the six-point first appeared. Obviously, $Y = X - 1$, and its sample space is $\Omega_Y = \{0, 1, 2, \ldots\}$. Find the PMF of Y.

Solution

For each roll, define S as the result of six points, and N as the results of the other points. It is obvious that $P(S) = 1/6$ and $P(N) = 5/6$. In this case, the die is rolled with a result of 1 to 5 point(s) in the first $X - 1$ times, and a result of 6 points in the last time. Therefore, the PMF of X can be given as:

$$P_X(x) = P(N)^{X-1} P(S) = \left(\frac{5}{6}\right)^{X-1} \left(\frac{1}{6}\right)$$

From the definition of X and Y, it is known that $g(X) = X - 1$ and $g^{-1}(Y) = Y + 1$. Therefore, the PMF of Y, according to Eq. (1.12), is

$$p_Y(y) = p_X\left(g^{-1}(y)\right) = \left(\frac{5}{6}\right)^Y \left(\frac{1}{6}\right)$$

When g is not a one-by-one function, there is no general mathematical method to obtain the PMF of Y. Instead, the PMF of Y is calculated according to the scenario itself. This is illustrated by the following example.

Example 1.7 In a Ludo game, to draw a token from the yard to the starting square, the player must roll a six in a die. When the game starts, all the players want to draw their first token to the starting square as soon as possible. If a player draws his first token within 3 rolls of a die, i.e., the number of 6 points appears within 3 rolls of a die, the player is in good luck, and denoted as $Y = 1$. Otherwise, the player is in bad luck and denoted as $Y = 0$. Calculate the PMF of Y.

Solution
Recall that in Example 1.6, X is denoted as the roll number of a die when a six-point appears for the first time. Thus, the random variable Y in this example can be related to X through the following relationship:

$$Y = g(X) = \begin{cases} 1 & X = 1, 2, 3 \\ 0 & X = 4, 5, 6, \ldots \end{cases}$$

The probability of $Y = 1$ is relatively easy to calculate as follows:

$$\begin{aligned} P(Y=1) &= P(X=1) + P(X=2) + P(X=3) \\ &= \sum_{x=1}^{3} P_X(x) \\ &= \frac{1}{6} + \frac{5}{6} \times \frac{1}{6} + \left(\frac{5}{6}\right)^2 \times \frac{1}{6} \\ &= 0.4213 \end{aligned}$$

The probability of $Y = 0$ becomes $P(Y = 0) = 1 - P(Y = 1) = 0.5787$.
Finally, the PMF of Y can be expressed as:

$$p_Y(y) = \begin{cases} 0.5787 & y = 0 \\ 0.4213 & y = 1 \end{cases}$$

For the continuous random variable X, suppose the PDF of X is $f_X(x)$ and g is a one-by-one and differentiable function with the inverse denoted by $X = g^{-1}(y)$ and $dx/dy = d[g^{-1}(y)]/dy$; then, the PDF of Y is obtained as:

$$f_Y(y) = f_X\left(g^{-1}(y)\right)\left|\frac{dx}{dy}\right| \tag{1.13}$$

The proof is as follows: Because $g(x)$ is one-by-one and continuous (since it is differentiable), it is either strictly monotonically increasing or decreasing. For a monotonically increasing g(x), the CDF of Y is expressed as:

$$F_Y(y) = P(Y \le y) = P(g(X) \le y) = P\left(X \le g^{-1}(y)\right) = F_X\left(g^{-1}(y)\right) \tag{1.14}$$

The PDF of Y is obtained by taking the derivative of the CDF:

$$f_Y(y) = \frac{d}{dy} F_Y(y) = f_X\left(g^{-1}(y)\right)\frac{dx}{dy} \tag{1.15}$$

For a monotonically decreasing g(x), the CDF of Y is obtained similarly as:

$$\begin{aligned} F_Y(y) &= P(Y \le y) = P(g(X) \le y) = P\left(X \ge g^{-1}(y)\right) \\ &= 1 - P\left(X \le g^{-1}(y)\right) = 1 - F_X\left(g^{-1}(y)\right) \end{aligned} \tag{1.16}$$

The PDF of Y is obtained as:

$$f_Y(y) = \frac{d}{dy} F_Y(y) = -f_X\left(g^{-1}(y)\right)\frac{dx}{dy} = f_X\left(g^{-1}(y)\right)\left(-\frac{dx}{dy}\right) \tag{1.17}$$

Equations (1.15) and (1.17) show high similarity except for a minus sign on dx/dy. In fact, for a monotonically increasing $g(x)$, dx/dy shows a positive value, while it is negative for a monotonically decreasing $g(x)$ with $-dx/dy = |dx/dy|$. $dx/dy = \left(\frac{d}{dy}\right)g^{-1}(y)$ is also referred to as the **Jacobian** of the transformation, denoted as $J = dx/dy$. Thus, Eqs. (1.15) and (1.17) can be combined for the expression of the PDF of Y as:

$$f_Y(y) = \frac{d}{dy} F_Y(y) = f_X\left(g^{-1}(y)\right)\left|\frac{dx}{dy}\right| \tag{1.18}$$

Example 1.8 Suppose there is a random variable X with PDF $f_X(x)$ defined as follows:

$$f_X(x) = \begin{cases} 1 & 0 < x < 1 \\ 0 & else \end{cases}$$

The random variable $Y = X^2$. Obtain the PDF of Y.

Solution

Since the effective range of X is (0, 1), the corresponding range of Y is also (0, 1). Since X is positive, the inverse of the transformation is simply $X = g^{-1}(Y) = \sqrt{Y}$. The Jacobian of this transformation is

$$J = \frac{dx}{dy} = \frac{1}{2\sqrt{y}}$$

The PDF of Y can thus be expressed as:

$$\begin{aligned} f_Y(y) &= f_X\left(g^{-1}(y)\right)|J| \\ &= \begin{cases} \frac{1}{2\sqrt{y}} & 0 < y < 1 \\ 0 & else \end{cases} \end{aligned}$$

1.2.4 Expectation

Assume that X is a discrete random variable and its realizations are $x_1, x_2, x_3, \ldots$ with probabilities $p_X(x_1), p_X(x_2), p_X(x_3), \ldots$, respectively. The **expectation** of the discrete random variable X is obtained by the following equation:

$$E(X) = \sum_{k=1}^{\infty} x_k p_X(x_k) \tag{1.19}$$

For a continuous random variable X, the expectation of X is calculated by the integration as shown below:

$$E(X) = \int_{-\infty}^{+\infty} x f_X(x) dx \tag{1.20}$$

where $f_X(x)$ is the PDF of X. It is common to use the **mean** of a random variable X, denoted as μ, to refer to its expectation.

According to the definition of random variables, the functions of random variables are also random variables. In reliability analysis, the system performance function composed of random variables is usually of greater concern. Assume that Y is a function of the random variable X, i.e., $Y = g(X)$. According to the different types of random variables, their expectation calculation formulas are different.

1. Discrete random variables

$$E(Y) = E[g(X)] = \sum_{k=1}^{\infty} g(x_k) p_X(x_k) \tag{1.21}$$

2. Continuous random variables

$$E(Y)=E[g(X)]=\int_{-\infty}^{+\infty}g(x)f_X(x)dx \tag{1.22}$$

where $p_X(x_k)$ and $f_X(x)$ are the PMF and PDF of X, respectively.

In practical engineering problems, the specific probability distribution of a random variable is usually unknown. Instead, several realizations, $x_1, x_2, x_3, \ldots,$ are obtained by experiments or measurement. The sample mean can be used as an approximation of the true mean of the random variable, which is calculated as follows:

$$\widehat{\mu}_X=\frac{1}{M}\sum_{k=1}^{M}x_k \tag{1.23}$$

where M is the number of realizations.

Furthermore, there are some fundamental properties of expectation as follows:

1. For a constant c, $E(c) = c$.
2. For a random variable X and constant c, $E(cX) = cE(X)$.
3. For random variables X and Y, $E(X + Y) = E(X) + E(Y)$.
4. For mutually independent random variables X and Y, $E(XY) = E(X)E(Y)$.

   Property 3 and Property 4 can be extended to the case of the sums and products of a finite number of random variables. With the above properties, the computation of expectation can be effectively reduced.

Example 1.9 Suppose that the random variable X follows a Poisson distribution with parameter λ, and its probability mass function is $p_X(x)=\frac{\lambda^x e^{-\lambda}}{x!}, x=0,1,2,3\ldots(\lambda>0)$, find the expectation of the random variable X.

Solution

Calculated from Eq. (1.19) we can obtain

$$E(x)=\sum_{x=0}^{+\infty}xp_X(x)=\sum_{x=0}^{+\infty}x\frac{\lambda^x e^{-\lambda}}{x!}=\sum_{x=0}^{+\infty}x\frac{\lambda^x e^{-\lambda}}{x!}=\lambda e^{-\lambda}\sum_{x=1}^{+\infty}\frac{\lambda^{x-1}}{(x-1)!}=\lambda e^{-\lambda}e^{\lambda}=\lambda$$

Example 1.10 Suppose that the random variable X follows an exponential distribution with parameter λ, and its probability mass function is $f_X(x)=\begin{cases}\frac{1}{\theta}e^{-x/\theta}, & x>0,\\ 0, & \text{otherwise}\end{cases}$, find the expectation of the random variable X.

Solution

Calculated from Eq. (1.20) we can obtain

$$E(X) = \int_{-\infty}^{+\infty} xf_X(x)dx = \int_0^{+\infty} x\frac{1}{\theta}e^{-x/\theta}dx = -\int_0^{+\infty} xde^{-x/\theta}$$
$$= -xe^{-x/\theta}\Big|_0^{+\infty} + \int_0^{+\infty} e^{-x/\theta}dx = -\theta e^{-x/\theta}\Big|_0^{+\infty} = \theta$$

1.2.5 *Variance*

In Sect. 1.2.4, we introduced the expectations of random variables, which are used to reflect the average condition of a random variable. However, in real engineering applications, the expectation alone is not sufficient to describe the uncertainty. Obviously, the *dispersion* of the random variable values with respect to its mean is needed to help fully describe the uncertainty.

Variance is defined as the deviation of a random variable from its mean. Assume X is a random variable with finite mean value μ; if $E[(X - \mu)^2]$ exists and is finite, then it is said to be the variance of X, and is denoted as $D(X)$ or $\mathrm{Var}(X)$, i.e.,

$$D(X) = \mathrm{Var}(X) = E\left[(x - \mu)^2\right] \tag{1.24}$$

Additionally, $\sqrt{D(X)}$ is referred to as the **standard deviation**, denoted as $\sigma(X)$.

A small of $D(x)$ value indicates that the values of X are more concentrated around $E(X)$ and vice versa. Using the definition of variance, it is known that the variance is actually the expectation of the function $g(X) = (X - E(X))^2$ of the random variable X. For discrete and continuous random variables, the formula to calculate variance is different, and is shown as follows:

1. Discrete random variable

$$D(X) = \sum_{k=1}^{\infty} [x_k - E(X)]^2 p_X(x_k) \tag{1.25}$$

2. Continuous random variable

$$D(X) = \int_{-\infty}^{+\infty} [x_k - E(X)]^2 f_X(x)dx \tag{1.26}$$

The variance of the random variable X can also be calculated by the following formula:

$$D(X) = E(X^2) - [E(X)]^2 \tag{1.27}$$

The above formula converts the calculation of variance into calculating the expectation of the square of the random variable and calculating the square of the expectation of the random variable.

In practical engineering problems, realizations of a random variable x_1, x_2, x_3, ... are generally obtained by experimental or simulation methods. The sample variance is calculated as follows:

$$\widehat{D}(X) = \frac{1}{M-1}\sum_{k=1}^{M}(x_k - \widehat{\mu}_X)^2 \tag{1.28}$$

where $\widehat{\mu}_X$ is the sample mean and M is the number of realizations.

In addition, the following fundamental properties of variance exist.

1. For a constant c, $D(c) = 0$.
2. For a random variable X and constant c, $D(cX) = c^2D(X)$, $D(c + X) = D(X)$.
3. For random variables X and Y, $D(X + Y) = D(X) + D(Y) + 2E\{[X - E(X)][Y - E(Y)]\}$. If X and Y are independent of each other, $D(X + Y) = D(X) + D(Y)$.

   Property 3 can be extended to the case of a finite number of random variables. With the properties above, the computations of variances can be effectively reduced.

Example 1.11 The probability density function of a random variable X is known to be

$$f_X(x) = \begin{cases} ax^2 + bx + c, & 0 < x \leq 1, \\ 0, & \text{otherwise} \end{cases}$$

Moreover, it is known that $E(x) = 0.5$ and $D(x) = 0.15$. Find the values of a, b, and c.

Solution

According to the probability density function property $\int_{-\infty}^{+\infty} f_X(x)dx = 1$, it is known that $\int_0^1 (ax^2 + bx + c)dx = \frac{a}{3} + \frac{b}{2} + c = 1$.

From the definition of $E(x)$, it follows that:

$$E(X) = \int_{-\infty}^{+\infty} x f_X(x)dx = \int_0^1 x(ax^2 + bx + c)dx = \frac{a}{4} + \frac{b}{3} + \frac{c}{2} = 1$$

The variance can be seen using Eq. (1.27)

$$D(X) = E(X^2) - [E(X)]^2 = \int_0^1 x^2(ax^2 + bx + c)dx - 0.5^2 = \frac{a}{5} + \frac{b}{4} + \frac{c}{3} - 0.25 = 0.15$$

By solving the above three equations, we obtain $a = 12$, $b = -12$, and $c = 3$.

1.2.6 Covariance

For random variables (X, Y), in addition to discussing their respective mathematical expectations and variances, the interrelationship between X and Y needs to be assessed. Property 3 of variance in Sect. 1.3.2 shows that two random variables are independent of each other if the random variables X and Y satisfy the following equation:

$$E\{[X - E(X)][Y - E(Y)]\} = 0 \tag{1.29}$$

The relationship between two random variables is expressed through the **covariance** and the **correlation coefficient**, which are defined as follows:

1. Covariance: $E\{[X\text{-}E(X)][Y\text{-}E(Y)]\}$ is called the covariance of the random variables X and Y, denoted as $\mathrm{Cov}(X,Y)$, i.e.,

$$\mathrm{Cov}(X, Y) = E\{[X - E(X)][Y - E(Y)]\} \tag{1.30}$$

2. Correlation coefficient: The covariance divided by the product of the two random variables' standard deviation is called the correlation coefficient of the random variables X and Y, denoted as ρ_{XY}, i.e.,

$$\rho_{XY} = \frac{Cov(X, Y)}{\sqrt{D(X)}\sqrt{D(Y)}} \tag{1.31}$$

The covariance of the random variables X and Y can be calculated by the following formula:

$$\mathrm{Cov}(X, Y) = E(XY) - E(X)E(Y) \tag{1.32}$$

The basic properties of covariance are listed as follows:

1. For random variables X and Y, $\mathrm{Cov}(X, Y) = \mathrm{Cov}(Y, X)$, and $\mathrm{Cov}(X, X) = D(X)$.
2. For random variables X and Y and constants a and b, $\mathrm{Cov}(aX, bY) = ab\mathrm{Cov}(X, Y)$.
3. For random variables X_1, X_2 and Y, $\mathrm{Cov}(X_1 + X_2, Y) = \mathrm{Cov}(X_1, Y) + \mathrm{Cov}(X_2, Y)$.

For random variables X and Y, the basic properties for the correlation coefficient are listed as follows:

1. $|\rho_{XY}| \leq 1$.
2. $|\rho_{XY}| = 1$ if and only if there are constants a and b such that $P[Y = a + bX] = 1$.

From the properties above, it can be seen that ρ_{XY} is the quantity that indicates the closeness of the linear relationship between random variables X and Y. When $|\rho_{XY}|$ is close to 1, X and Y have a good linear correlation. $|\rho_{XY}| = 1$ indicates that X and Y show a perfect linear relationship. A zero correlation coefficient, i.e., $\rho_{XY} = 0$, means that X and Y are not correlated at all.

Example 1.12 The random variable Z is known to follow a uniform distribution of $[-\pi, \pi]$, i.e., its probability density function is $f_Z(z) = \begin{cases} \frac{1}{2\pi}, & -\pi < z \leq \pi, \\ 0, & \text{otherwise} \end{cases}$. For two random variables, $X = \sin Z$ and $Y = \cos Z$, try to find the correlation coefficient ρ_{XY}.

Solution

The expectation of X and Y can be calculated as follows:

$$E(X) = E(\sin Z) = \int_{-\pi}^{+\pi} f_Z(z) \sin z dz = \int_{-\pi}^{+\pi} \frac{1}{2\pi} \sin z dz = 0$$

$$E(Y) = E(\cos Z) = \int_{-\pi}^{+\pi} f_Z(z) \cos z dz = \int_{-\pi}^{+\pi} \frac{1}{2\pi} \cos z dz = 0$$

The variance of X and Y can be calculated as follows:

$$E(X^2) = E(\sin^2 Z) = \int_{-\pi}^{+\pi} f_Z(z) \sin^2 z dz = \int_{-\pi}^{+\pi} \frac{1}{2\pi} \sin^2 z dz = \frac{1}{2}$$

$$E(Y^2) = E(\cos^2 Z) = \int_{-\pi}^{+\pi} f_Z(z) \cos^2 z dz = \int_{-\pi}^{+\pi} \frac{1}{2\pi} \cos^2 z dz = \frac{1}{2}$$

$$E(XY) = E(\sin Z \cos Z) = \int_{-\pi}^{+\pi} f_Z(z) \sin z \cos z dz = \int_{-\pi}^{+\pi} \frac{1}{2\pi} \sin z \cos z dz = 0$$

According to Eq. (1.32),

$$\text{Cov}(X, Y) = E(XY) - E(X)E(Y) = 0$$

According to Eq. (1.31),

$$\rho_{XY} = \frac{Cov(X, Y)}{\sqrt{D(X)}\sqrt{D(Y)}} = \frac{Cov(X, Y)}{\sqrt{E(X^2)}\sqrt{E(Y^2)}} = \frac{0}{\frac{1}{\sqrt{2}}\frac{1}{\sqrt{2}}} = 0$$

1.3 Probability Distribution

This section will introduce the commonly used distributions for both the discrete and continuous random variables, including the Bernoulli distribution, binomial distribution, and Poisson distribution in the discrete probability distribution, in addition to the uniform distribution, exponential distribution, normal distribution, and Weibull distribution in the continuous probability distribution [3].

1.3.1 Typical Distributions of Discrete Random Variables

1. Bernoulli distribution.
   A random experiment, such as a coin toss, with only two possible outcomes, A and $\overline{A}$, is defined as a **Bernoulli experiment**. Let us denote $X(A) = 1$ and $X(\overline{A}) = 0$; the PMF of X can be written as:

$$p(x) = p^x(1-p)^{1-x},\ x = 0, 1 \tag{1.33}$$

   where p is the probability of the outcome or random event A happening.
2. Binomial distribution.
   Suppose that the Bernoulli experiment is repeated n times independently, so that it is an n-fold Bernoulli experiment. The random variable X is the number of random events A occurring during the n-fold Bernoulli experiment regardless of their order of occurrence. Then, the random variable X with a sample space $\Omega = \{0, 1, 2, 3, \ldots, n\}$ follows a **binomial distribution** with parameters n, p; its probability mass function is expressed as follows:

$$p_X(x) = C_n^k p^x(1-p)^{n-x}, x = 0, 1, 2, \ldots, n (0 < p < 1) \tag{1.34}$$

   where C_n^k is the number of combinations, indicating the number of all combinations of k elements from n elements, which is calculated as

$$C_n^k = \frac{n!}{k!(n-k)!} \tag{1.35}$$

3. Poisson distribution.
   If the discrete random variable X follows a **Poisson distribution** with parameter λ and sample space $\Omega = \{0, 1, 2, 3, \ldots\}$, then its probability mass function is

$$p_X(x) = \frac{\lambda^x e^{-\lambda}}{x!}, x = 0, 1, 2, 3 \ldots (\lambda > 0) \tag{1.36}$$

   The Poisson distribution can be used as an approximate solution to the binomial distribution when n is large, p is small, and $np = \lambda$ (a constant):

$$C_n^k p^x(1-p)^{n-x} \approx \frac{(np)^x e^{-np}}{x!} \tag{1.37}$$

Example 1.13 Imagine a person who is shooting at targets with a hit rate of 0.01 per shot. Try to find the probability of hitting at least twice when he fires 500 shots independently.

Solution

Each shot is considered a random experiment. Suppose a random variable X represents the number of hits among this 500-shot process; then, the random variable X follows a binomial distribution with:

$p_X(x) = C_{500}^{k} 0.01^x 0.99^{500-x}, x = 0, 1, 2, \ldots, n$

$P\{X \geq 2\} = 1 - P\{X = 0\} - P\{X = 1\} = 1 - 0.99^{500} - 500 \times 0.01 \times 0.99^{499} \approx 0.96$

1.3.2 Typical Distributions of Continuous Random Variables

1. Uniform distribution.
   Suppose that the continuous random variable X follows a **uniform distribution** over the interval (a, b), denoted as $X \sim U(a, b)$, and its probability density function is

$$f_X(x) = \begin{cases} \dfrac{1}{b-a}, & a < x < b \\ 0, & \text{otherwise} \end{cases} \tag{1.38}$$

   The cumulative distribution function of the random variable X is

$$F_X(x) = \begin{cases} 0, & x < a \\ \dfrac{x-a}{b-a}, & a \leq x < b \\ 1, & x \geq b \end{cases} \tag{1.39}$$

   The probability density function and cumulative distribution function of uniform distribution under different parameters (a, b) are shown in Fig. 1.3.
2. Exponential distribution.
   If the continuous random variable X follows an exponential distribution with parameter μ ($\mu > 0$ and is a constant), its probability density function is

$$f_X(x) = \begin{cases} \dfrac{1}{\mu} e^{-x/\mu}, & x > 0, \\ 0, & \text{otherwise} \end{cases} \tag{1.40}$$

   The cumulative distribution function of the random variable X is

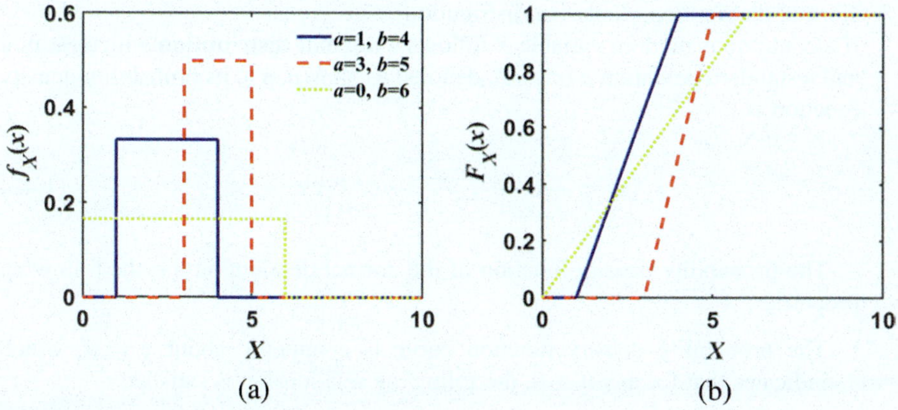

Fig. 1.3 Cases of uniform distribution: (**a**) Probability mass function and (**b**) Cumulative distribution function

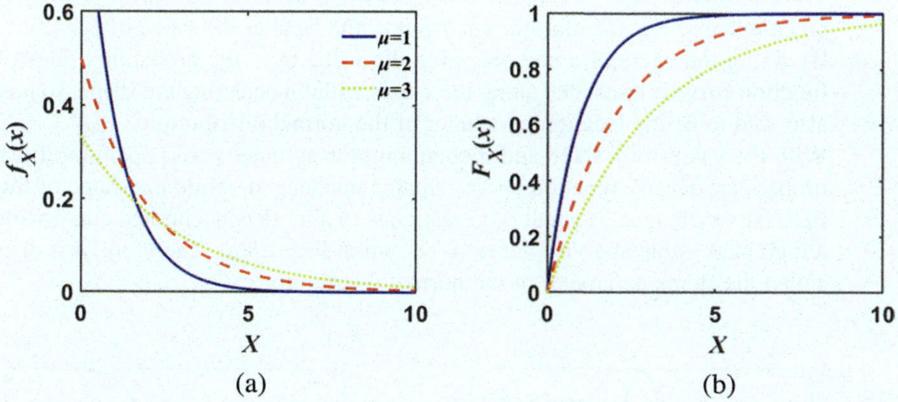

Fig. 1.4 Cases of exponential distribution: (**a**) Probability density function and (**b**) Cumulative distribution function

$$F_X(x) = \begin{cases} 1 - e^{-x/\mu}, & x > 0, \\ 0, & \text{otherwise} \end{cases} \tag{1.41}$$

The probability density function and cumulative distribution function of the exponential distribution with different parameters are shown in Fig. 1.4.

The exponential distribution is memoryless, i.e., $P(X > s + t | X > s) = P(X > t)$. In other words, the conditional probability that random variable X is larger than $s + t$, given that random variable X is known to be larger than s, is equal to the probability that random variable X is larger than t. This property allows the exponential distribution to be used in reliability theory to model the life distribution of an engineering system at a constant failure rate.

3. Normal distribution (Gaussian distribution).
   If a continuous random variable X follows a **normal distribution** with a mean μ and a standard deviation σ ($\sigma > 0$), denoted as $X \sim N(\mu, \sigma^2)$, its probability density function is

$$f_X(x) = \frac{1}{\sqrt{2\pi}\sigma} e^{-\frac{(x-\mu)^2}{2\sigma^2}}, -\infty < x < +\infty \tag{1.42}$$

The probability density function of the normal distribution has the following properties:

- The probability density function curve is symmetric about $x = \mu$, which indicates that for any $h > 0$, the following relationship is satisfied:

$$P\{\mu - h < X \leq \mu\} = P\{\mu < X \leq \mu + h\} \tag{1.43}$$

- The maximum value of the probability density function $f_X(\mu) = 1/(\sqrt{2\pi}\sigma)$ is taken when $x = \mu$. The farther x is from μ, the smaller the value of $f_X(x)$ is.
- By fixing the value of σ and changing the value of μ, the probability density function curve is translated along the x-axis without changing the shape, so μ is also said to be the location parameter of the normal distribution.
- With the values of μ fixed and σ changing, the symmetry axis position of the probability density function curve remains unchanged, while the shape of the PDF curve changes. A small σ corresponds to a more concentrated curve with a high peak value and vice versa, as shown in Fig. 1.5(a). Therefore, σ is also called the shape parameter of the normal distribution.

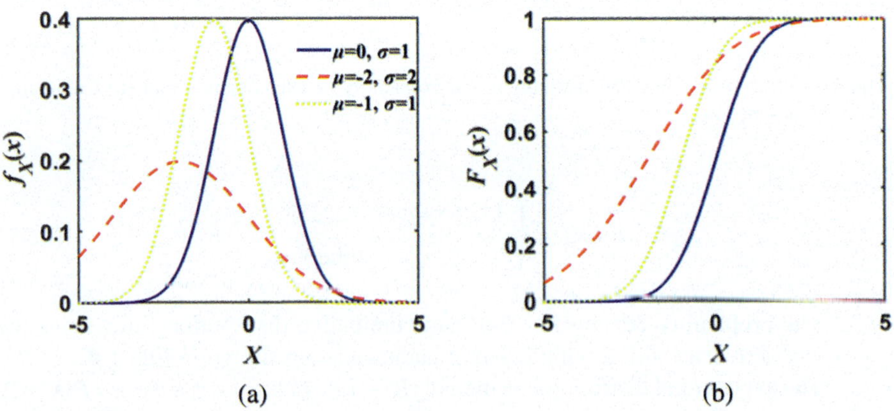

Fig. 1.5 Cases of normal distribution: (**a**) Probability density function and (**b**) Cumulative distribution function

The cumulative distribution function of the random variable X is

$$F_X(x)=\frac{1}{\sqrt{2\pi}\sigma}\int_{-\infty}^{x} e^{-\frac{(t-\mu)^2}{2\sigma^2}}\mathrm{d}t \tag{1.44}$$

The probability density function and cumulative distribution function of the normal distribution with different parameters are shown in Fig. 1.5(b).

When $\mu = 0$ and $\sigma^2 = 1$, the random variable X is said to follow the **standard normal (Gaussian) distribution**. Its probability density function and cumulative distribution function are denoted by $\varphi(x)$ and $\Phi(x)$, respectively, and expressed as follows:

$$\varphi(x)=\frac{1}{\sqrt{2\pi}}e^{-\frac{t^2}{2}} \tag{1.45}$$

$$\Phi(x)=\frac{1}{\sqrt{2\pi}}\int_{-\infty}^{x} e^{-\frac{t^2}{2}}\mathrm{d}t \tag{1.46}$$

Since the standard normal distribution curve is symmetric about $x = 0$, it is easy to obtain that $\Phi(-x) = 1 - \Phi(x)$.

For a general normal distribution $X \sim N(\mu, \sigma^2)$, we can transform it into a standard normal distribution by linearly varying $Z = (X - \mu)/\sigma$.

Below, we explain the transformation of a continuous random variable X and a standard normal variable U. For a continuous random variable X with CDF $F_X(x)$ and a standard normal random variable U with CDF $\phi(u)$, the transformation between X and U is explained as follows.

From the definition of CDF, it is known that

$$F_X(x)=P[X\le x]\triangleq p \tag{1.47}$$

p, the probability of $X \le x$, can be related to u with

$$\phi(u)=p \tag{1.48}$$

then, the mapping between X and U can be expressed as follows:

$$\begin{aligned}\phi(u)&=F_X(x)\\ u&=\phi^{-1}(F_X(x))\\ x&=F_X^{-1}(\phi(u))\end{aligned} \tag{1.49}$$

Consider an n-dimensional random vector $\boldsymbol{X}=[x_1\ x_2\ \ldots\ x_n]$ with a generic joint PDF $F_{\boldsymbol{X}}(\boldsymbol{x})$. $T : \boldsymbol{X} \to \boldsymbol{U}$ is a transformation between $\boldsymbol{x}$- and $\boldsymbol{u}$-space, and $\boldsymbol{U}=[u_1\ u_2\ \ldots\ u_n]$ is an n-dimensional standard Gaussian random vector. Rosenblatt [5] gives the transformation as follows:

$$T:\begin{cases} u_1=\phi^{-1}[F_{X_1}(x_1)] \\ u_2=\phi^{-1}[F_{X_2}(x_2|x_1)] \\ \cdots \\ u_n=\phi^{-1}[F_{X_n}(x_n|x_1,x_2,\ldots,x_{n-1})] \end{cases} \tag{1.50}$$

where $F_{X_i}(x_i|x_1,x_2,\ldots,x_{i-1}), i=1,2,\ldots,n$ is the CDF of X_i given the condition that $X_1=x_1$, $X_2=x_2$, …, $X_{i-1}=x_{i-1}$ and is calculated as:

$$F_{X_i}(x_i|x_1,x_2,\ldots,x_{i-1})=\frac{\int_{-\infty}^{x_i} f_{X_1,X_2,\ldots,X_i}(x_1,x_2,\ldots,x_{i-1},\xi)d\xi}{f_{X_1,X_2,\ldots,X_{i-1}}(x_1,x_2,\ldots,x_{i-1})} \tag{1.51}$$

where $f_{X_1,X_2,\ldots,X_{i-1}}(x_1,x_2\ldots,x_{i-1})$ is the joint PDF of X_1, X_2, …, X_{i-1} expressed as:

$$f_{X_1,X_2,\ldots,X_{i-1}}(x_1,x_2,\ldots,x_{i-1}) = \frac{\partial^{i-1}}{\partial x_1 \partial x_2 \ldots \partial x_{i-1}} F_{X_1,X_2,\ldots,X_{i-1}}(x_1,x_2,\ldots,x_{i-1}) \tag{1.52}$$

where $F_{X_1,X_2,\ldots,X_{i-1}}(x_1,x_2,\ldots,x_{i-1})$ is the joint CDF of X_1, X_2, …, X_{i-1} as:

$$F_{X_1,X_2,\ldots,X_{i-1}}(x_1,x_2,\ldots,x_{i-1})=P\left(\bigcap_{k=1}^{i-1} X_k \le x_k\right) \tag{1.53}$$

The inverse transformation of T is

$$T^{-1}:\begin{cases} x_1=F_{x_1}^{-1}[\phi(u_1)] \\ x_2=F_{x_2}^{-1}[\phi(u_2|x_1)] \\ \cdots \\ u_n=F_{x_n}^{-1}[\phi(u_n|x_1,x_2,\ldots,x_{n-1})] \end{cases} \tag{1.54}$$

4. Weibull distribution.
   If the continuous random variable X follows a Weibull distribution with parameters λ and k ($\lambda > 0$, $k > 0$ and is constant), its probability density function is

$$f_X(x) = \begin{cases} \frac{k}{\lambda}\left(\frac{x}{\lambda}\right)^{k-1} e^{-\left(\frac{x}{\lambda}\right)^k} & x \geq 0 \\ 0 & x < 0 \end{cases} \tag{1.55}$$

The cumulative distribution function of the random variable X is

$$F_X(x) = \begin{cases} 1 - e^{-\left(\frac{x}{\lambda}\right)^k} & x \geq 0 \\ 0 & x < 0 \end{cases} \tag{1.56}$$

If the Weibull distribution is often used to model the distribution of failure times, the probability of failure can be calculated by the following equation.

$$h_X(x) = \frac{f_X(x)}{1 - F_X(x)} = \frac{k}{\lambda}\left(\frac{x}{\lambda}\right)^{k-1} \tag{1.57}$$

where x denotes time ($x > 0$). When $k = 1$, the failure probability is constant, and the Weibull distribution becomes exponential at that time.

The probability density function and cumulative distribution function of the Weibull distribution with different parameters are shown in Fig. 1.6.

Example 1.14 The life of an electronic component (unit: hour) follows an exponential distribution with parameter $\lambda = 1/1000$. Determine the probability that at least one of the three electronic components survives after 1000 hours of usage.

Solution

Let the lifetime of the electronic component be a random variable X, and its cumulative distribution function is

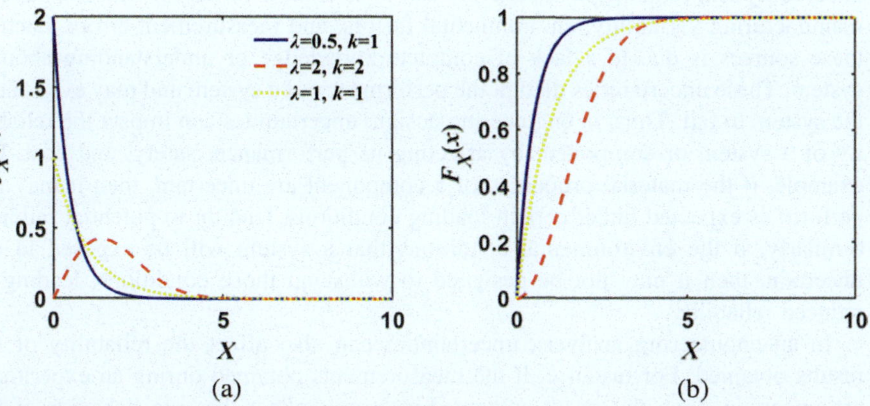

Fig. 1.6 Cases of Weibull distribution: (a) Probability mass function and (b) Cumulative distribution function

$$F_X(x) = \begin{cases} 1 - e^{-x/1000}, & x > 0, \\ 0, & \text{otherwise} \end{cases}$$

Therefore, the probability that the life of the component is greater than 1000 hours is

$$P\{X > 1000\} = 1 - P\{X \leq 1000\} = 1 - F_X(1000) = e^{-1}$$

Since the lives of different electronic components are independent of each other, a random variable Y can be used to represent the number of components that survive after 1000 hours of usage. Then, Y follows a binomial distribution with the parameters $n = 3$, $p = 1 - e^{-1}$). The probability that at least one of the three electronic components survives after 1000 hours of usage is calculated as follows

$$P\{Y \geq 1\} = 1 - P\{Y = 0\} = 1 - C_3^0 \left(1 - e^{-1}\right)^0 \left(e^{-1}\right)^3 = 1 - e^{-3}$$

The following table offers a comprehensive summary of the PDF and CDF of commonly used distributions. In addition, it also provides the relationship between the distribution parameters (a, b) and the mean μ and the standard deviation σ, as well as the transformation from closed forms of the transformation, from the standard Gaussian variable U with mean zero and a unit variance, to the random variable X [6] (Table 1.1).

1.4 Reliability

It is of critical importance to have a clear understanding of the relationship between uncertainty and reliability. Uncertainty can arise from a variety of sources, such as manufacturing variability, environmental factors, and measurement errors. Each of these sources is due to a lack of complete knowledge or understanding about a system. These uncertainties disrupt the performance of a system and may even cause the system to fail. Thus, in engineering design, uncertainties can impact the reliability of a system or component by affecting its performance, safety, and cost. For example, if the material properties of a component are uncertain, then it may not perform as expected under certain loading conditions, leading to potential failures. Similarly, if the environmental conditions that a system will be exposed to are uncertain, then it may not be designed to withstand those conditions, leading to reduced reliability.

In an engineering analysis, uncertainties can also affect the reliability of the results obtained. For instance, if the measurements obtained during an experiment are uncertain, then the results obtained may not reflect the true behavior of the

Table 1.1 A comprehensive summary of the PDF, CDF, distribution parameters, and U-X transformations of commonly used distributions

Distribution	PDF $f_X(x)$	CDF $F_X(x)$	Parameters	Range of a	Range of b	Transformation from U to X
Normal	$\frac{1}{b\sqrt{2\pi}}\exp\left[-\frac{1}{2}\left(\frac{x-a}{b}\right)^2\right]$	$\Phi\left(\frac{x-a}{b}\right)$	$a=\mu, b=\sigma$	$(-\infty, \infty)$	$(0, \infty)$	$X=a+bU$
Lognormal	$\frac{1}{bx\sqrt{2\pi}}\exp\left[-\frac{1}{2}\left(\frac{\ln x-a}{b}\right)^2\right]$	$\Phi\left(\frac{\ln x-a}{b}\right)$	$b=\sqrt{\ln\left[1+\left(\frac{\sigma}{\mu}\right)^2\right]}, a=\ln(\mu)-\frac{1}{2}b^2$	$(0,\infty)$	$(0,\infty)$	$X=\exp(a+bU)$
Exponential	$f(x)=\frac{1}{b}\exp(-\frac{x-a}{b}), x\geq a$	$1-\exp(-\frac{x-a}{b}), x\geq a$	a is the shift of distribution $b=\sigma=\mu-a$	$(-\infty, \infty)$	$(0, \infty)$	$X=a-b\ln(\Phi(-U))$
Uniform	$f(x)=\frac{1}{b-a}, a\leq x\leq b$	$\frac{x-a}{b-a}, a\leq x\leq b$	$a=\mu-\frac{\sqrt{12}}{2}\sigma, b=\mu+\frac{\sqrt{12}}{2}\sigma$	$(-\infty, \infty)$	$(-\infty, \infty)$	$X=a+(b-a)\Phi(U)$
Rayleigh	$f(x)=\frac{x}{b^2}\exp\left[-\frac{1}{2}\left(\frac{x}{b}\right)^2\right], x\geq a$	$1-\exp\left(\frac{x^2}{2b^2}\right), x\geq a$	a is the shift of distribution $b=\frac{\sigma}{\sqrt{2-\frac{\pi}{2}}}=\frac{\mu-a}{\sqrt{\frac{\pi}{2}}}$	$(-\infty, \infty)$	$(0, \infty)$	$X=a+\sqrt{-2b^2\ln(\Phi(-U))}$
Weibull[a]	$\frac{b}{a}\left(\frac{x}{a}\right)^{b-1}\exp\left[-\left(\frac{x}{a}\right)^b\right]$	$1-\exp\left[-\left(\frac{x}{a}\right)^b\right]$	b is the solution of $h(b\vert\mu,\sigma)=0$, where $h(b\vert\mu,\sigma)=\mu^2\left(\frac{\Gamma\left(1+\frac{2}{b}\right)}{\Gamma^2\left(1+\frac{1}{b}\right)}-1\right)-\sigma^2,$ $a=\frac{\mu}{\Gamma\left(1+\frac{1}{b}\right)}$	$(0,\infty)$	$(0, \infty)$	$X=a[-\ln(\Phi(-U))]^{\frac{1}{b}}$
Gamma[a]	$\frac{1}{\Gamma(a)b^a}x^{a-1}\exp(-x/b), x\geq 0$	$\frac{1}{\Gamma(a)}\gamma(a, \frac{x}{b}), x\geq 0$	$a=\left(\frac{\mu}{\sigma}\right)^2, b=\frac{\mu}{a}$	$(-\infty, \infty)$	$(0, \infty)$	Not available in closed-form
Extreme type-I[b]	$\frac{1}{b}\exp\left[-\frac{x-a}{b}-\exp(-\frac{x-a}{b})\right]$	$\exp\left[-\exp(-\frac{x-a}{b})\right]$	$b=\frac{\sqrt{6}\sigma}{\pi}, a=\mu-\eta b$	$(-\infty, \infty)$	$(0, \infty)$	$X=a-b\ln[-\ln(\Phi(U))]$

(continued)

Table 1.1 (continued)

Distribution	PDF $f_X(x)$	CDF $F_X(x)$	Parameters	Range of a	Range of b	Transformation from U to X
Extreme type-II	$\frac{a}{b}\left(\frac{b}{x}\right)^{a+1}\exp\left[-\left(\frac{b}{x}\right)^{a}\right]$	$\exp\left[-\left(\frac{b}{x}\right)^{a}\right]$	a is the solution of $h(a\mid\mu,\sigma)=0$, where $h(a\mid\mu,\sigma)=\mu\frac{\sqrt{\Gamma(1-2/a)-\Gamma^2(1-1/a)}}{\Gamma(1-1/a)}-\sigma$, $b=\frac{\mu}{\Gamma\left(1-\frac{1}{a}\right)}$	$(-\infty, \infty)$	$(0, \infty)$	$X=\frac{b}{[-\ln(\Phi(U))]^{\frac{1}{a}}}$
Beta	$f(x)=\frac{x^{a-1}(1-x)^{b-1}}{B(a,b)}, 0\leq x\leq 1$ where $B(a,b)=\frac{\Gamma(a)\Gamma(b)}{\Gamma(a+b)}$	/	$a=\frac{(1-\mu)\mu^2}{\sigma}, b=\frac{(1-\mu)^2\mu}{\sigma}$	$(0, \infty)$	$(0, \infty)$	Not available in closed-form

[a]$\Gamma(\cdot)$ is the complete gamma function of a, $\gamma(\cdot,\cdot)$ is the lower incomplete gamma function;
[b]η is the Euler–Mascheroni constant ($\eta \approx 0.57721$)

system being studied. This can lead to incorrect conclusions and ultimately inaccurate predictions.

Therefore, it is essential for engineers to understand and manage uncertainties when analyzing and designing engineering systems. This involves identifying sources of uncertainty, quantifying their impact on reliability, and developing strategies to mitigate them. This can be accomplished through the use of statistical methods, probabilistic analysis, and sensitivity analysis, to name a few.

1.4.1 Basic Concepts of Reliability

An event where a system performs its required function under specified conditions and within a specified time is a random phenomenon, which can be described by using probabilistic and statistical methods. The probability of occurrence of this preferred event is called the **reliability** of the system. The complementary event is called the **failure** of the system.

Based on the definition above, a general mathematical expression of reliability is derived here. The **performance** of a system Z is related to a series of random variables, such as geometric parameters, structural properties, and applied loads. These random variables are usually expressed as a vector $\boldsymbol{X} = [x_1 \ x_2 \ \dots \ x_n]$. Z and $\boldsymbol{X}$ are related through a **performance function**: $Z = G(X)$. The failure of the system can then be defined as $Z < 0$. To obtain the reliability of a system, its probability of failure, P_F, can be calculated as [7]:

$$P_F = \int \dots \int_{G(\mathbf{X})<0} f(\mathbf{X}) d\mathbf{X} \tag{1.58}$$

where $f(X)$ is the joint PDF of random variables $\boldsymbol{X}$. Then, the reliability of the system is simply $1 - P_F$. In a reliability analysis of engineering systems, two commonly used disciplines, namely, probability theory and statistics, are needed. Since the concepts of these two disciplines are similar and easily confused, the two concepts are explained in detail for a better understanding [8].

1. Probability theory: Probability theory is mainly a theory-oriented branch of mathematics. It is an axiomatic pure mathematical theory with a strict axiomatic basis, and the relevant conclusions are obtained by requiring rigorous mathematical derivations. Probability consists of calculating the possibility of a given event, and is the basis of statistics, which allows us to calculate the results of an analysis under strict assumptions.
2. Statistics: Statistics is an application-oriented branch of mathematics. Statistics focuses on analyzing the frequency of past events or quantifying uncertainty based on real-world observations. Statistics is an application of probability theory by thinking backward about the process of data generation based on observed

data. Statistics is a discipline that emphasizes statistical inference, including issues such as statistical determination, estimation, and testing.

An engineering system designed by reliability theory should not only ensure that the system does not fail when performing its function at a certain instant but should also ensure that the system functions normally without catastrophic failure within a specified time period [9]. According to its temporal characteristics, reliability analysis can be divided into two categories [10]:

1. Time-invariant reliability analysis estimates the probability that a system is able to perform normal operational functions *statically* under uncertainty.
2. Time-variant reliability analysis estimates the probability that a system is able to perform intended functions *over a specified period of time*.

In addition, there are relevant optimization methods that explicitly incorporate system reliability into the design algorithm, whereas reliability is usually not considered in traditional deterministic designs. The design with the lowest probability of failure is usually selected in practical engineering. Reliability-based design optimization (RBDO) has been widely investigated to design cost-effective and reliable systems.

1.4.2 Importance of Reliability Assessment

In practical engineering, systems are usually designed to ensure the maximum level of safety and the minimum level of cost. However, existing deterministic optimal design approaches, which consider uncertainty through safety factors, usually do not obtain sufficient information to achieve the best use of the available resources to maximize safety [11]. Using the reliability analysis approach, more information can be obtained about the system behavior, the effects of different uncertain variables on the system performance and the interactions between different system components, providing more information for optimal design [12]. For example, there are many uncertainties in engineering systems, such as loads, material density, stiffness, and mass properties, which affect the performance response of the structure (e.g., stresses at critical locations, deformations, resonant frequencies, etc.). Reliability quantification is performed for the uncertainty factors, and the influence of the uncertainty factors on the structural performance is determined by reliability analysis [13]. Finally, the design variables could be optimized using reliability-based design optimization methods to maximize the safety level and minimize the cost. The reliability quantification, reliability analysis, and reliability-based optimization methods will be described in detail in the following chapters.

Exercises

1.1 Write the sample space of random variables based on the following experiments:

(a) A coin is tossed 10 times consecutively. Write the sample space of the number of heads being up.
(b) There are 5 balls in a box, numbered 1–5, from which 2 balls are randomly taken out at once. Write the sample space of the sum of the numbers of the 2 balls.
(c) Consider a square with the side length 1 and its center located at the origin of a 2-dimensional coordinate system. Write the sample space of the coordinate of an arbitrary point inside the square.

1.2 State the relationship between the following random events *A* and *B*:

(a) Roll a die, observe the number of points. Define the random event $A = \{$the number of points is even$\}$ and the random event $B = \{2, 4, 6\}$. Find the relationship between A and B.
(b) Turn 2 coins consecutively. Define the random event $A = \{$both coins are heads$\}$ and the random event $B = \{$different results for both coins$\}$. Find the relationship between A and B.
(c) Record the life of a light bulb. Define the random event $A = \{$light bulb life greater than 1000 hours$\}$ and the random event $B = \{$light bulb life greater than 1500 hours$\}$. Find the relationship between A and B.

1.3 Suppose the probability that a bearing has a service life of at least 6500 hours is 70% and the probability that it has a service life of at least 10,000 hours is 45%. Calculate the probability that a bearing can be used for 10,000 hours if it has already been used for 6500 h.

1.4 There are 5 balls in a box, numbered 1 to 5, from which 2 balls are taken at random. Assume that *X* is the sum of the 2 balls' numbers. Calculate the following probabilites: (1) $P\{3 < X < 7\}$, (2) $P\{X \le 5\}$, (3) $P\{X > 6\}$, and (4) the cumulative distribution function $F_X(x)$ of *X*.

1.5 The probability density function of a random variable *X* is defined as

$$f_X(x) = \begin{cases} 2 \times (1 - x) & 0 < x < 1 \\ 0 & \text{otherwise} \end{cases}$$

Calculate the following probabilities: (1) $P\{0.1 < X \le 0.6\}$, (2) $P\{X > 0.8\}$, and (3) the cumulative distribution function of *X*.

1.6 Consider a cantilever beam with length *L*, width *b*, and height *h*, which is under a tip concentrated force *P*. It is known based on the structural analysis theory that the tip deflection $\delta = PL^3/3EI$, where *E* is the Young's module of the beam and *I* is the moment of inertia. *I* is calculated as $I = bh^3/12$ in this case. Assuming that only the length *L* is uncertain and follows $N \sim (\mu_L, \sigma_L)$, calculate the PDF and CDF of δ.

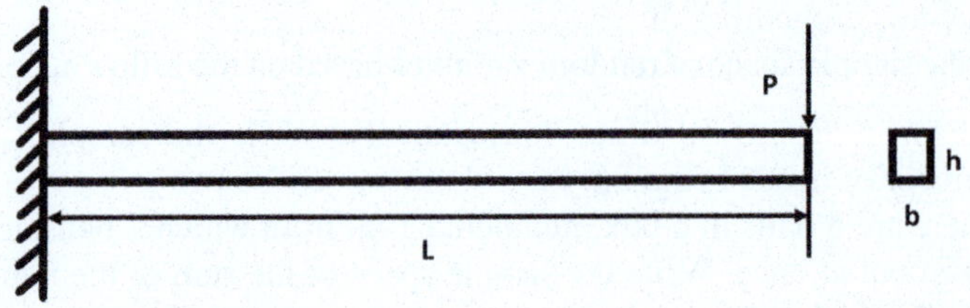

1.7 The probability density function of the random variable X is known as

$$f_X(x) = \begin{cases} \frac{x}{2} & 0 < x < 2 \\ 0 & \text{otherwise} \end{cases}$$

Calculate (1) the probability density function of the random variable $Y = 3X + 2$ and (2) the probability density function of the random variable $Z = X^2$.

1.8 Suppose there are 4 kinds of products in a batch, namely, grades 1, 2, 3, and substandard products. Their proportions are 40%, 30%, 20%, and 10%, respectively, and the selling prices of the products are 100 yuan, 80 yuan, 60 yuan, and 0 yuan, respectively. Calculate the mean and variance of the selling price of the products.

1.9 The probability density function of the random variable X is known as

$$f_X(x) = \begin{cases} \frac{1}{b-a} & a \le x \le b \\ 0 & \text{otherwise} \end{cases}$$

Calculate the expectation and variance of the random variable X.

1.10 Assume the substandard rate of a mechanical part in the machining process is 0.2%. Assuming that the examination of an individual part does not influence that of the others, what is the probability that there are at least 3 substandard products among 1500 products?

1.11 Suppose the length (cm) of bolts follows the normal distribution $N(10.1, 0.05^2)$, and the length of bolts within 10.1 ± 0.1 is qualified. Calculate the probability of a randomly selected bolt being unqualified.

1.12 The temperature of a liquid at a processing site is adjusted by a thermostat. When the thermostat is set at a(°C), the liquid temperature X(°C) is a random variable and $X \sim N(a, 0.2^2)$. Solve the following questions: (1) If $a = 75$, what is the probability that X is less than 74? (2) To ensure that the probability of the liquid temperature being at least 75 °C is not less than 0.99, what is the range of a?

1.13 The damage probability of an electronic component is 0.02, 0.001, 0.05, if its supply voltage does not exceed 200 V, the supply voltage is between 200 and 240 V, and the supply voltage is over 240 V, respectively. Assume that the supply voltage follows a normal distribution $N(220,25^2)$. (1) Calculate the probability of the electronic component being damaged; (2) Given the electronic component is damaged, calculate the probability of the voltage being within the range of 200–240 V.

References

1. Schmetterer, L. (2012). *Introduction to mathematical statistics*. Springer.
2. Grimmett, G., & Welsh, D. (2014). *Probability: An introduction*. Oxford University Press.
3. Haldar, A., & Mahadevan, S. (2000). *Probability, reliability, and statistical methods in engineering design*. Wiley.
4. Hausdorff, F. (2005). *Set theory*. American Mathematical Soc.
5. Rosenblatt, M. (1952). Remarks on a multivariate transformation. *The Annals of Mathematical Statistics, 23*, 470–472.
6. Hu, W., Jiang, Z., & Wang, Y. *Reliability analysis of offshore wind turbines using copula*. In ASME 2018 power conference collocated with the ASME 2018 12th international conference on energy sustainability and the ASME 2018 nuclear forum 2018.
7. Jiang, Z., Hu, W., Dong, W., Gao, Z., & Ren, Z. (2017). Structural reliability analysis of wind turbines: A review. *Energies, 10*, 2099.
8. Giri, N. C. (2019). *Introduction to probability and statistics*. CRC Press.
9. Hu, W., Yan, J., Zhao, F., Jiang, C., Liu, H., Cho, H., et al. (2023). Surrogate-based time-dependent reliability analysis for a digital twin. *Journal of Mechanical Design, 2023*, 1–36.
10. Cheng, K., & Lu, Z. (2019). Time-variant reliability analysis based on high dimensional model representation. *Reliability Engineering & System Safety, 188*, 310–319.
11. Hu, W., Choi, K., & Cho, H. (2016). Reliability-based design optimization of wind turbine blades for fatigue life under dynamic wind load uncertainty. *Structural and Multidisciplinary Optimization, 54*, 953–970.
12. Royset, J., Kiureghian, A. D., & Polak, E. (2001). Reliability-based optimal design of series structural systems. *Journal of Engineering Mechanics, 127*, 607–614.
13. Teixeira, R., Nogal, M., & O'Connor, A. (2021). Adaptive approaches in metamodel-based reliability analysis: A review. *Structural Safety, 89*, 102019.

Chapter 2
Uncertainty Modeling

Nomenclature

$Bel(\cdot)$	The belief function
$\underline{\boldsymbol{F}}()$	The lower boundary of the cumulative distribution function (CDF)
$\bar{\boldsymbol{F}}()$	The upper boundary of the CDF
$\boldsymbol{F_X(x)}$	The CDF of X
$f_X(x,\boldsymbol{\theta})$	The probability density function (PDF) of X
$G(\cdot)$	The P-box of Z
$g(\cdot)$	A known continuous function
H	The Shannon entropy
$L(\cdot)$	The likelihood function
$l(\cdot)$	The log likelihood function
$\boldsymbol{m}(\cdot)$	The mapping function
n	The number of samples
$\underline{\boldsymbol{P}}()$	The derivatives of $\underline{\boldsymbol{F}}()$
$\bar{\boldsymbol{P}}()$	The derivatives of $\bar{\boldsymbol{F}}()$
$Pl(\cdot)$	The plausibility function
p_i	The probability associated with the i-th sample
X	A given random variable
$\underline{\boldsymbol{x}}$	The lower bound of interval x^I
$\bar{\boldsymbol{x}}$	The upper bound of interval x^I
x^A	The maximum absolute difference between the midpoint and the bounds of the interval set
x^I	The interval variable
x^M	The average of the upper and lower bounds of the interval set
x^R	Half of the width of the interval set
x^W	The difference between the upper and lower bounds of the interval set
$[x]$	An interval vector
y^I	The interval variable

W. Hu, *Design Optimization Under Uncertainty*,
https://doi.org/10.1007/978-3-031-49208-2_2

z^I	The interval variable
θ	The estimated parameter
$\widehat{\theta}$	The maximum likelihood estimator
μ	The mean of the normal distribution random variable
σ	The standard deviation of the normal distribution random variable
$\varnothing$	The empty set
Θ	The identification frame

2.1 Introduction

Uncertainty is a fundamental concept in the field of probability, statistics, and decision-making [1]. It expresses the inability to precisely determine an outcome based on available information. Uncertainty in design optimization is often related to reliability and robustness. The goal of design engineering is to design reliable systems and components that can perform optimally under a variety of conditions. To achieve this goal, engineers must be aware of the sources and types of uncertainty, and use some methods to quantify uncertainty.

In general, during the process of design optimization under uncertainty (DOUU), engineers need to consider various uncertainties at different phases of the equipment's whole lifecycle [2]. For example, throughout the lifecycle of a tunnel boring machine, the sources of uncertainty can be divided into the following categories: (i) in the mission analysis phase, the main sources of uncertainty are the needs and requirements of the mission, relevant national support policies and institutional regulations, funding, mission schedules, and variability in political and cultural factors, etc.; (ii) in the design phase, uncertainties arise primarily from computational simulation. There are three main sources, namely uncertainty in the model input (external uncertainty), uncertainty in the model (uncertainty in the model parameters), and model error, as shown in Fig. 2.1; (iii) in the manufacturing phase, sources of uncertainty are manual operation errors, tolerances in manufacturing, and so on; and (iv) in the actual operation phase, uncertainties arise from actual environmental influences. These uncertainties are shown in Fig. 2.2.

According to their different characteristics, uncertainty can be classified into two categories: aleatory uncertainty and epistemic uncertainty [3]. Aleatory uncertainty is caused by probabilistic variability and is also referred to as statistical uncertainty. It is inherently uncertain and may be characterized by a probability distribution. In contrast, epistemic uncertainty arises from a lack of knowledge such as an incomplete understanding of underlying processes, phenomena, or imprecise evaluation of the system performance. This type of uncertainty is also known as systematic uncertainty and may be reduced through the acquisition of more information about the system [4].

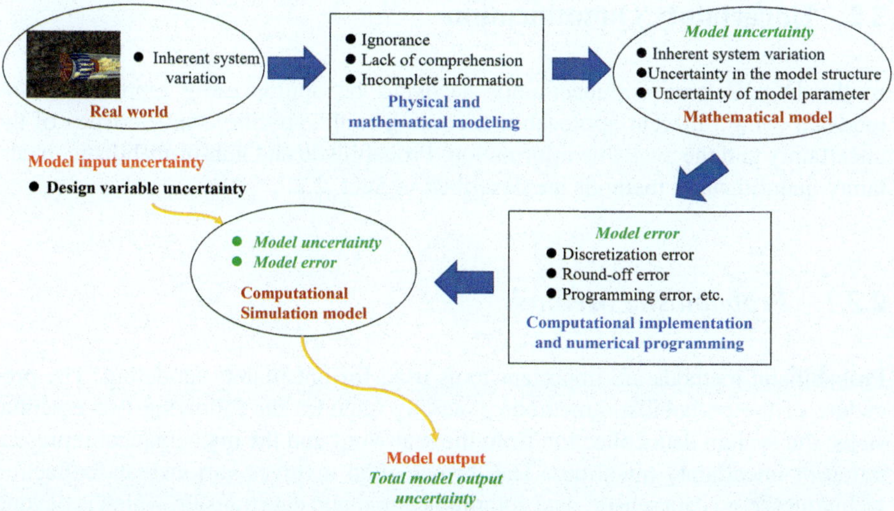

Fig. 2.1 Uncertainty sources in the simulation-based design process

Fig. 2.2 Uncertainty sources throughout the equipment's lifecycle

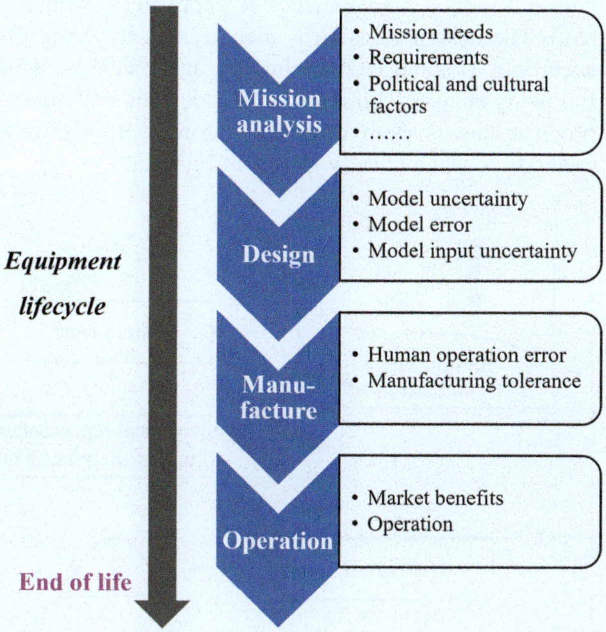

2.2 Uncertainty Quantification

In the DOUU process, uncertainty needs to be appropriately represented and modeled using different approaches according to the specific characteristics of the uncertainty and the available information. Probabilistic and non-probabilistic uncertainty quantification methods are described in Sect. 2.2.

2.2.1 *Probabilistic Methods*

Probabilistic methods are important tools used for uncertainty modeling. The procedure of the probabilistic methods generally includes the following two essential steps: the system data collection from the real word and the mathematical representation of uncertainty quantities. The former often involves various data collection techniques, e.g., data sensor, data communication, and data transfer, which is beyond the scope of this chapter. The latter in probabilistic methods often requires investigation in the two aspects: histogram or frequency determination (e.g., probability density function) and statistics or parameters estimation (e.g., distribution parameters). The design decision is then created by using these two aspects information according to some DOUU theories and methods, which will be elaborated in the following chapters. Finally, the new designs will affect the system in real world and produce new system data. The aforementioned procedure of the probabilistic methods is schematically shown in Fig. 2.3.

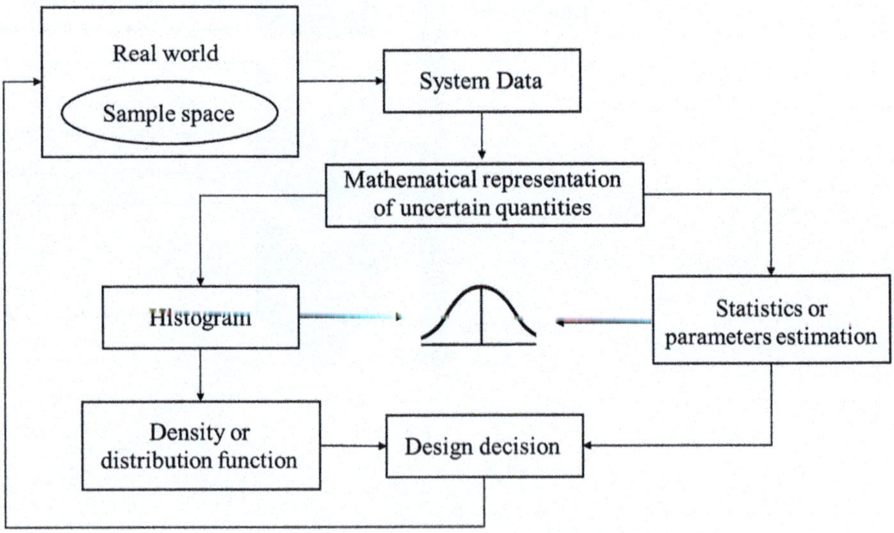

Fig. 2.3 Steps in probabilistic methods for uncertainty modeling. (Adapted from [5])

In practical engineering applications, determining the potential distribution of random variables is a major challenge. In general, it is necessary to estimate the distribution and parameters of a random variable using the measurement data. Once the type of distribution of a random variable has been chosen, the specific distribution needs to be uniquely defined by evaluating its parameters. In this section, the maximum entropy distribution and parameter estimation methods will be presented for determining the type of distribution and parameters of a random variable, respectively.

2.2.1.1 Maximum Entropy Distribution

Currently, in the processing of measurement data, the probability distribution of the data relies on certain specific assumptions. In contrast, the probability distribution function obtained by the maximum entropy distribution method is objective and can improve the accuracy of the results.

$$H = -\sum_i p_i \ln(p_i) \tag{2.1}$$

where H is the Shannon entropy (so named after its original proponent) and p_i is the probability associated with the i-th sample. It is easily shown that the maximum entropy corresponds to a uniform distribution, where all samples are equally likely.

Any other distribution would have a concentration of probability away from the extreme values, leading to a reduction of uncertainty and hence a reduction of entropy. The principle of maximum entropy aims to select a probability density function (PDF) that maximizes entropy while adhering to the given constraints. By utilizing all available information and satisfying the known constraints, this approach minimizes uncertainty without making unnecessary assumptions. Consequently, ignorance is preserved, ensuring maximal uncertainty regarding unknown information. Harr [6] provides a discussion on how the maximum entropy principle aids in assigning probability distributions based on known constraints, as summarized in Table 2.1.

Table 2.1 Maximum entropy distributions

Constraint	Assigned PDF
Upper bound, lower bound	Uniform
Mean, standard deviation	Normal
Range, mean, standard deviation	Beta
Mean occurrence rate	Poisson

2.2.1.2 Method of Moments

In Chap. 1, it is explained that the mean or expected value of a random variable represents its first moment, while the variance represents its second moment. The principal idea behind the method of moments is that all parameters of a distribution can be estimated by using its moments' information. If a distribution has only one parameter, then one piece of information needs to be extracted from the samples; typically, this would be the first moment, which is the mean of the random variables. On the other hand, if there are two parameters in the distribution, two pieces of information need to be extracted from the samples, which would most likely be the first two moments, namely the mean and variance of the random variable. As the name suggests, the parameters of a distribution have some correlation with the moments of the random variable. Most common relationships between the distribution parameters and the moments for continuous and discrete random variables are shown in Table 1.1 of Chap. 1.

Although the method of moment estimation is a simple and easy way to implement, it has its limitations. For example, it may not produce accurate estimates when the sample size is small or poorly distributed. In such cases, other methods such as maximum likelihood estimation (MLE) may be more suitable.

2.2.1.3 Maximum Likelihood Estimation

The maximum likelihood estimation is based on the idea that, for a given random variable X with n sample observations $x_1, x_2, ..., x_n$, the parameter estimate should be the value that maximizes the probability of obtaining those observations. In other words, the MLE is the set of parameter values that make the observed data most likely to occur, assuming a particular distribution type of the random variable. This technique is widely used in statistical inference and has numerous applications in fields such as engineering, finance, and biology. Consider the probability density function (PDF) of X to be $f_X(x,\theta)$, where θ is the parameter that needs to be estimated. The basis of our inferential procedures is the likelihood function given by

$$L(\theta; X) = \prod_{i=1}^{n} \mathbf{f}(x_i; \theta), \theta \in \Omega \tag{2.2}$$

In this section, L is considered as a function of the parameter θ. To achieve this, the order of x_i and θ in the argument of the likelihood function is reversed. The likelihood function is conventionally represented as $L(\theta)$, representing the probability of observing the sample data given a specific value of θ. In practice, it is often more convenient to work with the natural logarithm of this function, which is denoted as $\log L(\theta)$.

$$l(\theta) = \log L(\theta) = \sum_{i=1}^{n} \log f(x_i;\theta), \theta \in \Omega \tag{2.3}$$

Using the log likelihood function $l(\theta)$ does not result in any loss of information, as the logarithm is a one-to-one function. This means that the probability of observing the sample data given a parameter value θ can be equivalently expressed using either the likelihood function or the log likelihood function. Additionally, our discussion in this section is mostly applicable even if X is a random vector rather than a scalar random variable. The parameter θ is estimated using a point estimator denoted by $\widehat{\theta}$. The value of $\widehat{\theta}$ is chosen to maximize the likelihood function $L(\theta)$ based on the observed data $X_1, ..., X_n$. This estimator is commonly referred to as the maximum likelihood estimator (MLE) and is denoted as $\widehat{\theta}$.

To simplify the mathematics involved in maximizing the likelihood function, it is often convenient to work with the logarithm of the likelihood function. This approach is based on the observation that the likelihood function is a product of probabilities, and taking the logarithm transforms the product into a summation, which is often simpler to work with. The benefits of using the logarithmic transformation are demonstrated in the following examples.

Example 2.1: Normal Distribution The random variable X follows a normal distribution $N(\mu, \sigma^2)$, find the maximum likelihood estimate of μ.

Solution
The likelihood function of n observations of a normal random variable X is

$$L = \prod_{i=1}^{n} \frac{1}{\sqrt{2\pi}\sigma_X} e^{-\frac{1}{2}\left(\frac{x_i - \mu_X}{\sigma_X}\right)^2} = \left(\frac{1}{\sqrt{2\pi}\sigma_X}\right)^n \times e^{-\frac{1}{2\sigma_X^2}\sum_{i=1}^{n}(x_i - \mu_X)^2}$$

Logarithm:

$$l = \log L = n \log \frac{1}{\sqrt{2\pi}\sigma_X} - \frac{1}{2\sigma_X^2}\sum_{i=1}^{n}(X_i - \mu_X)^2$$

$$\frac{\partial l}{\partial \mu_X} = \frac{1}{2\sigma_X^2}\sum_{i=1}^{n}(x_i - \mu_X) = 0$$

or

$$\widehat{\mu}_X = \frac{1}{n}\sum_{i=1}^{n} x_i = \overline{X}$$

2.2.2 Non-probabilistic Methods

2.2.2.1 Probability Box

In the field of uncertainty analysis, the stochastic model plays an important role and has obtained a lot of research results [7]. However, it is always difficult to obtain enough test data to verify the probability density of random variables in pratical engineering problems, and the probability method may produce a large deviation in uncertainty modeling [8]. In addition, practical engineering problems often suffer from both aleatory uncertainty and epistemic uncertainty, requiring a model that can incorporate multiple existing uncertainty models [9]. In view of these needs, a hybrid model combining probability and interval is proposed, called probability box (P-box).

P-box is a representation that describes aleatory uncertainty and epistemic uncertainty. It defines the uncertainty of the model by using the upper and lower cumulative distribution functions, CDF, as bounds. Figure 2.4 illustrates a basic P-box case with an uncertain parameter X. The probability distribution of X consists of upper and lower bounds. For X, the bounds for its values less than 0 and greater than 30 overlap. The bounds, whatever their shape, are sufficient as long as they satisfy monotonicity and do not cross each other. In the P-box case, the aleatory uncertainty of the parameter X can be expressed in terms of the overall slant of the CDF curve, and its epistemic uncertainty can be expressed in terms of the breadth between the left and right edges of the P-box. The P-box is a method of constraining the mean and

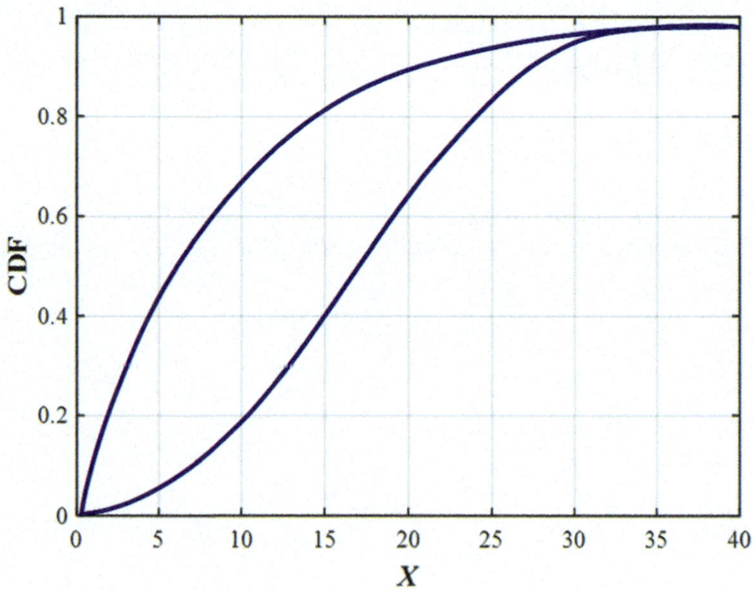

Fig. 2.4 Basic representation of P-box

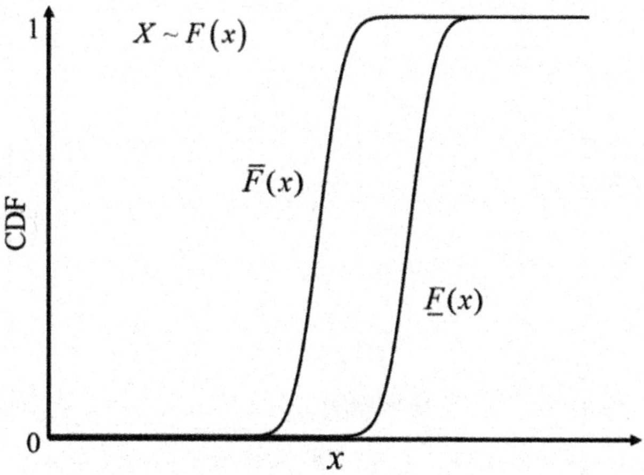

Fig. 2.5 A probability box defined by a normal distribution

variance of a variable to a specified interval specified by the left and right bounds of the distribution function of that variable and the type of distribution of the variable (monotonicity, symmetry, etc.).

The definition is as follows: Suppose that $\underline{F}(x)$ and $\overline{F}(x)$ are non-decreasing functions ranging over the real between the real numbers $\mathbb{R} \rightarrow [0, 1]$, and have $\underline{F}(x) \leq \overline{F}(x)$, $x \in \mathbb{R}$, and $\mathbb{R}$ is the real number field. Define $\left[\underline{F}, \overline{F}\right]$ as the set containing all non-decreasing functions $F \in [0, 1]$, with $\underline{F}(x) \leq F(x) \leq \overline{F}(x)$. When expressing an uncertain probability distribution in terms of $\underline{F}(x)$ and $\overline{F}(x)$, the area contained by $\underline{F}(x)$ and $\overline{F}(x)$ is called the P-box, namely: $F(x) = \left[\underline{F}(x), \overline{F}(x)\right]$. Its form is shown in Fig. 2.5. P-boxes are a flexible representation of intervals, probability distributions, and scalars. The relationship between these factors is represented by one coordinate system, as shown in Fig. 2.6.

In addition, P-box can be used for representation of multi-type input uncertainties. In practical engineering problems, uncertainties in mechanical system often have multi-types input uncertainties. Suppose the uncertain inputs of mechanical system consist of stochastic variables $\boldsymbol{X}$ and interval variables $\boldsymbol{Y}$. The output is variable Z. These vectors are defined as:

$$\boldsymbol{X} = [X_1, X_2, X_3, \cdots, X_m]^T \tag{2.4}$$

$$\boldsymbol{Y} = [Y_1, Y_2, Y_3, \cdots, Y_n]^T \tag{2.5}$$

The lower bound of P-box is $\underline{F}(Z)$ and the upper bound of P-box is $\overline{F}(Z)$. The process of P-box can be described with stochastic variables $\boldsymbol{X}$ and interval variables $\boldsymbol{Y}$.

$$\left[\underline{F}(Z), \overline{F}(Z)\right] = G(\boldsymbol{X}, \boldsymbol{Y}) \tag{2.6}$$

where $G(\boldsymbol{X}, \boldsymbol{Y})$ is the P-box of Z.

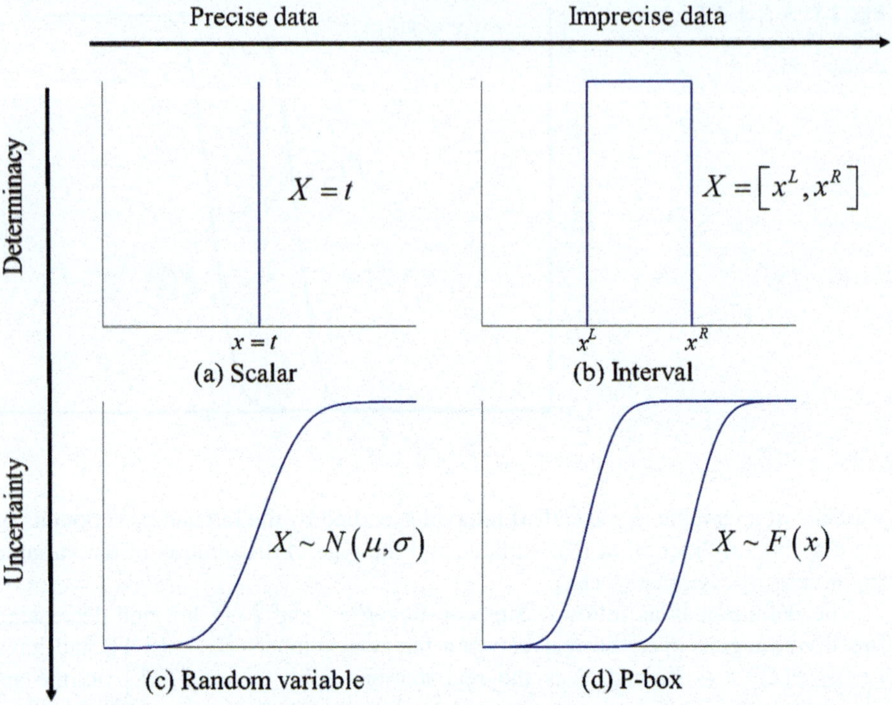

Fig. 2.6 The relationship between P-box and other factors about uncertainty

According to the results of the P-box, the expected value of Z can also be represented by an interval variable which can be obtained by

$$[\underline{E}(Z), \overline{E}(Z)] = \left[\int z \cdot \underline{P}(z)dz, \int z \cdot \overline{P}(z)dz\right] \tag{2.7}$$

where $\underline{P}(z)$ and $\overline{P}(z)$ are obtained by the derivatives of $\underline{F}(Z)$ and $\overline{F}(Z)$, respectively.

Example 2.2 Let $y{\sim}N(x, \sigma^2)$ and y is a probabilistic quantity that is normally distributed with imprecise parameter x as the mean and the known σ^2 as the variance. For example, $x = [50, 60]$ and $\sigma^2 = 5$. Express the uncertainty of y with the P-box method.

Solution

The lower and upper boundaries of this model, i.e., $\underline{F}(x)$ and $\overline{F}(x)$, are needed. First, the corresponding cumulative distribution function curves are calculated from the probability density function curves of $x^L{\sim}N(50, 5)$ and $x^R{\sim}N(60, 5)$.We get that the lower bound of P-box is $\underline{F}(x) = F(x^R)$ and the upper bound of P-box is $\overline{F}(x) = F(x^L)$. In this way, the uncertainty in y with the P-box is presented in Fig. 2.7a–c.

Fig. 2.7 The uncertainty of y with the P-box method

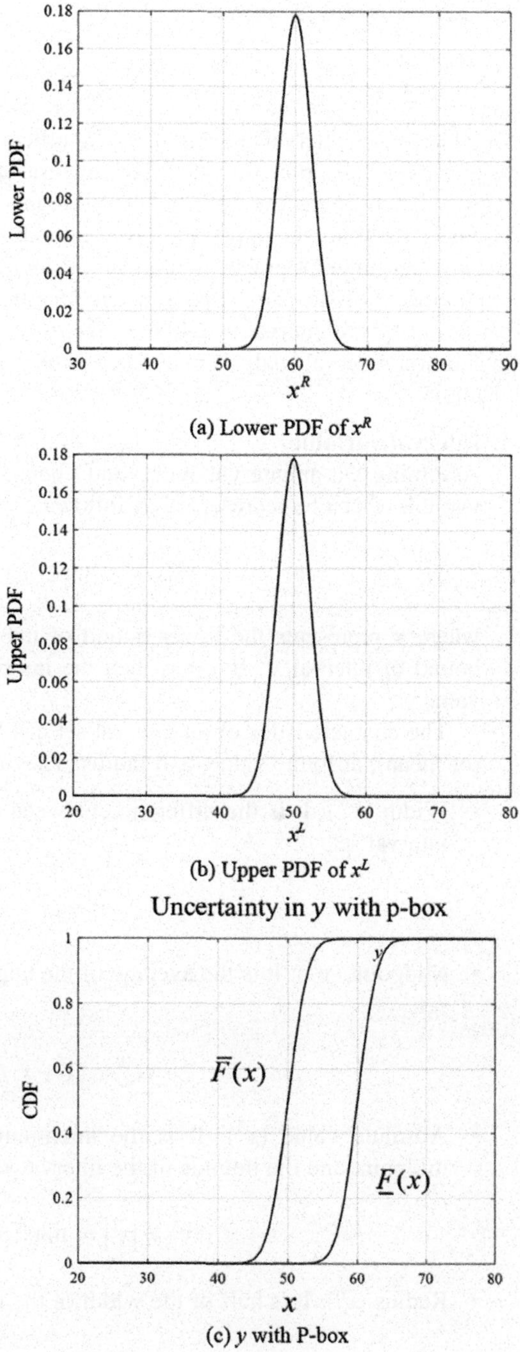

(a) Lower PDF of x^R

(b) Upper PDF of x^L

(c) y with P-box

2.2.2.2 Interval Analysis

Interval analysis is a technique used to represent uncertainty in models by defining intervals with upper and lower bounds for variable values. By replacing uncertain variables with intervals, the aim is to efficiently express and compute various responses of uncertain models. The interval analysis method [10, 11] does not require determining the specific distribution of uncertain sample data, but only involves obtaining the upper and lower bounds of the data. Interval analysis method has a wider range of practical applications and can theoretically meet the reliability requirements of structures operating in uncertain environments. In this section, the basic concepts of interval variable and the rules of interval operations are introduced. In addition, interval analysis methods will be explained through a specific numerical example.

1. **Interval variable.**
   Assuming X is an interval, with $\underline{x}$ and $\overline{x}$ being real numbers, and $\underline{x} < \overline{x}$, the interval variable x^I can be represented as follows:

$$x^I = [\underline{x}, \overline{x}] = \{x \in R | \underline{x} \leq x \leq \overline{x}\} \tag{2.8}$$

   where $\underline{x}$ represents the lower bound of interval x^I, and $\overline{x}$ represents the upper bound of interval x^I. If $\underline{x} = \overline{x}$, then the interval variable x^I represents a specific value.

   The characteristics of an interval set can be described by its width, midpoint (or mean), absolute value, and radius. They are represented as follows:

   - Width (x^W): It is the difference between the upper and lower bounds of the interval set.

$$x^W = \overline{x} - \underline{x} \tag{2.9}$$

   - Midpoint (x^M): It is the average of the upper and lower bounds of the interval set.

$$x^M = (\overline{x} + \underline{x})/2 \tag{2.10}$$

   - Absolute value (x^A): It is the maximum absolute difference between the midpoint and the bounds of the interval set.

$$x^A = |x^I| = \max(|\overline{x}|, |\underline{x}|) \tag{2.11}$$

   - Radius (x^R): It is half of the width of the interval set.

$$x^R = (\overline{x} - \underline{x})/2 \tag{2.12}$$

If there is a situation where the interval x^I exists such that $\underline{x} = -\overline{x}$, then in this case, the interval represents a symmetric interval. For any interval x^I, if there exists an interval vector $[\boldsymbol{x}]$ containing an ordered array of n intervals, it can be represented as follows:

$$[x] = ([x_1], [x_2], \cdots, [x_n]) \tag{2.13}$$

Similarly, the width $[\boldsymbol{x}]^W$, midpoint $[\boldsymbol{x}]^M$, and norm $\|[\boldsymbol{x}]\|$ of an interval can be represented as follows:

$$[x]^W = \max_{1 \le i \le n}\left(x_i^W\right) \tag{2.14}$$

$$[x]^M = \left([x_1]^M, [x_2]^M, \cdots, [x_n]^M\right) \tag{2.15}$$

$$\|[x]\| = \max_{1 \le i \le n}(|x_i|) \tag{2.16}$$

2. **Interval operations.**
For any intervals x^I, y^I, under the assumption of independence, the basic arithmetic rules are as follows:

$$x^I + y^I = \left[\underline{x} + \underline{y}, \overline{x} + \overline{y}\right] \tag{2.17}$$

$$x^I - y^I = \left[\underline{x} - \underline{y}, \overline{x} - \overline{y}\right] \tag{2.18}$$

$$x^I \times y^I = \left[\min\left(\underline{x} \bullet \underline{y}, \underline{x} \bullet \overline{y}, \overline{x} \bullet \underline{y}, \overline{x} \bullet \overline{y}\right), \max\left(\underline{x} \bullet \underline{y}, \underline{x} \bullet \overline{y}, \overline{x} \bullet \underline{y}, \overline{x} \bullet \overline{y}\right)\right] \tag{2.19}$$

$$x^I \div y^I = [\underline{x}, \overline{x}] \bullet \left[\frac{1}{\underline{y}}, \frac{1}{\overline{y}}\right] \tag{2.20}$$

In addition to the basic arithmetic operations mentioned above, the associative and commutative properties can also be applied to interval operations, and they are applicable to all interval operations.

Let x^I, y^I, and z^I be three interval variables. The associative property of interval operations is as follows:

$$x^I + \left(y^I + z^I\right) = \left(x^I + y^I\right) + z^I \tag{2.21}$$

$$x^I \times \left(y^I \times z^I\right) = \left(x^I \times y^I\right) \times z^I \tag{2.22}$$

The commutative property of interval operations is as follows:

$$x^I + y^I = y^I + x^I \tag{2.23}$$

$$x^I \times y^I = y^I \times x^I \tag{2.24}$$

The operation of an interval with a constant a is as follows:

$$x^I + a = [\underline{x} + a, \overline{x} + a] \tag{2.25}$$

$$x^I \times a = \begin{cases} [\underline{x} \times a, \overline{x} \times a], a \geq 0 \\ [\overline{x} \times a, \underline{x} \times a], a < 0 \end{cases} \tag{2.26}$$

Example 2.3 Consider an input $x^I = [2, 5]$, and solve for the distribution of the output $y = ax + b$, where $a = [0.8, 1.2]$, $b = [-1, 1]$ are constants.

Solution

Step 1: Within different variable intervals, randomly generate 10,000 samples using the Monte Carlo simulation method for each interval.

Step 2: Calculate the output data y based on the generated sample data.

Step 3: Plot the distribution of the output data y, as shown in Fig. 2.8.

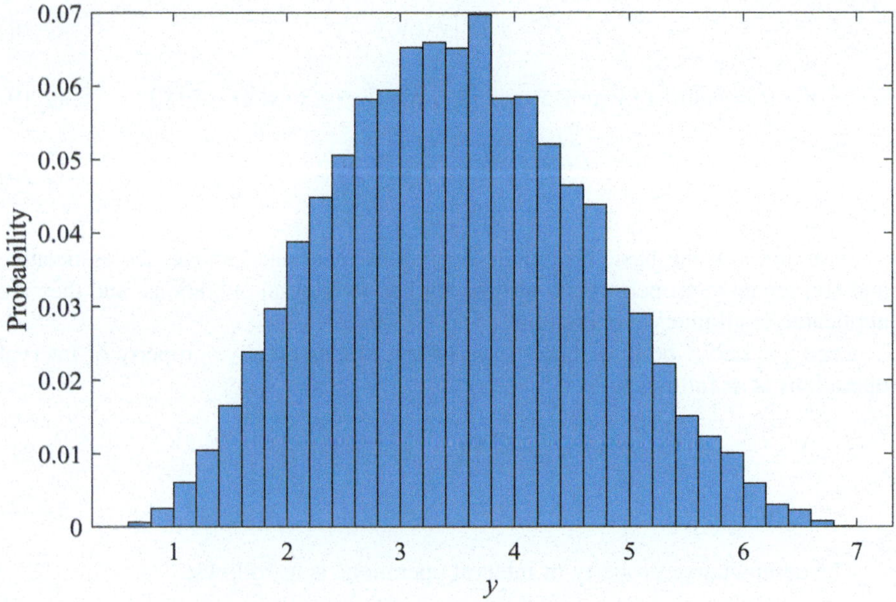

Fig. 2.8 The uncertainty of y with the interval analysis method

2.3 Uncertainty Propagation

DOUU requires the quantification of uncertainties in system output performances which propagate from uncertain inputs, often known as uncertainty propagation [12]. Simulation-based methods [13] for uncertainty propagation have been widely studied, and in this section the distribution function transformation method, the Monte Carlo simulation method, and evidence theory method will be used to describe how input uncertainty can be transformed into model-predicted uncertainty.

2.3.1 *Distribution Function Transformation Method*

In Chap. 1, the basic concepts of probabilistic, such as PDF $f_X(x)$ and CDF $F_X(x)$, have been introduced. However, in some experiments, the random variable of interest is often not directly measurable, but rather a function of another random variable that can be measured directly. For example, we can directly measure the diameter d of a circular section, but what we are interested in is the cross-sectional area S. Here, the random variable S is a function of the random variable d.

In this section, based on Sect. 1.2.3 of Chap. 1, the determination of the probability distribution of a function $Y = g(X)$ based on the known probability distribution of a random variable X will be discussed. Here, $g(\cdot)$ is a known continuous function that relates the values of X to the values of Y. Specifically, when a value x is taken by X, the value $g(x)$ is assumed by Y.

Example 2.4 For a linear function, let's assume that the random variable X has the following probability density function. To determine the probability density function of the random variable Y.

$$f_X(x) = \begin{cases} \frac{x}{8}, 0 < x < 4 \\ 0, \text{ otherwise} \end{cases}$$

$$Y = 2X + 8$$

Solution
First, define and solve for the distribution function of Y, $F_Y(y)$.

$$\begin{aligned} F_Y(y) &= P\{Y \leq y\} = P\{2X + 8 \leq y\} \\ &= P\left\{X \leq \frac{y-8}{2}\right\} = F_X\left(\frac{y-8}{2}\right) \end{aligned}$$

Next, calculate the derivative of $F_Y(y)$ with respect to Y. Then the PDF $f_Y(y)$ is obtained as:

$$f_Y(y) = f_X\left(\frac{y-8}{2}\right) \times \left(\frac{y-8}{2}\right) = \begin{cases} \frac{1}{8}\left(\frac{y-8}{2}\right) \times \frac{1}{2}, 0 < \frac{y-8}{2} < 4 \\ 0, \text{otherwise} \end{cases}$$

$$= \begin{cases} \frac{y-8}{32}, 8 < y < 16 \\ 0, \text{otherwise} \end{cases}$$

Example 2.5 For the random variable X have PDF $f_X(x)$, where $-\infty < x < \infty$. Find the PDF of $Y = X^2$.

Solution
As with the previous example, define and solve for the distribution function of Y, $F_Y(y)$. Calculate the derivative of $F_Y(y)$ with respect to Y. Then the PDF $f_Y(y)$ is obtained as:

$$\begin{aligned} F_Y(y) &= P\{Y \le y\} = P\{X^2 \le y\} \\ &= P\{-\sqrt{y} \le X \le \sqrt{y}\} \\ &= F_X(\sqrt{y}) - F_X(-\sqrt{y}) \end{aligned}$$

$$f_Y(y) = \begin{cases} \frac{1}{2\sqrt{y}}\left[f_X(\sqrt{y}) + f_X(-\sqrt{y})\right], y > 0 \\ 0, \qquad y \le 0 \end{cases}$$

For example, assume X follow a normal distribution, $X \sim N(0,1)$. Then its PDF $f_X(x)$ is expressed as follows:

$$f_X(x) = \frac{1}{\sqrt{2\pi}} e^{-\frac{1}{2}x^2}, -\infty < x < \infty$$

The PDF $f_Y(y)$ is obtained as follows:

$$f_Y(y) = \begin{cases} \frac{\sqrt{y}}{\sqrt{2\pi}} e^{-\frac{y}{2}}, y > 0 \\ 0, \qquad y \le 0 \end{cases}$$

From these two examples, the universal theorem for the calculation of PDF $f_Y(y)$ can be concluded where $g(\cdot)$ is a strictly monotonic function.

Example 2.6 The random variable X follows a normal distribution $N(\mu, \sigma^2)$, find the PDF $f_Y(y)$ of $Y = aX + b$.

Solution
The PDF $f_X(x)$ of X is defined as follows:

$$f_X(x) = \frac{1}{\sqrt{2\pi}\sigma} e^{-\frac{(x-\mu)^2}{2\sigma^2}}, -\infty < x < \infty$$

Based on Eq. (1.12) in Sect. 1.2.3, it is necessary to solve for the inverse function $h(y)$.

$$x = h(y) = \frac{y-b}{a}, \quad h'(y) = \frac{1}{a}$$

From Eq. (1.12), the PDF $f_Y(y)$ of Y is

$$\begin{aligned} f_Y(y) &= \frac{1}{|a|} f_X\left(\frac{y-b}{a}\right) \\ &= \frac{1}{|a|}\frac{1}{\sqrt{2\pi}\sigma} e^{-\frac{\left(\frac{y-b}{a}-\mu\right)^2}{2\sigma^2}} \\ &= \frac{1}{|a|\sigma\sqrt{2\pi}} e^{-\frac{[y-(b+a\mu)]^2}{2(a\sigma)^2}}, \quad -\infty < y < \infty \end{aligned}$$

The calculation shows that $Y = aX + b$ also obeys a normal distribution, i.e., $Y \sim N(a\mu + b, (a\sigma)^2)$.

2.3.2 *Monte Carlo Simulation Method*

Monte Clarlo simulation method is a compuational algorithm that rely on repeated random sampling to obtain numerical results. In this section, the uncertainty propagation method based on Monte Carlo simulation (MCS) with sampling is explained in detail. This process entails generating multiple sets of samples with random values assigned to model parameters. It involves selecting an appropriate number of runs to accurately estimate the uncertainty in model outcomes. The specific steps are shown in Fig. 2.9.

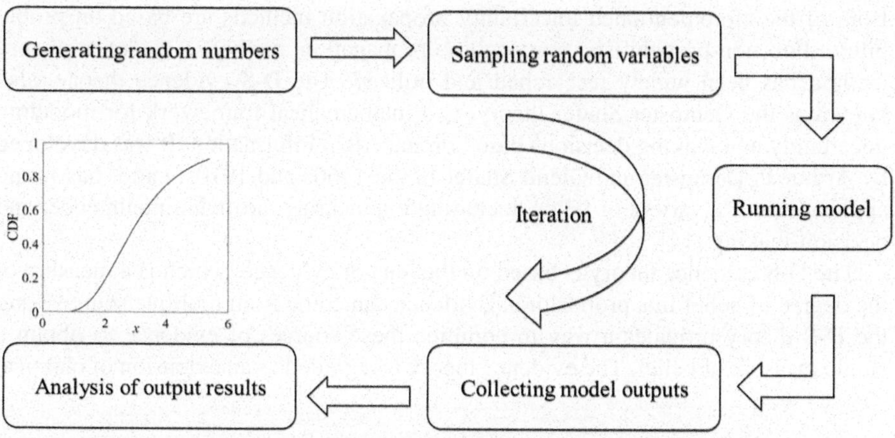

Fig. 2.9 Flowchart of the MCS and sampling method for uncertainty propagation

MCS is a widely used numerical method based on probability and statistical theory. It relies on repeated random sampling to obtain numerical approximations for analysis results that are difficult to solve analytically. Due to its intuitive nature and ease of implementation, MCS is widely applied in engineering simulations to propagate uncertainty. A detailed description of the MCS method is described in Chap. 5. In practical applications, to ensure the reliability of uncertainty quantification results, MCS typically requires a sufficiently large number of samples. However, simple random sampling, due to the need for a larger number of samples to achieve convergence of sampling results, increases the computational time cost. In this section, Latin hypercube sampling (LHS) is chosen as an alternative method to address the issue effectively. Similarly, the detailed description of the LHS method is described in Chap. 5.

Example 2.7: Uncertainty Propagation Based on MCS and LHS Methods For a variable $Y = X_1 + X_2$, where the random variables X_1 and X_2 follow the normal distribution $N(10, 1^2)$ and $N(5, 0.5^2)$, respectively. Calculate the probability distribution of the random variable Y by the MCS and LHS methods.

Solution

Step 1: Within the design space, a large number (about 10,000) of random samples of the MCS and LHS methods are generated and the probability distributions of input variables X_1 and X_2 are plotted, as shown in Fig. 2.10a–d.

Step 2: Using the sampled input data of X_1 and X_2, the output values of the random variable $Y = X_1 + X_2$ are calculated.

Step 3: Plot the probability distribution of the random variable Y to obtain the normal distribution parameters of Y, as shown in Fig. 2.11a, b.

2.3.3 *Evidence Theory*

Both of the aforementioned uncertainty propagation methods are based on probability. For non-probabilistic uncertainty propagation methods, evidence theory method has been widely researched and utilized. The D-S evidence theory, also known as the Dempster-Shafer theory, is a mathematical framework for modeling uncertainty and making decisions based on uncertain information. It was developed by Arthur P. Dempster and Glenn Shafer in the 1960s and 1970s, and it has found applications in a variety of fields, including engineering, artificial intelligence, and decision-making [14].

The D-S evidence theory is based on the idea of evidence, which is a measure of the degree of belief in a proposition. Evidence can come from multiple sources, and the D-S theory provides a way to combine these sources of evidence to obtain a single measure of belief. The evidence theory is regarded as an extension of classical

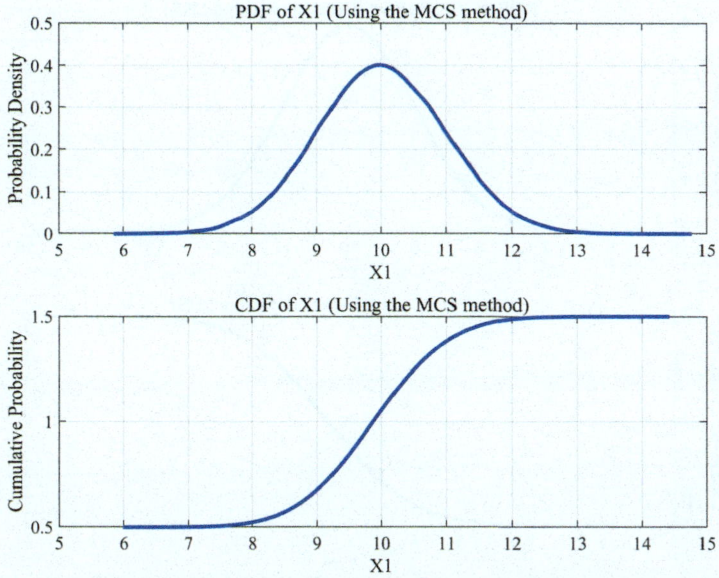

(a) the probability distributions and the cumulative probability distributions of input variables X_1 using the MCS method (100000 samples)

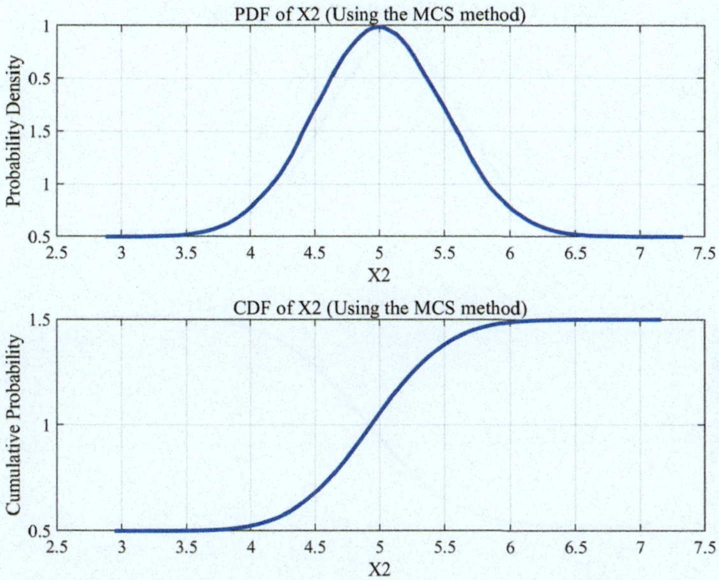

(b) the probability distributions and the cumulative probability distributions of input variables X_2 using the MCS method(100000 samples)

Fig. 2.10 Probability distributions of input variables

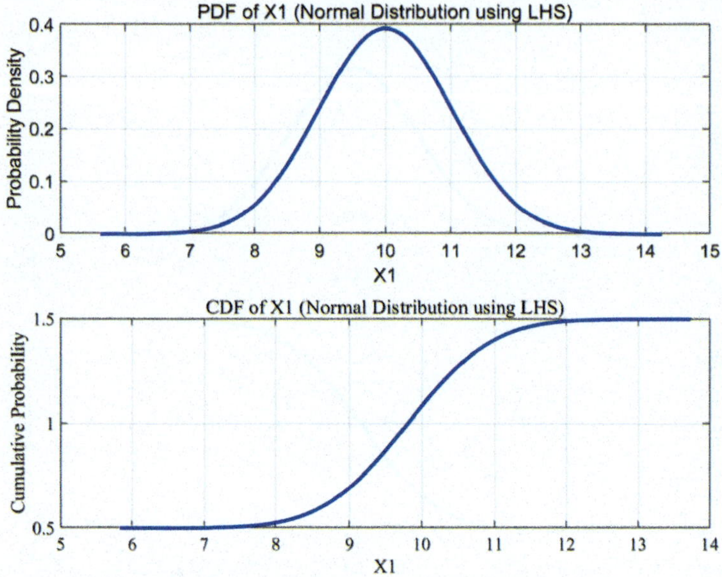

(c) the probability distributions and the cumulative probability distributions of input variables X_1 using the LHS method(10000 samples)

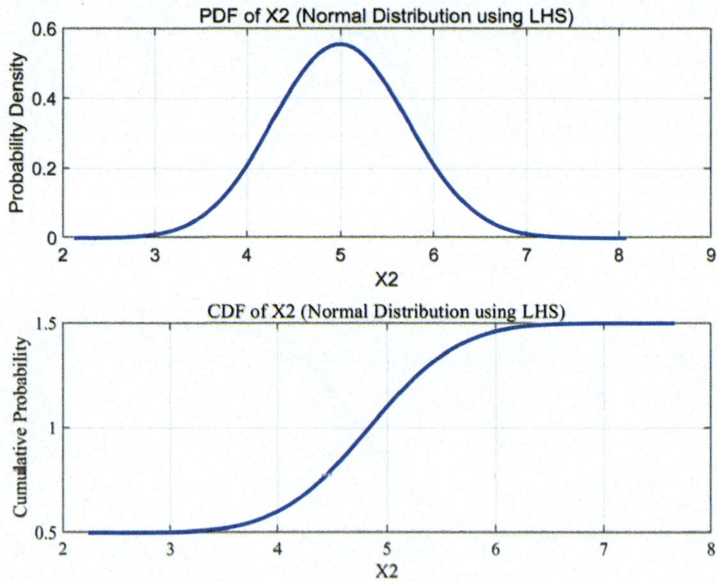

(d) the probability distributions and the cumulative probability distributions of input variables X_2 using the LHS method(10000 samples)

Fig. 2.10 (continued)

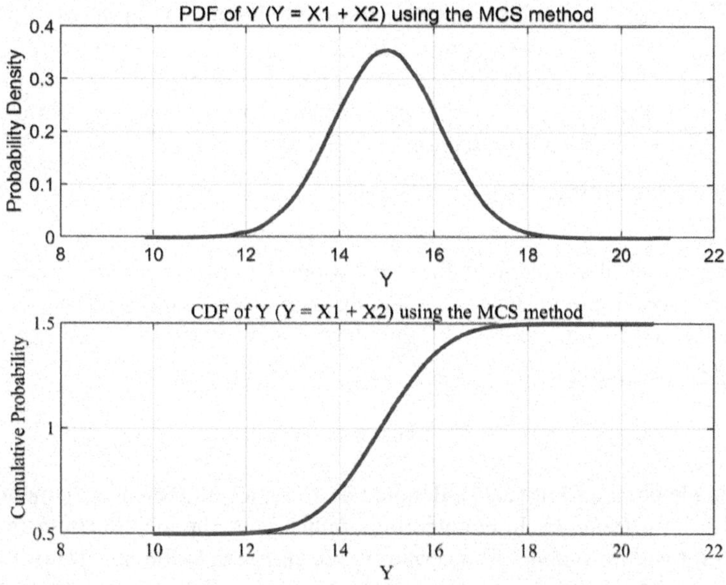

(a) the probability distributions and the cumulative probability distributions of output variable *Y* using the MCS method

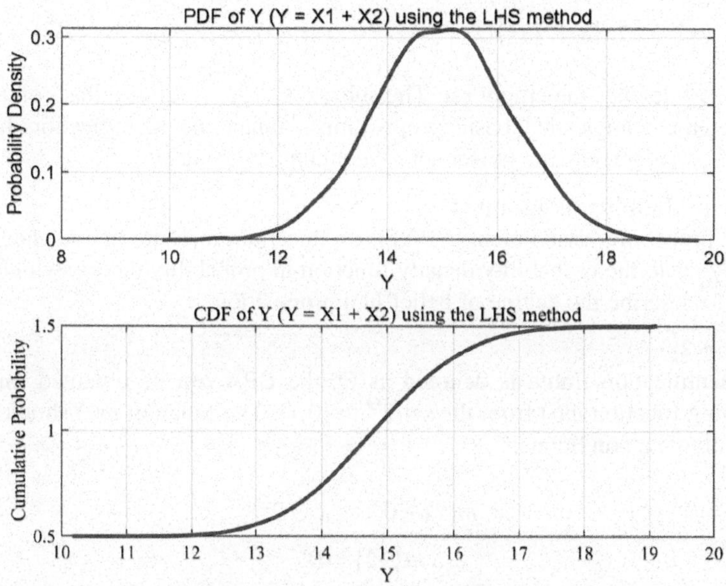

(b) the probability distributions and the cumulative probability distributions of output variable *Y* using the LHS method

Fig. 2.11 Probability distributions of output variable

probability theory, allowing for the allocation of belief to any set of possible events. The evidence theory constructs an uncertain modeling framework through three fundamental concepts, establishing a one-to-one correspondence between propositions and sets, thereby transforming the uncertainty of propositions into the uncertainty of sets. Its detailed introduction is as follows.

1. **Identification frame.**

Definition 1
The set of all possible outcomes for a given problem or event is referred to as the identification frame Θ. Θ is a collection of pairwise exclusive and exhaustive elements, and for finite sets only, the identification frame Θ can be expressed as follows:

$$\Theta = \{x_1, x_2, \cdots, x_n\} \tag{2.27}$$

The elements in Eq. (2.27), denoted as x_i, represent the basic elements of the identification frame Θ. In engineering applications, the identification frame can be interpreted as a range of uncertainty, which refers to the interval of uncertain parameters.

The power set of the identification frame Θ, denoted by 2^Θ, is a collection of all possible subsets of Θ. Specifically, it can be expressed as:

$$2^\Theta = \{\varnothing, \{x_1\}, \{x_2\}, \{x_3\}, \cdots, \{x_n\}, \{x_1, x_2\}, \{x_1, x_3\}, \{x_2, x_3\}, \cdots, \Omega\} \tag{2.28}$$

where $\varnothing$ denotes the empty set. The subsets x_1, x_2, ..., and x_n that contain only one element are called basic propositions. When the identification frame Θ contains n elements, the power set 2^Θ contains $2n$ elements.

2. **Basic probability assignment.**
Basic probability assignment (BPA) is an uncertain measure in evidence theory, analogous to the probability density function in probability theory, which can be used to describe the degree of belief in a proposition.

Definition 2
If the identification frame is denoted as Θ, the BPA can be obtained through a mapping function $m(\cdot)$ from the set 2^Θ to [0,1]. The function $m(\cdot)$ should satisfy the following conditions:

$$m(x) \geq 0, \quad x \in 2^\Theta \tag{2.29}$$

$$m\{\varnothing\} = 0 \tag{2.30}$$

$$\sum m(x) = 1, \quad x \in 2^\Theta \tag{2.31}$$

where x represents any subset of the power set 2^Θ, which can be interpreted as a proposition or event. $m(x)$ represents the degree of belief in proposition or event x, and it reflects the confidence level of relevant and feasible evidence in x. Propositions with $m(x) > 0$ are called focal elements. Equation (2.30) indicates that no degree of belief is assigned to the empty set (empty proposition), while Eq. (2.31) states that although any degree of belief can be assigned to a proposition, the sum of all degrees of belief assigned to all propositions must be equal to 1.

In practical engineering applications, the identification frame can be understood as an interval of uncertain parameters. Focal elements can be interpreted as small intervals within the identification frame, and the BPA is the degree of belief assigned to each small interval. Compared with probability theory, evidence theory's BPA satisfies weaker axioms, making it more intuitive, flexible, and effective for quantitatively describing the degree of belief in propositions in information-scarce situations. The flexibility in uncertainty measurement provided by evidence theory will be further illustrated through several common BPA structures shown in Fig. 2.12.

(a) **Independence of BPA.**
In the power set, the degree of belief in each proposition is obtained based on the known information and is not influenced by other propositions. For example, in Fig. 2.12, $m(\{x_4\}) = 0.1$ indicates that the degree of belief in the true value of parameter x_4 falling within the interval x_4 is 0.1. If there is evidence or an authoritative expert appointed to assign degrees of belief to propositions in the power set 2^{Θ}, as shown in Fig. 2.9a, b, and c, the degrees of belief in other propositions are determined by the known information and are independent of $m(\{x_4\})$.

(b) **Non-additivity of BPA.**
In evidence theory, additivity does not necessarily hold due to the uncertainty of information, which is present in probability theory. For example, in Fig. 2.12c, $m(\{x_1\}) = 0.1$, $m(\{x_2\}) = 0.3$, but $m(\{x_1,x_2\}) = 0.1$.

(c) **Non-monotonicity of BPA.**
In evidence theory, BPA does not satisfy monotonicity, which prevents the unnecessary assumption during the allocation of degrees of belief in the case of incomplete information. In probability theory, it is always the case that $p(\{x_1\}) \leq p(\{x_1,x_2\})$. However, in evidence theory, the degree of belief in proposition $\{x_1\}$ is determined by the known information and evidence, which cannot be directly used to support proposition $\{x_1,x_2\}$. Similarly, the evidence supporting $\{x_1,x_2\}$ cannot be allocated to the corresponding subset propositions $\{x_1\}$ or $\{x_2\}$. Therefore, situations may arise, such as in Fig. 2.12b, where $m(\{x_4\}) > m(\{x_3,x_4\})$.

3. **Belief and plausibility functions.**
Belief function and plausibility function are among the fundamental concepts in evidence theory, which are often used to describe the possibility of proposition validity through the probability interval constructed.

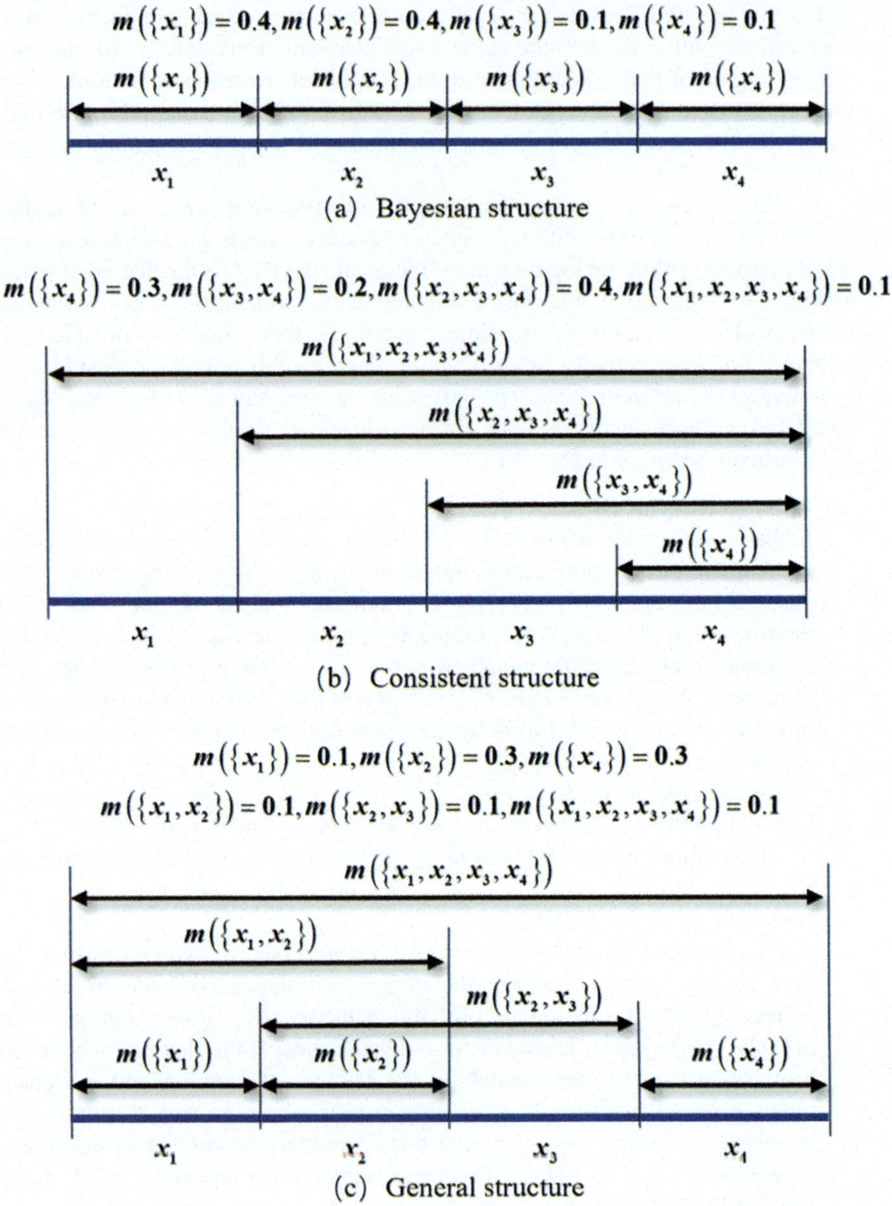

Fig. 2.12 Common BPA structures

Definition 3

Let Θ be the identification framework, and Belief function (*Bel*) be a mapping from 2^{Θ} to [0,1]. Then, the *Bel*(*A*) for proposition *A* is defined as:

$$Bel(A) = \sum_{x|x \subseteq A} m(x) \tag{2.32}$$

The *Bel*(*A*) describes the degree of belief in the truth of proposition *A*, and is obtained by summing up the probabilities of all subsets x that provide complete support for the validity of *A*. It is a lower bound of uncertainty measure.

Definition 4

In the identification framework where Θ is defined, the plausibility function (*Pl*) is also a mapping from 2^{Θ} to [0,1]. Then, the *Pl*(*A*) for proposition *A* is defined as follows:

$$Pl(A) = \sum_{x|x \cap A \neq \varnothing} m(x) \tag{2.33}$$

The *Pl*(*A*) describes the degree to which proposition *A* is not false, indicating the level of confidence in proposition *A*. It is obtained by summing up the probabilities of all subsets x that provide complete and partial support for the validity of *A*. *Pl*(*A*) is a more relaxed estimate than *Bel*(*A*) and can be viewed as an upper bound of uncertainty measure. The relationship between plausibility and belief functions is as follows:

$$Pl(A) = 1 - Bel(\overline{A}) \tag{2.34}$$

here, $\bar{A}$ represents the negation of proposition *A*.

The evidence theory constructs a general framework for uncertain reasoning, establishing a one-to-one correspondence between propositions and sets. Therefore, the uncertainty of propositions can be transformed into the uncertainty of sets through the evidence framework. In the field of structural engineering, the uncertain evaluation problem of system response can also be transformed into the corresponding uncertain problem of sets by constructing an evidence framework for solving. This section next describes the evidence-theoretic uncertainty propagation process through a two-dimensional system functional function.

Example 2.8 Consider a two-dimensional performance equation consisting of two random variables a and b, defined as follows:

$$y = g(a, b)$$

In the equation, $a \in [0, 2]$ and $b \in [-1, 1]$ are two independent evidential variables. Assuming that an authoritative expert has provided a BPA structure based on limited information as shown in Fig. 2.13, an evidence framework will be constructed using the BPA structure to evaluate the uncertainty characteristics of the response y.

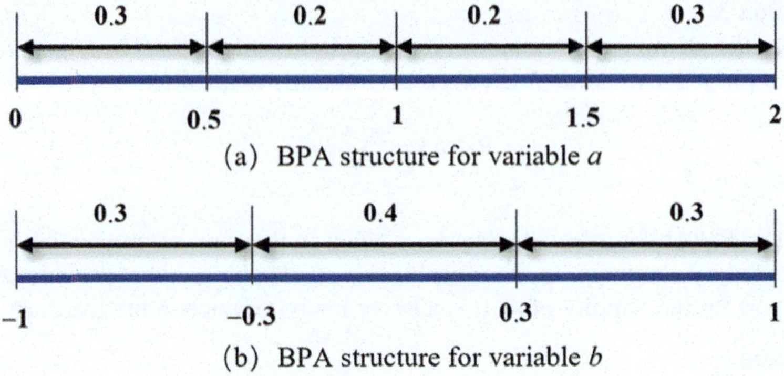

Fig. 2.13 BPA structures for variables *a* and *b*

Solution

Step 1: Constructing a joint BPA structure.

As the evidence variables *a* and *b* are mutually independent, we can define the joint recognition framework *D* for the above problem through the Cartesian product:

$$D = A \times B = \left\{ d^I(i,j) = a^I(i) \times b^I(j), a^I(i) \in A, b^I(i) \in B \right\} \tag{2.35}$$

where *A* and *B* are the recognition frameworks of the evidence variables *a* and *b*, $a^I(i)$ represents the *i*-th focal element of variable *a*, which is an interval, $b^I(j)$ represents the *j*-th focal element of variable *b*, and $d^I(ij)$ is a joint focal element within the joint framework *D*.

Correspondingly, based on the independence of the evidence variables, the BPA on the joint focal elements can be obtained as follows:

$$m(d^I(i,j)) = \begin{cases} m_a(a^I(i)) \cdot m_b(b^I(j)) & a^I \in A, b^I \in B \\ 0 & \text{other} \end{cases} \tag{2.36}$$

Then, the joint BPA structure for the above example is shown in Fig. 2.14. For example, the basic belief assignment of $m(d^I(2,2)) = 0.2 \times 0.04 = 0.08$ corresponds to the joint focal element $d^I(2,2) = a^I(2) \times b^I(2)$.

Step 2: Computing *Bel* and *Pl*.

For the functional function shown in Eq. (2.18), the event domain corresponding to a certain event *t* can be defined as follows:

$$G = \{(a,b) | g(a,b) \le g_0\} \tag{2.37}$$

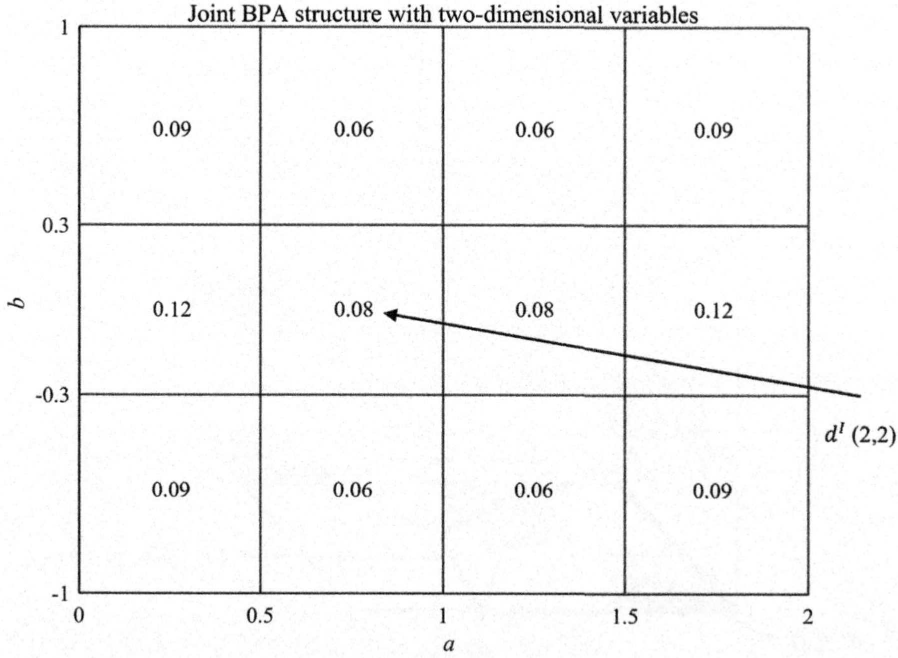

Fig. 2.14 Joint BPA structure for variables *a* and *b*

In the equation, g_0 is the threshold. After obtaining the joint focal elements and the corresponding basic BPA, an evidence uncertain inference framework is constructed, and the original problem can be transformed into a set membership judgment problem. Through the evidence framework, a one-to-one correspondence between propositions and sets is established, therefore the measurement of the original event is converted to the measurement of the corresponding event domain. In the context of uncertainty, both the event domain *G* and the joint focal elements can be viewed as sets. By determining the membership relationship between the focal elements and the event domain *G*, the *Bel* and *Pl* of the event under the threshold g_0 can be obtained.

$$Bel(G) = \sum_{d^I | d^I \subseteq G} m(d^I) \tag{2.38}$$

$$Pl(G) = \sum_{d^I | d^I \cap G \neq \varnothing} m(d^I) \tag{2.39}$$

If the joint focal element d^I is fully contained within the event domain *G*, then $d^I \subseteq G$, and the probability of this focal element should be included in $Bel(G)$ and $Pl(G)$. If the joint focal element d^I is fully or partially outside the event domain *G*,

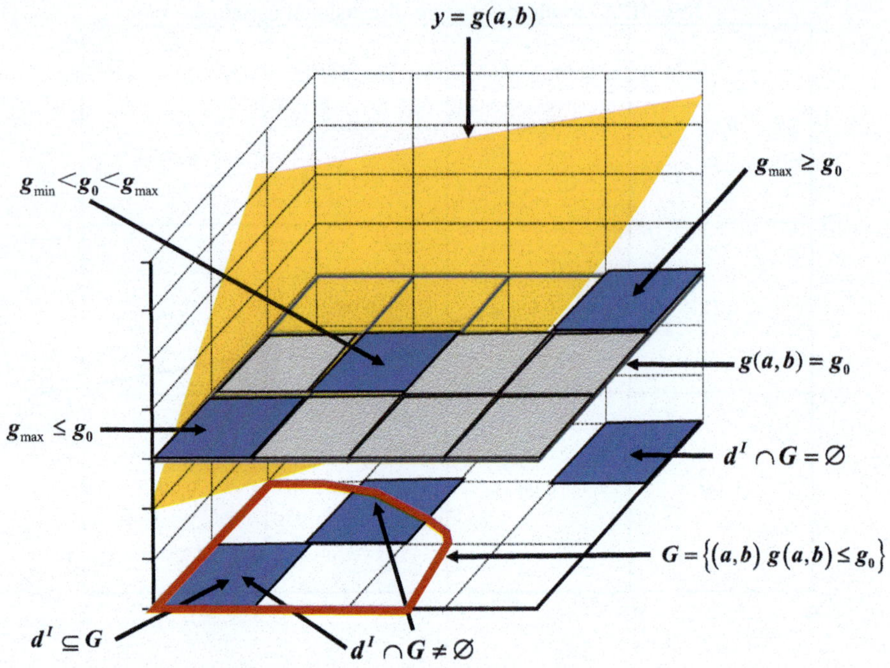

Fig. 2.15 Joint focal element attribution

then $d^I \cap G \neq \varnothing$, and the probability of this focal element should be included in $Pl(G)$. If the joint focal element d^I falls outside the event domain G, then $d^I \cap G = \varnothing$, and the probability of this focal element does not contribute to either $Bel(G)$ or $Pl(G)$.

As mentioned above, the key to determining $Bel(G)$ and $Pl(G)$ lies in effectively judging the relationship between the joint focal element d^I and the event domain G. By analyzing the extreme values of the system functional function within the joint focal element d^I, we can determine the positioning of the joint focal element as follows:

$$[g_{\min}, g_{\max}] = \left[\min_{a,\, b \in d^I} g(a,b),\ \max_{a,\, b \in d^I} g(a,b)\right] \tag{2.40}$$

As shown in Fig. 2.15, if $g_{\max} \leq g_0$ within the joint focal element, then the entire joint focal element falls within the event domain, that is, $d^I \subseteq G$, and the probability of this focal element should be included in $Bel(G)$ and $Pl(G)$. If $g_{\min} \leq g_0 \leq g_{\max}$, then only a part of the joint focal element is within the event domain, and the probability of this focal element should only be included in $Pl(G)$. If $g_{\min} \geq g_0$, it indicates that the focal element falls outside the event domain, and the corresponding probability does not contribute to either $Bel(G)$ or $Pl(G)$.

By analyzing the extreme values of the joint focal element, it is possible to obtain the extreme values of the system response on the joint focal element, and then determine the association between the focal element and the event domain, as well as obtain the corresponding *Bel* and *Pl* measures for the event. MCS sampling or optimization methods are generally used to obtain the extreme values of the focal element response. For cases with low nonlinearity, it is approximately assumed that the function within each joint focal element is monotonic, so the vertex method can be used to quickly obtain the extreme values of the response within the focal element. The analysis process above is carried out under the fixed threshold g_0. By changing the threshold g_0 in the sequence, the cumulative belief function (CBF) and cumulative plausibility function (CPF) of the system response can be obtained to characterize the uncertainty characteristics of the system response within the recognition framework.

Exercises

2.1 Briefly describe the sources and classification of different uncertainties, and explain different uncertainties with specific cases.

2.2 Briefly describe the key idea of the Maximum entropy distribution method to quantify uncertainty.

2.3 Let X be the life of a mechanical gearbox bearing that obeys an exponential distribution with the following probability density

$$f(x) = \begin{cases} \frac{1}{\theta}\exp[-(x-\mu)/\theta], & x \geq \mu \\ 0, & \text{elsewhere} \end{cases}$$

where θ, μ are unknown parameters, n pieces are randomly selected from this batch of parts for the life test, let their failure times be $x_1, x_2, \ldots, x_n$, respectively. Find the moment estimates of the two parameters θ, μ.

References

1. Aien, M., Hajebrahimi, A., & Fotuhi-Firuzabad, M. (2016). A comprehensive review on uncertainty modeling techniques in power system studies. *Renewable Sustainable Energy Reviews, 57*, 1077–1089.
2. Yao, W., Chen, X., Luo, W., Van Tooren, M., & Guo, J. (2011). Review of uncertainty-based multidisciplinary design optimization methods for aerospace vehicles. *Progress in Aerospace Sciences, 47*(6), 450–479.
3. Acar, E., Bayrak, G., Jung, Y., Lee, I., Ramu, P., & Ravichandran, S. S. (2021). Modeling, analysis, and optimization under uncertainties: a review. *Structural and Multidisciplinary Optimization, 64*(5), 2909–2945.

4. Cheng, J., Lu, W., Hu, W., Liu, Z., Zhang, Y., & Tan, J. (2019). Hybrid reliability-based design optimization of complex structures with random and interval uncertainties based on ASS-HRA. *IEEE Access, 7*, 87097–87109.
5. Haldar, A., & Mahadevan, S. (2000). *Probability, reliability, and statistical methods in engineering design*. Wiley.
6. Harr, M. E. (1987). Reliability-based design in civil engineering.
7. Ferson, S., & Ginzburg, L. R. (1996). Different methods are needed to propagate ignorance and variability. *Reliability Engineering System Safety, 54*(2–3), 133–144.
8. Faes, M. G., Daub, M., Marelli, S., Patelli, E., & Beer, M. (2021). Engineering analysis with probability boxes: A review on computational methods. *Structural Safety, 93*, 102092.
9. Peng, X., Ye, T., Hu, W., Li, J., Liu, Z., & Jiang, S. (2022). Construction of adaptive Kriging metamodel for failure probability estimation considering the uncertainties of distribution parameters. *Probabilistic Engineering Mechanics, 70*, 103353.
10. Chaturvedi, A., Prasad, K., & Ranjan, R. (2006). Use of interval arithmetic to incorporate the uncertainty of load demand for radial distribution system analysis. *IEEE Transactions on Power Delivery, 21*(2), 1019–1021.
11. Noghabi, A. S., Mashhadi, H. R., & Sadeh, J. (2010). Optimal coordination of directional overcurrent relays considering different network topologies using interval linear programming. *IEEE Transactions on Power Delivery, 25*(3), 1348–1354.
12. Lee, S. H., & Chen, W. (2009). A comparative study of uncertainty propagation methods for black-box-type problems. *Structural and Multidisciplinary Optimization, 37*(3), 239–253.
13. Madsen, H. O., Krenk, S., & Lind, N. C. (2006). *Methods of structural safety*. Courier Corporation.
14. Mourelatos, Z. P., & Zhou, J. (2005). *A design optimization method using evidence theory*. In Procedings of the international design engineering technical conferences and computers and information in engineering conference, pp 1153–1161.

Chapter 3
Surrogate Modeling

Nomenclature

$\mathbf{A}$	An initial matrix
$\boldsymbol{a}$	The regression coefficients vector
$\boldsymbol{a}_{ij}$	The member of the initial matrix $\mathbf{A}$
b	The value that minimize the sum of residuals
C	A set of Voronoi cells
$\boldsymbol{c}_i$	The i-th Voronoi cell
c_s	The most sensitive region
$\mathrm{CDM}(\cdot)$	A crowding distance metric
D	The set of training data
$\boldsymbol{Dens}_i$	The spatial density metric
$\boldsymbol{d}_{n_i}{}^{+}$	The distance of z_{n_i} to the ideal solution
$\boldsymbol{d}_{n_i}{}^{-}$	The distance of z_{n_i} to the negative ideal solution
$\boldsymbol{dom}(\boldsymbol{x}_i, \boldsymbol{x}_j)$	The closed half-plane bounded by the perpendicular bisector of $\boldsymbol{x}_i$ and $\boldsymbol{x}_j$
$\boldsymbol{d}_{sj}^{t}$	The Euclidean distance between the candidate point $\boldsymbol{x}_s^t$ and the existing sample $\boldsymbol{x}_{sj}$
$\boldsymbol{e}_i^{LOO}$	Leave-one-out cross-validation error at sample $\boldsymbol{x}_i$
e_j	The entropy of metrics
$\boldsymbol{e}_{st}^{LOO}$	The estimated error of the candidate sample $\boldsymbol{x}_s^t$
$f_{near}(\cdot)$	The output value predicted by the linear least squares fit model
$\widehat{f}(\cdot)$	The prediction function
$G(\cdot)$	An ordinary Kriging model
$Grad(\cdot)$	A local gradient estimation
k	The number of neighboring samples
$\boldsymbol{k}_i$	The number of random samples falling within the cell $\boldsymbol{c}_i$
M	A black-box function
m	The dimension of design variables

W. Hu, *Design Optimization Under Uncertainty*,
https://doi.org/10.1007/978-3-031-49208-2_3

N	The number of Monte Carlo simulation samples
N_D	The number of training points
$\boldsymbol{NonL_i}$	The nonlinear at cell $\boldsymbol{c_i}$
n	The number of input variables
$\boldsymbol{n_i}$	The number of samples that fall in the Voronoi cell $\boldsymbol{c_i}$
$\mathbf{P}$	The normalized matrix
$\boldsymbol{p_{ij}}$	The member of the normalized matrix $\mathbf{P}$
q	The Pareto optimal samples
$R(\cdot)$	A correlation function in Kriging method
S	The comprehensive evaluation score
s_{n_i}	The close degree
$s^2(\cdot)$	The estimated covariance function
$\boldsymbol{t}(\cdot)$	A linear fit of the points around $\boldsymbol{x_i}$
$U(\cdot)$	The learning function
$\boldsymbol{Var_{st}}$	The variance of the candidate sample
w	The weight value that minimize the sum of residuals
w_0	A scaling factor
w_{global}	The weight of exploration
w_{local}	The weight of exploitation
$\boldsymbol{w_i}$	The corresponding weight in SVR method
w_j	The weight of metrics
$\boldsymbol{X_0}$	An initial design of experiments
$\boldsymbol{X_i^{near}}$	The neighboring samples set
$\boldsymbol{X_s^{near}}$	The neighboring samples set of the sensitive sample $\boldsymbol{x_s}$
$\boldsymbol{x}$	The input variable
$\boldsymbol{x_{new}}$	The new sample
$\boldsymbol{x_s}$	The most sensitive sample
$\boldsymbol{x_{sj}}$	The existing sample
$\boldsymbol{x_s^t}$	The candidate sample
$\boldsymbol{Y_0}$	An output data
y	The real output response
$\overline{y}$	The mean of real output response
$\widehat{y}$	The predicted output response
$\mathbf{Z}$	The decision matrix
z_j^+	The ideal solution
z_j^-	The negative ideal solution
z_{nij}	The member of the decision matrix $\mathbf{Z}$
$\boldsymbol{\beta}$	The regression coefficients in Kriging method
$\boldsymbol{\delta}$	A user-defined tolerance parameter in SVR method
$\boldsymbol{\zeta}$	A vector in Kriging method
$\boldsymbol{\eta}(\boldsymbol{x})$	The stochastic process
λ	The weights in RBF method
$\boldsymbol{\mu}$	The base term in SVR method
$\boldsymbol{\rho}(\boldsymbol{x})$	A vector

σ^2	The process variance
$\phi(\cdot)$	The selected basis function in RBF method
ψ	The basis function in SVR method
$\|\cdot\|_2$	The Euclidean norm

3.1 Introduction

Surrogate modeling is a kind of approach used to approximate the characteristics of functions or systems based on the knowledge extracted from training samples. Common techniques such as the response surface model (RSM) [1], radial basis function (RBF) [2], support vector regression (SVR) [3], and Kriging model [4] have been extensively applied, which have their own characteristics and provide different model used for various applications (e.g., design optimization [5–7], reliability analysis [8, 9], and digital twin construction [10, 11]). Other state-of-the-art surrogate models, e.g., multi-fidelity surrogate model [12, 13], are often improving the prediction accuracy and efficiency based on the multi-level or hybrid utilization of these fundamental models. Thus, it is very important to understand the basics of these commonly used surrogate models. Recently, artificial neural network (ANN) becomes an emerging new type of surrogate models, which show promising prediction capability despite its expensive cost of ANN training. An advanced type of ANN, named physics-informed neural network (PINN), incorporates prior knowledges into the ANN construction and is utilized as a special surrogate model for design optimization under uncertainty. Details of PINN is explained in Chap. 9.

After determining the type of surrogate model, the initial data samples for building a surrogate model must be selected. This is usually called the sampling process, also referred to as the generation of design of experiments (DOE). Generally, the prediction quality of the surrogate model highly depends on the size and distribution of the given samples. The process of sampling points involves solve the governor equations, e.g., by computationally expensive simulation. As a result, obtaining the sampling results is a significant source of computational cost. However, by using appropriate sampling strategies, a high efficiency of the surrogate modeling can be achieved using acceptable sampling cost.

Sampling strategies can be generally divided into two categories: one-shot sampling and sequential sampling. The one-shot sampling strategy mainly considers the filling of the design space, and all samples are generated at one time such as the orthogonal array [14] and Latin hypercube sampling [15]. However, the main shortcoming is that a large sample size is required to explore the design space completely. To circumvent the limitations of one-shot sampling, sequential sampling (also called adaptive sampling) techniques have been proposed since the 1950s. In adaptive sampling, new samples are designed based on information extracted during previous iterations to place them in locations of high interest.

In this chapter, four surrogate model construction methods, RSM, RBF, SVR, and Kriging, are briefly introduced in Sect. 3.2. To reduce the computational cost required for surrogate modeling methods, evaluation metrics for global exploration and local exploitation of samples in surrogate model construction are proposed to select new samples accordingly in Sect. 3.3. Finally, in Sect. 3.4, a specific adaptive sampling method is applied to illustrate the overall process of surrogate model construction.

3.2 Surrogate Modeling Methods

Although there are various surrogate modeling methods, four commonly used ones are briefly explained: RSM, RBF, SVR, and Kriging.

3.2.1 Response Surface Method

To implement the RSM method, a surrogate model is constructed as a linear combination of polynomial functions of input variables, typically including linear and nonlinear terms. The order of RSM equals to the highest order of the polynomial consisting of input variables. Specifically, a first-order RSM can be expressed as follows:

$$\widehat{f}(\boldsymbol{x}, \boldsymbol{a}) = a_0 + a_1 x_1 + \cdots + a_n x_n \quad (3.1)$$

where the vector $\boldsymbol{a} = (a_0, \ldots, a_n)^T$ in the RSM method is determined by minimizing the sum of squared errors between the predicted values and the actual ones. Here, $\boldsymbol{x} = [\,x_1 \quad x_2 \quad \ldots \quad x_n\,]$ is the input variable, and n is the number of input dimensions.

Higher-order polynomials are sometimes employed in RSM, particularly when the problem exhibits a degree of nonlinearity. A second-order polynomial model is an example which can be written as follows:

$$\widehat{f}(\boldsymbol{x}, \boldsymbol{a}) = a_0 + \sum_{i=1}^{n} a_i x_i + \sum_{i=1}^{n} a_{ii} x_i^2 + \sum_{i=1}^{n-1} \sum_{j=i+1}^{n} a_{ij} x_i x_j \quad (3.2)$$

3.2.2 Radial Basis Function

RBF is a surrogate model that combines local univariate functions with weights determined by the distance measures from a reference point or specified center. Specifically, for n training points, an RBF surrogate can be expressed as:

$$\hat{f}(\boldsymbol{x}) = \sum_{i=1}^{n} \lambda_i \phi(\|\boldsymbol{x} - \boldsymbol{x}_i\|_2) \tag{3.3}$$

where the weights λ_i and $i = 1, \cdots, n$ in RBF are determined using the least-squares method. Here, $\phi(\cdot)$ denotes the selected basis function, and $\|\cdot\|_2$ represents the Euclidean norm.

3.2.3 Support Vector Regression

SVR employs the support vector machine (SVM) theory and utilizes basis functions similar to RBF. The corresponding formulation is expressed as follows:

$$\hat{f}(\boldsymbol{x}) = \mu + \sum_{i=1}^{n} w_i \psi(x, \boldsymbol{x}_i) \tag{3.4}$$

where the basis function ψ takes the i-th training point, x_i as input, and w_i represents its corresponding weight. The base term is denoted by μ.

These weights can be obtained by solving the following optimization problem:

$$\begin{cases} \min & \frac{1}{2}\|\mathbf{w}\|^2 \\ \text{s.t.} & \left|y_i - \hat{f}(\boldsymbol{x}_i)\right| \le \delta, i = 1, \ldots, n \end{cases} \tag{3.5}$$

SVR offers a user-defined tolerance parameter δ, which can be used to specify the acceptable error margin within the input data.

3.2.4 Kriging

Kriging is a commonly studied surrogate modeling method in engineering applications due to its strong interpolation capabilities. Based on Gaussian processes, it provides highly accurate predictions with zero error at each training point while also allowing for the estimation of predictive uncertainty. A typical Kriging model formulation is presented below [16]:

$$G(\boldsymbol{x}) = \boldsymbol{\rho}^T(\boldsymbol{x})\boldsymbol{\beta} + \eta(\boldsymbol{x}) \tag{3.6}$$

The Kriging model involves an n-dimensional input vector $\boldsymbol{x} \in \mathbb{R}^n$, where the user selects k basis functions represented by vector $\boldsymbol{\rho}(\boldsymbol{x}) = [\rho_1(\boldsymbol{x}), \cdots, \rho_k(\boldsymbol{x})]^T$, typically in polynomial form. The regression coefficients are given by $\boldsymbol{\beta} = [\beta_1, \cdots, \beta_k]^T$, and the response mean structure is represented by $\boldsymbol{\rho}^T(\boldsymbol{x})\boldsymbol{\beta}$. The covariance of the stochastic process $\eta(\boldsymbol{x})$, such as a Gaussian process, has zero mean and is expressed as follows:

$$Cov(\eta(\boldsymbol{x}_i),\ \eta(\boldsymbol{x}_j)) = \sigma^2 R(\zeta, \boldsymbol{x}_i,\ \boldsymbol{x}_j) \tag{3.7}$$

The covariance of the stochastic process, denoted by $\eta(\boldsymbol{x}_i)$ and $\eta(\boldsymbol{x}_j)$, is given by a correlation function $R(\boldsymbol{\zeta}, \boldsymbol{x}_i, \boldsymbol{x}_j)$ that depends on the inputs $\boldsymbol{x}_i$ and $\boldsymbol{x}_j$, as well as an unknown process variance σ^2 and correlation parameters included in vector $\boldsymbol{\zeta} = [\zeta_1, \cdots, \zeta_n]^T$ to be determined. A Gaussian correlation function takes the following form:

$$R(\zeta, \boldsymbol{x}_i,\ \boldsymbol{x}_j) = \prod_{k=1}^{n} \exp\left(-\zeta_k (x_{ik} - x_{jk})^2\right) \tag{3.8}$$

To fit a Kriging model, N_D training points $x_D = [x_1, \cdots, x_{N_D}]^T$ and their output responses $\boldsymbol{y}_D = [f(\boldsymbol{x}_1), \cdots, f(\boldsymbol{x}_{N_D})]^T$ are used. Maximum likelihood estimation (MLE) is commonly employed to estimate the regression coefficients $\boldsymbol{\beta}$ and the process variance σ^2. The parameters can be estimated by setting the gradients of the log-likelihood function with respect to $\boldsymbol{\beta}$ and σ^2 to zero [17]:

$$\widehat{\boldsymbol{\beta}} = \left(\boldsymbol{\rho}_D^T \mathbf{R}^{-1} \boldsymbol{\rho}_D\right)^{-1} \boldsymbol{\rho}_D^T \mathbf{R}^{-1} \mathbf{y}_D \tag{3.9}$$

$$\widehat{\sigma}^2 = \frac{1}{N_D}\left(\mathbf{y}_D - \boldsymbol{\rho}_D \widehat{\boldsymbol{\beta}}\right)^T \mathbf{R}^{-1} \left(\mathbf{y}_D - \boldsymbol{\rho}_D \widehat{\boldsymbol{\beta}}\right) \tag{3.10}$$

where $\boldsymbol{\rho}_D = [\boldsymbol{\rho}(\boldsymbol{x}_1), \cdots, \boldsymbol{\rho}(\boldsymbol{x}_{N_D})]^T$ is an $N_D \times k$ regression matrix; and $\mathbf{R}$ is a matrix consisting of the correlation functions $R(\boldsymbol{\zeta}, \boldsymbol{x}_i, \boldsymbol{x}_j)$.

Besides, the estimation $\widehat{\boldsymbol{\zeta}}$ of the optimal correlation parameters can be obtained by solving the following optimization problem

$$\widehat{\zeta} = \arg\max_{\zeta} \{\psi(\zeta)\} = -\frac{N_D}{2}\ln(\widehat{\sigma}^2) - \frac{1}{2}\ln(|\mathbf{R}|) \tag{3.11}$$

Once the model parameter estimates have been obtained, it is possible to calculate the expected value $\widehat{f}(\boldsymbol{x}_0)$ for an unobserved point $\boldsymbol{x}_0$ using the Kriging predictor.

$$\widehat{f}(\boldsymbol{x}_0) = \boldsymbol{\rho}^T(\boldsymbol{x}_0)\widehat{\boldsymbol{\beta}} + \mathbf{r}_0^T \mathbf{R}^{-1}\left(\mathbf{y}_D - \boldsymbol{\rho}_D \widehat{\boldsymbol{\beta}}\right), \tag{3.12}$$

The correlations between an unobserved point $\boldsymbol{x}_0$ and the N_D sampled points $\boldsymbol{x}_D = [\boldsymbol{x}_1, \cdots, \boldsymbol{x}_{N_D}]^T$ are represented by $\boldsymbol{r}_0 = [R(\boldsymbol{\zeta}, \boldsymbol{x}_1, \boldsymbol{x}_0), \ldots, R(\boldsymbol{\zeta}, \boldsymbol{x}_{N_D}, \boldsymbol{x}_0)]^T$. One of the distinctive features of the Kriging model is its ability to provide prediction variance for any unobserved point. The prediction variance can be calculated using the estimated covariance function, which takes into account the uncertainty in both the model and the input data.

$$s^2(\boldsymbol{x}_0) = \widehat{\sigma}^2\left[1 - \mathbf{r}_0^T\mathbf{R}^{-1}\mathbf{r}_0 + \left(\boldsymbol{\rho}_D\mathbf{R}^{-1}\mathbf{r}_0 - \boldsymbol{\rho}(\boldsymbol{x}_0)\right)^T\left(\boldsymbol{\rho}_D^T\mathbf{R}^{-1}\boldsymbol{\rho}_D\right)^{-1}\left(\boldsymbol{\rho}_D\mathbf{R}^{-1}\mathbf{r}_0 - \boldsymbol{\rho}(\boldsymbol{x}_0)\right)\right] \tag{3.13}$$

3.2.5 *Performance Evaluation of Surrogate Models*

The goal of performance evaluation is to assess the prediction accuracy and reliability of a surrogate model. Commonly used metrics for evaluating surrogate model performance include root mean square error (RMSE), mean absolute error (MAE), mean relative error (MRE), and coefficient of determination (R^2). The formulas of these metrics are shown in Table 3.1, where y_i represents real value, $\widehat{y}_i$ denotes predicted value, $\overline{y}$ is the mean of real values, and N is the number of samples.

It should be noted that the performance evaluation of surrogate models should not only consider their prediction accuracy, but also take into account their applicability range, computational efficiency, robustness, and interpretability. Only by comprehensively considering different metrics and selecting suitable surrogate models based on specific problems, can efficient, accurate, and reliable surrogate models be constructed.

Example 3.1 Use the four surrogate modeling methods to fit the following two functions expressed in Eqs. (3.14 and 3.15), as shown in Fig. 3.1. For Function 1, suppose there are 11 samples uniformly distributed in the design space, i.e., $x = [-1, -0.7, \ldots, 2]$; for Function 2 suppose there are 11 samples uniformly distributed in the design space, i.e., $x = [0, 0.1, \ldots, 1]$.

$$f(x) = (x-1)^2, \quad x \in [-1, 2] \tag{3.14}$$

Table 3.1 Performance evaluation metrics and formulas for surrogate models

Metric	Formula	Value	Note
RMSE	$\sqrt{\frac{1}{N}\sum_{i=1}^{N}(y_i - \widehat{y}_i)^2}$	$[0, +\infty)$	Best possible score is 0.0, smaller value is better.
MAE	$\frac{1}{N}\sum_{i=1}^{N}\lvert y_i - \widehat{y}_i\rvert$	$[0, +\infty)$	Best possible score is 0.0, smaller value is better.
MRE	$\frac{1}{N}\sum_{i=1}^{N}\frac{\lvert y_i - \widehat{y}_i\rvert}{\lvert y_i\rvert}$	$[0, +\infty)$	Best possible score is 0.0, smaller value is better.
R^2	$1 - \frac{\sum_{i=1}^{N}(y_i - \widehat{y}_i)^2}{\sum_{i=1}^{N}(y_i - \overline{y})^2}$	$(-\infty, 1]$	Best possible score is 1.0, bigger value is better.

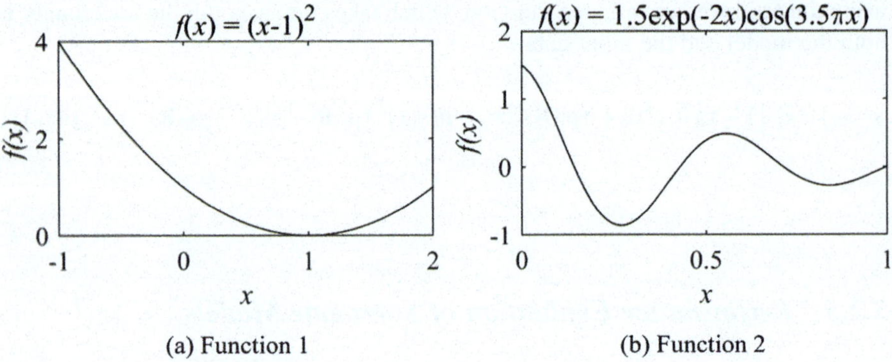

Fig. 3.1 Two numerical functions

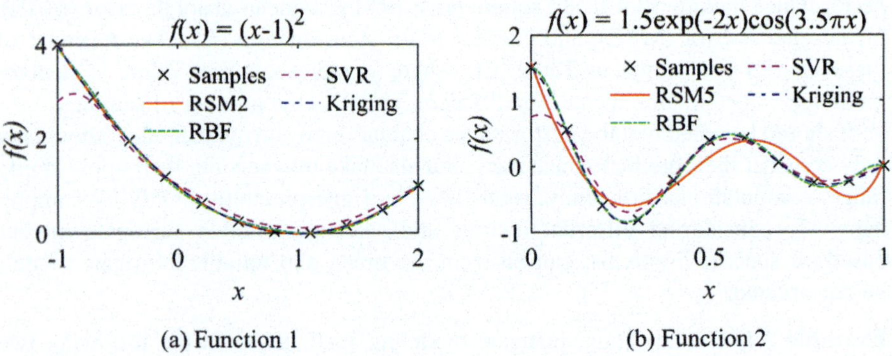

Fig. 3.2 The fitting results for two numerical functions under four surrogate models

$$f(x) = 1.5\exp(-2x)\cos(3.5\pi x), \quad x \in [0, 1] \tag{3.15}$$

Solution
Fit the two functions using the four surrogate models, the results are shown in Fig. 3.2, where the Gaussian kernel function is used in SVR. For Function 1, RSM is second order; for Function 2, the second order RSM is not enough to fit the function, so a fifth order RSM is used.

Example 3.2 Calculate the error and accuracy metrics of each surrogate model for the two functions, including root mean square error (RMSE) and determination coefficient (R^2).

Solution
The results are shown in Table 3.2.

This table presents the performance evaluation results of four commonly used surrogate models RSM, RBF, SVR, and Kriging on two test functions. The data

Table 3.2 Error and accuracy metrics of each surrogate model for Functions 1 and 2

Functions	Metric	RSM	RBF	SVR	Kriging
Function 1	RMSE	0.0000	0.0414	0.2205	0.0000
	R^2	1.0000	0.9986	0.9598	1.0000
Function 2	RMSE	0.1901	0.0318	0.1712	0.0033
	R^2	0.8768	0.9966	0.9002	1.0000

includes two evaluation metrics: root mean square error (RMSE) and coefficient of determination (R^2).

For Function 1, which has lower complexity, all four surrogate models show excellent performance. Specifically, both RSM and Kriging have very low RMSEs (0.0000), and RSM achieves an R^2 of 1.0000, indicating high prediction accuracy. On the other hand, SVR has a higher RMSE but still acceptable prediction accuracy, with an R^2 of 0.9598.

For Function 2, which is more complex and has two peaks, Kriging exhibits the best predictive performance, with an RMSE of only 0.0033 and an R^2 of 1.0000, the highest among the four surrogate models. Additionally, RBF also has a very low RMSE (0.0318) and perfectly fits the target function with an R^2 of 0.9966. In contrast, RSM's predictive performance is poor, with a higher RMSE (0.1901) and an R^2 of 0.8768.

Overall, the data in this table demonstrates that different surrogate models have different predictive performances on different problems. To choose the most suitable surrogate model, different evaluation metrics need to be considered in combination with the problem features and experimental results.

3.3 Adaptive Sampling Methods

It is expected that an efficient surrogate model requires as few samples as possible, in order to reduce computationally expensive simulations or experiments. From an initial design, an exact response is evaluated on design points, so information is available and employed to build the surrogate model. That model is intrinsically uncertain, due to a lack of knowledge which can be reduced by gaining more information. Thus, the current model is analyzed to identify where further experiments should be performed to benefit the most from supplementary information to reduce the lack of knowledge. In this loop process, points are gradually added in the under-sampling region until the surrogate model is sufficiently accurate.

3.3.1 Steps of Adaptive Sampling Methods

The general algorithm flowchart of an adaptive sampling method for global surrogate modeling is depicted in Fig. 3.3. Consider an initial design of

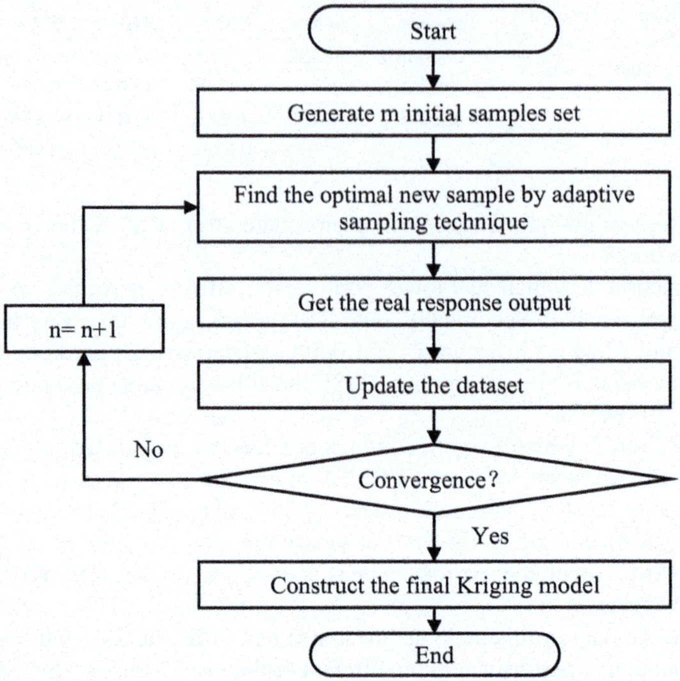

Fig. 3.3 General flowchart for building a surrogate model based on an adaptive sampling of experiments

experiments $\boldsymbol{X}_0 = [\, x_1 \quad x_2 \quad \ldots \quad x_n \,]$ associated with an output dataset as $\boldsymbol{Y}_0 = [\, y_1 \quad y_2 \quad \ldots \quad y_n \,]$. The creation of the surrogate model begins by fitting a black-box function M from this data. Then, supplementary samples are included in the set of training data D through successive iterations until reaching a convergence or stopping criterion.

To start the adaptive sampling procedure, an initial dataset is required for constructing the first surrogate model. Latin hypercube sampling (LHS) is a very classical technique for generating the initial dataset, which can effectively fill the design space with few samples. For m-dimensional design variables, LHS divides it into n mutually exclusive and exhaustive spaces with equal probability, as shown in the following formula:

$$\forall(i,k) \in [1,n]^2, \forall j \in [1,m], x_j^i \neq x_j^k, \text{ if } \; i \neq k \tag{3.16}$$

which means distinction and space-filling distribution of samples are ensured separately for every dimension. The basic idea of the Latin hypercube sampling (LHS) method is to partition the parameter space into equally probable sub-intervals and ensure that there is only one sample point within each sub-interval. This approach eliminates the correlation between parameter values and achieves a more uniform sampling result.

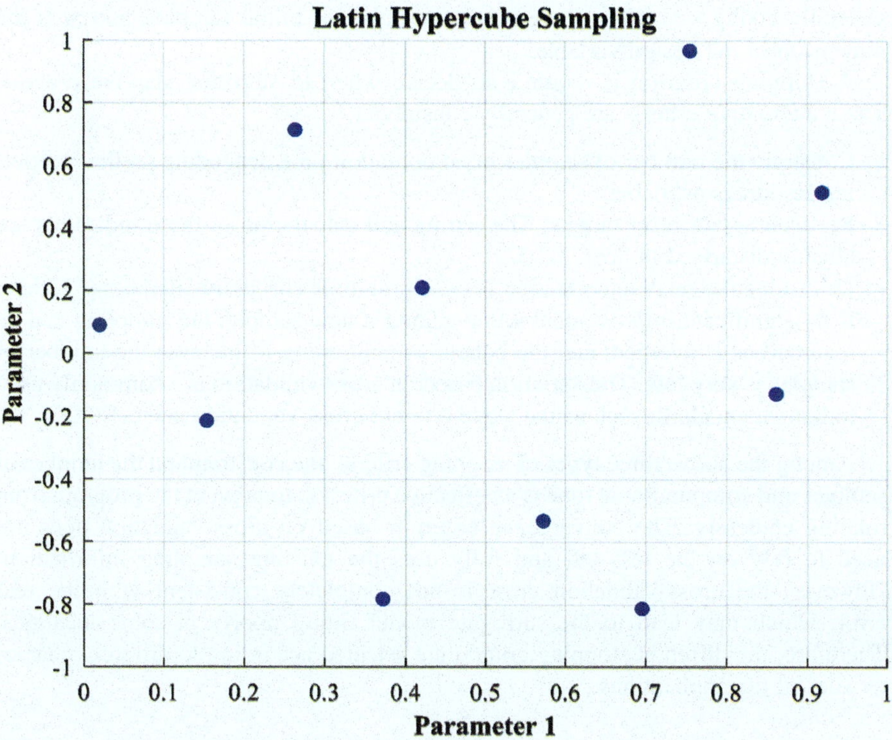

Fig. 3.4 LHS sampling results for the two-dimensional parameter space

Taking a two-dimensional parameter space as an example, suppose we want to perform LHS sampling on two parameters, generating 10 samples. The range of one parameter is [0, 1], while the range of the other parameter is [−1, 1]. Firstly, the range of each parameter is divided into equal sub-intervals, and then a random point is selected within each sub-interval as a sample, as shown in Fig. 3.4.

Since the LHS enforces differences between samples in all dimensions, the non-collapsing strategy is guaranteed [18]. In addition, the space-filling character of the initial design is particularly crucial when lack of priori knowledge of the mapping behavior. To enhance the space-filling character, it can be combined with maximizing the minimum distance criterion, i.e., Maximin LHS [19].

The number of samples included in the initial design also needs to be considered. A small number of initial samples leads to starting metamodels associated with a large lack of prior knowledge, which could mislead the first steps of the adaptive procedure. On the contrary, large numbers of initial designs may cause high computational costs due to numerous evaluations, and weaken the significance of adaptive sampling. Jones et al. [20] have given some suggestions on the relationship between the size of the initial sample set and the dimension of the design variables, generally taking $n = 10m$ where n is the number of initial samples, and m is the dimension of the design variable.

A stopping criterion is required to decide when to stop the adaptive process. Three alternative criteria are generally considered:

- Constrain the number of samples or computation time depending on the computing resources available.
- Set accuracy or error targets. The setting and calculation of these indicators are already discussed in Sect. 3.2.5.
- Determine the relative correction between multiple successive iterations. If there is no significant improvement while adding a new sample, the sampling can be regarded as converged, and the benefit of continuing to increase the number of samples is very low. The variation based on cross-validation is a commonly used judgment standard, such as the leave-one-out cross-validation error (LOOCV).

Among the above three types of stopping criteria, the constraint on the number of samples and computational time is an effective criterion to prevent the program from running endlessly. The convergence criterion based on cross-validation does not need to rely on the test set, and fully uses the existing sampling information. However, the cross-validation error is not completely consistent with the real error, which may lead to the surrogate model fitting poorly in some localities. Therefore, the different stopping criteria are selected according to different characteristics of the applications.

3.3.2 General Features of Adaptive Sampling Schemes

The samples are designed through an adaptive sampling strategy by solving an optimization problem. In the case of single-selection schemes, only one sample $\boldsymbol{x}_{n+1}$ is added per iteration, which is defined as the point that maximizes the refinement criterion (RC) given below:

$$\boldsymbol{x}_{n+1} = \arg \max_{\boldsymbol{x}^* \in X} RC(\boldsymbol{x}^*) \tag{3.17}$$

where the superscript * emphasizes the feasible solution of the optimization process.

Generally, there are two categories for adaptive sampling algorithms, namely the global exploration method and the local exploitation method. The global exploration method aims to comprehensively scan the entire design space and prevent local aggregation of samples. For example, Crombecq et al. [21] developed a type of global exploration method based on Monte Carlo simulation. Among the large number of candidate samples provided by Monte Carlo sampling. They calculate the minimum distance between the candidate sample and the existing sample, and select the candidate point sample that maximizes the minimum distance as the new sample, which ensures the good space filling and projection characteristics of the sampling point.

On the contrary, local exploitation makes full use of the output spatial features, finding the sub-regions that have the largest impact on the output, and improving the prediction accuracy of some parts of interest. For example, Jin et al. [22] proposed a sequential sampling method based on leave-one-out cross-validation, and the new samples focus on local areas with large leave-one-out errors. However, accurate estimates for some regions will inevitably increase prediction errors for others.

Moreover, it is challenging to select the most beneficial regions since the true model is unknown, as shown in Fig. 3.5. Consider the scenario depicted in Fig. 3.5, where an initial surrogate model is based on a dataset of seven samples shown in Fig. 3.5a. In order to explore and exploit the design space, an adaptive process is employed that stops after adding seven new observations. Using a local exploitation adaptive scheme, as shown in Fig. 3.5b, non-local behavior is identified and further investigated by adding supplementary samples near the initial outstanding sample. This approach allows for a precise description of the identified fluctuation, but fails to detect the second local nonlinearity of the true function. On the other hand, using global exploration as shown in Fig. 3.5c, the seven supplementary samples are evenly distributed across the design space, allowing for the identification of the second nonlinear region but providing less accuracy for both nonlinear local behaviors.

While exploration and exploitation appear to be opposing strategies for building an adaptive dataset, advanced adaptive procedures can be built by combining both methods to benefit from their features simultaneously. Hybrid adaptive sampling approaches can thus be developed by combining exploration and exploitation to yield a global refinement criterion, as follows:

$$RC(\boldsymbol{x}) = w_{\text{local}} local(\boldsymbol{x}) + w_{\text{global}} global(\boldsymbol{x}) \tag{3.18}$$

The literature proposes three general trade-off strategies for balancing exploration and exploitation, including the decreasing strategy and switch strategy. In the decreasing strategy, the weight of exploration w_{global} decreases while the weight of exploitation w_{local} increases. This enables the surrogate model to focus on exploration at the beginning and gradually shift its focus toward exploitation. On the other hand, in the switch strategy, the weights for local and global exploration are

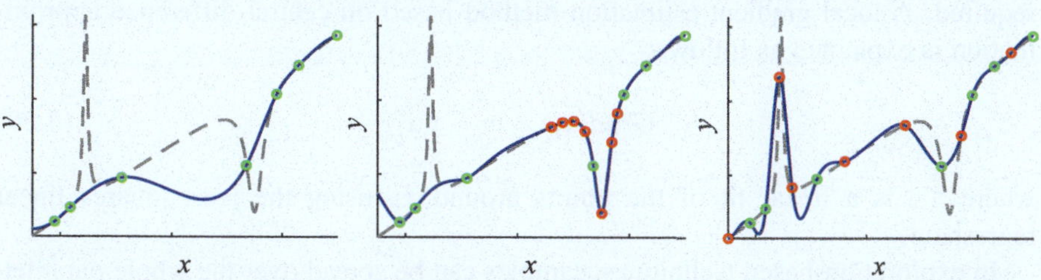

(a) Initial sampling (b) Exploitation strategy (c) Exploration strategy

Fig. 3.5 Local exploitation and global exploration strategy for adaptive sampling adapted from [23]

dynamically switched based on previous iterations. Greedy strategies are also employed in the switch method to estimate the weights and exploit information efficiently.

3.3.3 Techniques for Exploitation and Exploration

In exploitation-based techniques, samples are placed in areas of specific interest, and cross-validation error and gradient are commonly interesting information. In general, a larger cross-validation error or gradient at a certain region indicates that more information is contained there, and so more supplementary samples are required.

Cross-validation-based adaptive sampling is a strategy based on the analysis of the surrogate model accuracy with regard to unknown data. Algorithms generally rest on cross-validation error, including the k-fold cross-validation error, and leave-one-out cross-validation (LOOCV) error. Actually, the latter is a special case of the general k-fold cross-validation error.

The calculation process of the LOOCV error is shown below. Consider a surrogate model with n existing samples. For every $i = 1, 2, \ldots, n$, an auxiliary surrogate model $\widehat{M}_{-i}$ is trained on $n - 1$ observations consisting of the reduced set $D_{-i} = D \backslash (\boldsymbol{x}_i, y_i)$. Then, the cross-validation error e_i^{LOO} denoted at point $\boldsymbol{x}_i$ is calculated as follows:

$$\mathrm{e}_i^{LOO} = \left| y_i - \widehat{M}_{-i}(\boldsymbol{x}_i) \right|, \forall i \in [1, n] \tag{3.19}$$

where y_i is the real response value at sample $\boldsymbol{x}_i$ and \denotes the relative complement set between two sets. For two sets B, C, B\C is the objects that belong to set B and not to set C. To evaluate the local error for every sample, surrogate model needs to be reevaluated n times, which may be a numerically demanding task.

Gradient-based techniques are built around the premise that an accurate surrogate model requires less samples in subdomains corresponding with low gradient of the response surface than in subdomains with large gradient. For surrogate modeling, the gradient information is not directly available and so its numerical approximation is required. A local gradient estimation method based on central difference approximation is explained as follows:

$$Grad(\boldsymbol{x}_i) = |y_i - t(\boldsymbol{x}_i)| \tag{3.20}$$

where $t(\cdot)$ is a linear fit of the points around $\boldsymbol{x}_i$, using the least-squares linear regression.

In exploration-based techniques, samples can be spread over the whole parametric domain to unveil regions with high local nonlinearity. While the scale of current obversion is very small, the exploration feature is particularly important. Another interest of exploration is preventing local clustering of points, which will increase the sampling scale. There are conceptually two different tools to create an exploration component, including distance-based and variance-based exploration (if the surrogate model provides a prediction variance).

Distance-based exploration either generates a criterion to sparsely sample regions or sets restrictions. The preferred points have a large cumulative distance from existing samples. A crowding distance metric denoted CDM is defined by Garud et al. [24] as

$$\mathrm{CDM}(\boldsymbol{x}_i) = \sum_{j=1, j \neq i}^{n} \left\| \boldsymbol{x}_i - \boldsymbol{x}_j \right\|^2 \tag{3.21}$$

where n is the number of existing samples, and the notation $\|\cdot\|^2$ denotes the second-order norm operator.

The variance-based adaptive sampling approach is founded on the concept that large errors in the surrogate model approximation are likely to be concentrated in regions with higher predicted variances. This is because the variance can be easily computed as a byproduct of the Kriging surrogate model, and thus serves as an indicator of the proximity to potential areas of improvement for the surrogate.

3.4 An Effective Strategy for Adaptive Sampling

In this section, an efficient multi-metrics adaptive sampling approach for surrogate modeling based on Voronoi tessellation and dynamic weight trade-off strategy is illustrated. In the proposed method, the sampling space is partitioned into Voronoi cells according to existing samples. To evaluate the local exploitation and global exploration of sampling, three metrics (the cross-validation error, local nonlinearity, and crowding density) are calculated for each cell. Based on the entropy method, a dynamic weight decision strategy for the three metrics is developed to make a trade-off between the local exploitation and global exploration of sampling. Then, the most sensitive cell is identified based on the Technique for Order Preference by Similarity to Ideal Solution (TOPSIS) method. Finally, a new learning function based on the estimated leave-one-out error and variance is proposed to determine the new sampling point.

3.4.1 Voronoi Tessellation

The first step of the proposed methodology is to partition the entire design space in order to help designers to concentrate on local regions rather than the whole design space. Assume a set of points $\boldsymbol{X} = \{\boldsymbol{x}_1, \boldsymbol{x}_2, \cdots, \boldsymbol{x}_n\} \subseteq D$ represent the current samples. The Voronoi tessellation approach is employed to partition the entire design space into a set of Voronoi cells $C = \{c_1, c_2, \cdots, c_n\}$. The i-th Voronoi cell c_i represents the surrounding region of point $\boldsymbol{x}_i$, and the cell c_i can be defined as the following two formulas.

$$dom(\boldsymbol{x}_i, \boldsymbol{x}_j) = \left\{ \boldsymbol{x} \in D \middle| \|\boldsymbol{x} - \boldsymbol{x}_i\|_2 \le \|\boldsymbol{x} - \boldsymbol{x}_j\|_2 \right\} \tag{3.22}$$

$$c_i = \bigcap_{\boldsymbol{x}_j \in \boldsymbol{X} \backslash \boldsymbol{x}_i} dom(\boldsymbol{x}_i, \boldsymbol{x}_j) \tag{3.23}$$

where $dom(\boldsymbol{x}_i, \boldsymbol{x}_j)$ is a closed half-plane bounded by the perpendicular bisector of $\boldsymbol{x}_i$ and $\boldsymbol{x}_j$. The bisector separates all points of the plane closer to $\boldsymbol{x}_i$ from those closer to $\boldsymbol{x}_j$. As shown in Fig. 3.6, a 2D example partitioned by the Voronoi diagram approach is illustrated.

A Voronoi cell is a polyhedron with irregular boundaries in high dimensions, so it is difficult to describe the cell mathematically. Hence, the Monte Carlo method is adopted to generate a large number of random points to describe the Voronoi cells approximately. By computing the distances to all existing samples, each random point is assigned to its closest point. Thus, the Voronoi cell c_i can be approximately described by a set of random points, as

$$\boldsymbol{x}^i_{rand} = \left\{ \boldsymbol{x}^i_{r1}, \boldsymbol{x}^i_{r2}, \cdots \boldsymbol{x}^i_{rk_i} \right\} \tag{3.24}$$

where k_i is the number of random points falling within the cell c_i. If the total number of random points generated by Monte Carlo simulation is N, then $N = \sum_{i=1}^{n} k_i$. Generally, the more samples N generated by Monte Carlo simulation, the simulated Voronoi cell is closer to the real one. To avoid excessive consumption of computing resources, the total number of random points is generally determined by $N = m \times n \times w_0$, where m is the dimension of design variable and w is a scaling factor.

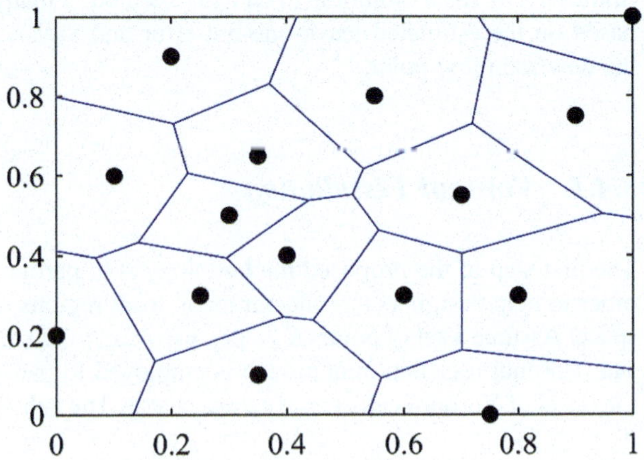

Fig. 3.6 A set of 2D samples and the corresponding Voronoi cells

3.4.2 Metrics for Evaluating the Existing Samples

Based on the partitioned Voronoi cells, three metrics are proposed to evaluate the local and global characteristics of the surrogate model respectively, including the cross-validation error, local nonlinearity, and crowding density.

Firstly, the leave-one-out cross-validation (LOOCV) error of each sample is calculated to evaluate the local prediction error of the surrogate model. The LOO error of point $\boldsymbol{x}_i$ can be calculated as

$$\mathrm{e}_i^{LOO} = \left| y_i - \widehat{M}_{-i}(\boldsymbol{x}_i) \right|, \forall i \in [1, n] \tag{3.25}$$

where y_i denotes the real response of $\boldsymbol{x}_i$, $\widehat{M}_{-i}(\boldsymbol{x}_i)$ denotes the predicted response of $\boldsymbol{x}_i$ by the surrogate model constructed on existing points without $\boldsymbol{x}_i$. The above process is repeated for all existing points by leaving them out one at a time and finally, the prediction errors of all existing samples will be obtained.

Secondly, a method of calculating the local nonlinearity of the sample is proposed to reflect the high-order change characteristics of the response. The algorithm is divided into two parts: (1) select the neighboring set of the sample based on Voronoi cells; (2) use the sample and its neighboring set to estimate the nonlinearity of the local region.

As shown in Fig.3.7, each Voronoi cell shares a common edge with its neighbors, thus the samples at the center of the neighbor Voronoi cells are regarded as neighboring points. According to the correspondence between Delaunay triangles and Voronoi cells, the neighboring point set of a certain sample is determined indirectly by finding the triangles that share vertices. As shown in Fig. 3.7, the neighboring points of sample A are determined as samples B, D, F, and G.

After determining the neighboring points of each existing sample, the linear least squares method is applied to fit the sample $\boldsymbol{x}_i$ and its neighboring points set $\boldsymbol{X}_i^{near} = \{\boldsymbol{x}_{i1}, \boldsymbol{x}_{i2}, \cdots \boldsymbol{x}_{ik}\}$. The nonlinear $NonL_i$ at local cell c_i is calculated by the sum of the residuals of the true value and the fitted value of these local points, as follows:

$$NonL_i = \sum_{j=0}^{k} \left| f_{near}(\boldsymbol{x}_{ij}) - y_{ij} \right| \tag{3.26}$$

where k is the number of neighboring points; when j equals 0, $\boldsymbol{x}_{ij}$ is the sample $\boldsymbol{x}_i$ itself; y_{ij} is the real response at sample $\boldsymbol{x}_{ij}$; $f_{near}(\boldsymbol{x}_{ij})$ is the output value predicted by the linear least squares fit model at sample $\boldsymbol{x}_{ij}$, and its calculation formula is as follows:

$$f_{near}(\boldsymbol{x}) = w \cdot \boldsymbol{x} + b \tag{3.27}$$

where w and b are the values that minimize the sum of residuals, as

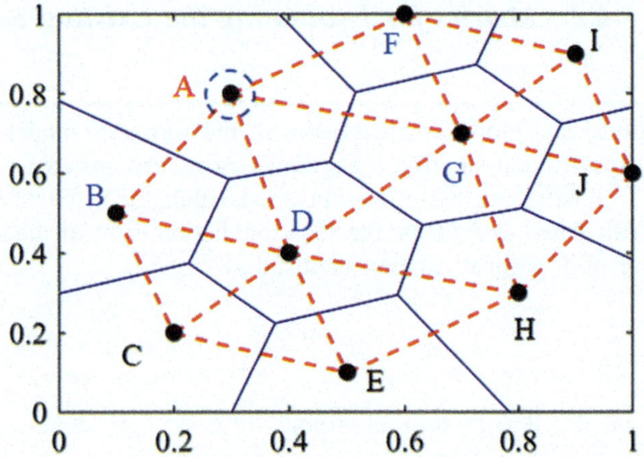

Fig. 3.7 Neighboring samples selection

$$\min\left(\sum_{j=0}^{k}\left|f_{near}(\boldsymbol{x}_{ij})-y_{ij}\right|\right) \tag{3.28}$$

The above process is repeated for all existing samples, and the local nonlinear at all samples can be obtained.

Both of the above two metrics reflect the local exploitation of the sampling process. If only considering these two metrics, the new samples that gradually added will appear in local aggregation. To avoid unreasonable local enrichment of samples, another metric considering global space exploration is proposed, as the spatial density of samples. The size of the Voronoi cell where each sample is located to represent the spatial density of the sample. Based on the Monte Carlo simulation, the spatial density $Dens_i$ at sample $\boldsymbol{x}_i$ is defined as the proportion of the number of samples falling in the Voronoi cell c_i, as:

$$Dens_i = n_i/N \tag{3.29}$$

where n_i is the number of samples that fall in the Voronoi cell c_i, and N is the total number of Monte Carlo simulation samples.

3.4.3 Identification of Sensitive Voronoi Cell

In the previous section, three metrics are proposed to evaluate the exploration and exploration of current samples, including leave-one-out cross-validation error, local nonlinearity, and spatial density of samples. Considering each metric separately, the larger the value, the more samples should be inserted in the Voronoi cell. Figure 3.8

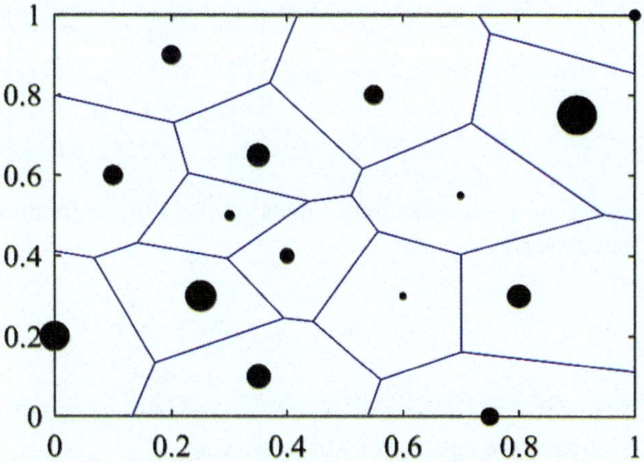

Fig. 3.8 Schematic diagram of an evaluation index for samples

shows a certain evaluation metric of each sample, where larger circles represent larger values.

However, the order of the evaluation metrics of samples is not consistent in the three dimensions. A certain sample is larger than another sample in some metrics, but the performance may be opposite in the remaining metrics. The Technique for Order Preference by Similarity to Ideal Solution (TOPSIS) based on entropy weight is a classic and effective multi-attribute decision-making method. Firstly, the entropy weight method is used to calculate the information entropy of each evaluation metric. After weighting, the weights of each index are assigned, and the weights of each index become skewed. Steps are as follows. Then, an improved TOPSIS method is applied to select the optimal solution (sensitive Voronoi cell). The specific steps are as follows:

Firstly, an initial matrix **A** is established, according to the calculated metrics of existing samples, as shown in the next matrix.

$$\mathbf{A} = \begin{bmatrix} a_{11} & a_{12} & \cdots & a_{1m} \\ a_{21} & a_{22} & \cdots & a_{2m} \\ \vdots & \vdots & a_{ij} & \vdots \\ a_{n1} & a_{n2} & \cdots & a_{nm} \end{bmatrix} \tag{3.30}$$

where n is the number of samples, m is the dimension of metrics, and m equals 3 for this method; a_{ij} is the initial of the j-th evaluation metric in the i-th sample. Among the matrix, the first column is the leave-one-out error (e^{LOO}), the second column is the local nonlinearity (*NonL*), and the third column is the spatial density (*Dens*). For example, a_{i1} is the leave-one-out error of the i-th sample.

The initial matrix **A** is normalized to matrix **P**, as:

$$\mathbf{P}=\begin{bmatrix} p_{11} & p_{12} & \cdots & p_{1m} \\ p_{21} & p_{22} & \cdots & p_{2m} \\ \vdots & \vdots & p_{ij} & \vdots \\ p_{n1} & p_{n2} & \cdots & p_{nm} \end{bmatrix} \tag{3.31}$$

where p_{ij} is the standard values of the j-th evaluation metric in the i-th sample, calculated by a_{ij}, as:

$$p_{ij}=\frac{a_{ij}}{\sum\limits_{i=1}^{n} a_{ij}} \tag{3.32}$$

Define the entropy of j-th metrics, as:

$$e_j=-\frac{1}{\ln(n)}\sum_{i=1}^{n} p_{ij}\ln\left(p_{ij}\right) \tag{3.33}$$

Since the smaller the entropy, the greater the weight, so the right of the j-th metrics is calculated as:

$$w_j=\frac{1-e_j}{\sum\limits_{j=1}^{m}\left(1-e_j\right)} \tag{3.34}$$

So far, each evaluation metric has been given a weight coefficient, and an improved TOPSIS method based on non-dominated sorting strategy is proposed to select the most sensitive sample and its Voronoi cell. The non-dominated sorting method is applied to select q Pareto optimal samples, as $P_{best}=\left[p_{n_1},\cdots,p_{n_q}\right]^T$, to reduce the selection range of sensitive cells, where $n_i\in[1,n],\ q\leq n$. As shown in Fig. 3.9, the candidate samples F, G, and H are abandoned to enter the competition for most sensitive sample.

The remain samples are used to establish the decision matrix $\mathbf{Z}$, as follows:

$$\mathbf{Z}=\begin{bmatrix} z_{n_1 1} & z_{n_1 2} & \cdots & z_{n_1 m} \\ z_{n_2 1} & z_{n_2 2} & \cdots & z_{n_2 m} \\ \vdots & \vdots & z_{n_i j} & \vdots \\ z_{n_q 1} & z_{n_q 2} & \cdots & z_{n_q m} \end{bmatrix} \tag{3.35}$$

where $z_{n_i j}$ is calculated as follows:

$$z_{n_i j}=w_j\cdot p_{n_i j} \tag{3.36}$$

Calculate the Euclidean distance of remain samples to the ideal solution and negative ideal solution, as follows:

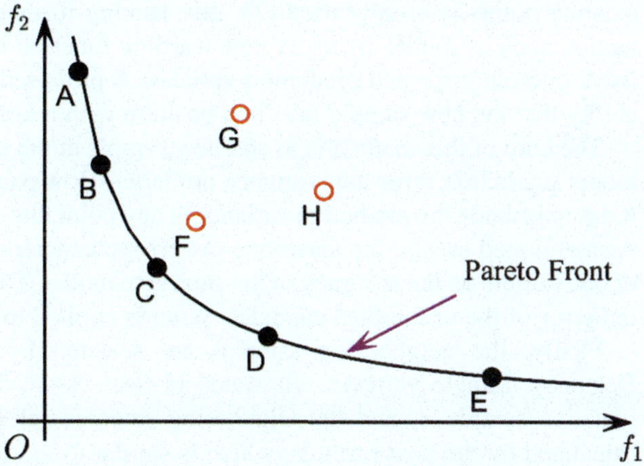

Fig. 3.9 Non-dominated sorting to select Pareto optimal samples

$$d_{n_i}{}^{+}=\sqrt{\sum_{j=1}^{m}\left(z_j{}^{+}-z_{n_ij}\right)^2},\ d_{n_i}{}^{-}=\sqrt{\sum_{j=1}^{m}\left(z_j{}^{-}-z_{n_ij}\right)^2} \tag{3.37}$$

where $d_{n_i}{}^{+}$ is the distance of remain sample metrics z_{n_i} to the ideal solution, and the $d_{n_i}{}^{-}$ is the distance of z_{n_i} to the negative ideal solution; the ideal solution $z_j{}^{+}$and negative ideal solution $z_j{}^{-}$ are calculated by the following formulas.

$$z_j{}^{+}=\max\left\{z_{n_1j},z_{n_2j},\cdots,z_{n_qj}\right\},z_j{}^{-}=\min\left\{z_{m_1j},z_{m_2j},\cdots,z_{m_qj}\right\} \tag{3.38}$$

Calculate the close degree s_{n_i} of the solution z_{n_i}, as follows:

$$s_{n_i}=\frac{d_{n_i}{}^{-}}{d_{n_i}{}^{+}+d_{n_i}{}^{-}} \tag{3.39}$$

The comprehensive evaluation score $S=\left\{s_{n_1},s_{n_2},\ldots,s_{n_q}\right\}$ of each sample is calculated. Select the sample $\boldsymbol{x}_{n_i}$ with the largest comprehensive evaluation score in the sample set (that is, point $\boldsymbol{x}_s$, $s=n_i$) as the sensitive sample, and the Voronoi cell c_s is the most suitable sensitive area for adding new samples.

3.4.4 Determination of the Location of New Sample Point

The most sensitive Voronoi cell for adding a new sample is identified now. Next, it is still necessary to use a certain criterion to finally determine the specific location of the new sample in the sensitive cell. The criterion of maximizing the distance from existing points is usually used [25, 26], lacking further mining of sampling information such as model errors. A new learning function that combines the estimated leave-one-out error and prediction variance is proposed, which increases the probability that the new sample falls in a position with a large predication error.

The core of this method is to add new sample at the position where the surrogate model prediction error and variance are large. However, surrogate models such as Kriging provide the prediction variance at any point, but without the prediction error. As mentioned earlier, the leave-one-out error characterizes the influence of the lack of one sample on the accuracy of the surrogate model. Thus, the estimated leave-one-out error of the unsampled candidate point is applied to evaluate its error property.

Firstly, the neighboring samples are selected by the criterion of common Delaunay triangle vertexes, illustrated in Sect. 3.4.2. The leave-one-out errors of sensitive sample $\boldsymbol{x}_s$ and its neighboring samples set $\boldsymbol{X}_s^{near} = \{\boldsymbol{x}_{s1}, \boldsymbol{x}_{s2}, \cdots, \boldsymbol{x}_{sk}\}$ is calculated by the next formula, which is similar to the formula (25).

$$\mathrm{e}_{sj}^{LOO} = \left| y_{sj} - \widehat{M}_{-sj}(\boldsymbol{x}_{sj}) \right|, \forall j \in [0, k] \tag{3.40}$$

where k is the number of neighboring points; when j equals 0, $\boldsymbol{x}_{sj}$ is the point $\boldsymbol{x}_s$ itself; y_{sj} is the real response at point $\boldsymbol{x}_{sj}$; $\widehat{M}_{-sj}(\boldsymbol{x}_{sj})$ is the output value predicted by the surrogate model constructed on all existing points without $\boldsymbol{x}_s$.

Monte Carlo simulation is used to generate a large number of candidate points in the sensitive Voronoi cell c_s. For any candidate point $\boldsymbol{x}_s^t$, the leave-one-out error of the sensitive sample $\boldsymbol{x}_s$ and its neighboring samples set $\boldsymbol{X}_s^{near}$ is used to estimate the leave-one-out error of the candidate point, as:

$$e_{st}^{LOO} = \frac{\sum_{j=0}^{k} \left(\frac{1}{d_{sj}^t} \cdot e_{sj}^{LOO} \right)}{\sum_{j=0}^{k} \frac{1}{d_{sj}^t}}, t \in [1, n_s] \tag{3.41}$$

where n_s is the total number of candidate points located in the sensitive Voronoi cell c_s; d_{sj}^t is the Euclidean distance between the candidate point $\boldsymbol{x}_s^t$ and the existing sample $\boldsymbol{x}_{sj}$, calculated by the next formula:

$$d_{sj}^t = \|\boldsymbol{x}_s^t - \boldsymbol{x}_{sj}\|_2 \tag{3.42}$$

where $\|\cdot\|_2$ denotes the second-order norm.

In some special cases, such as the candidate point coincides with the sensitive point $\boldsymbol{x}_s$, the leave-one-out error calculated by this method is equal to the one of the sample $\boldsymbol{x}_s$, which shows the rationality of the leave-one-out error estimated by this method. However, due to the existence of existing samples, it is necessary to control the new samples as far away from the existing samples as possible. In order to ensure that the new sampling point falls in a position with a large estimation error and at the same time stay away from the existing point, the estimated error e_{st}^{LOO} of the candidate point $\boldsymbol{x}_s^t$ is multiplied by the variance Var_{st} as a learning function, as follows:

$$U\left(\boldsymbol{x}_s^t\right) = e_{st}^{LOO} \times Var_{st} \tag{3.43}$$

where Var_{st} is the prediction variance of the candidate point $\boldsymbol{x}_s^t$, which is provided by the surrogate model.

Finally, select the candidate point with the maximum learning function calculation value in the sensitive cell c_s as the new sampling point, as follows:

$$\boldsymbol{x}_{new} = \arg\max_{\boldsymbol{x}_s^t \in c_s}\left\{U\left(\boldsymbol{x}_s^t\right)\right\} \tag{3.44}$$

Example 3.3 To test the effectiveness of the algorithms proposed in this section, some numerical benchmark cases are given, as shown in the following formulas. Try to program these numerical functions and draw the images of one-dimensional and two-dimensional cases.

$$\begin{aligned}\mathcal{M}_{Hump,1D}(x) &= 5x + \frac{0.05}{(x-0.45)^2+0.002} \\ &- \frac{0.5}{(x-3.5)^2+0.03} - 6,\ \ x \in [-0.5, 5.0]\end{aligned} \tag{Function 1}$$

$$\begin{aligned}\mathcal{M}_{Ea,2D}(\mathrm{x}) &= -\cos(x_1)\cdot\cos(x_2) \\ &\cdot \exp\left(-(x_1-\pi)^2-(x_2-\pi)^2\right),\ \ \mathrm{x} \in [0,7]^2\end{aligned} \tag{Function 2}$$

$$\begin{aligned}\mathcal{M}_{Peaks,\ 2D}(\mathrm{x}) &= 3(1-x_1)^2 e^{\left(-x_1^2-(x_2+1)^2\right)} - 10\left(\frac{x_1}{5}-x_1^3-x_2^5\right)e^{\left(-x_1^2-x_2^2\right)} \\ &- \frac{1}{3}e^{\left(-(x_1+1)^2-x_2^2\right)},\ \mathrm{x} \in [-44]^2\end{aligned} \tag{Function 3}$$

Solution

Function 1, named Hump function, is a one-dimensional case with strong nonlinear, shown in Fig. 3.10a. Function 2, named Eason function, is a two-dimensional case with a single peak in the center area, shown in Fig. 3.10b. Function 3, named Peaks function, is a two-dimensional case with multiple peaks and valleys, shown in Fig. 3.10c.

Example 3.4 Taking Function 1 in Example 3.3 as an example, define 5 initial samples, as $X = [0.5, 2, 3.4, 3.6, 4.5]$, select Kriging as the surrogate model, and use the method proposed in Sect. 3.4.2 to calculate the 3 metrics (e^{LOO}, *NonL*, and *Dens*) for each initial sample. Similarly, try to calculate the metrics matrix of Function 2 under 10 initial samples generated by TPLHS.

Fig. 3.10 Graphs of different functions

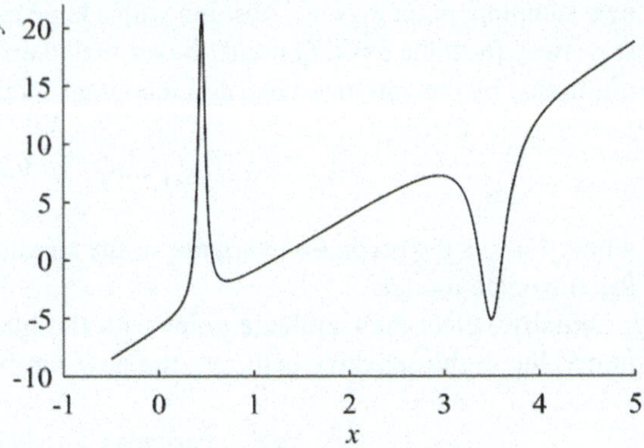

(a) Function 1: Hump function

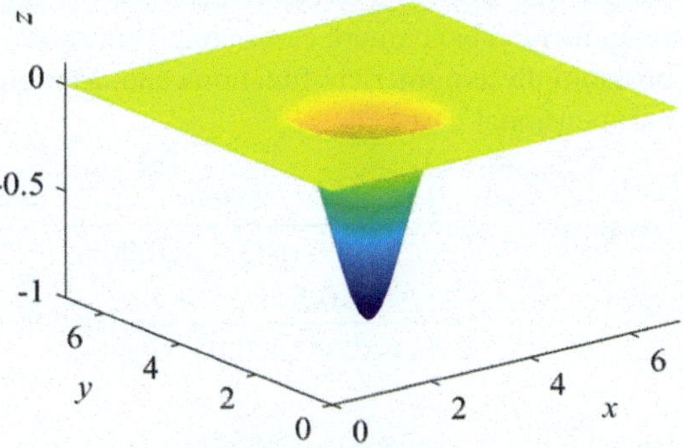

(b) Function 2: Eason function

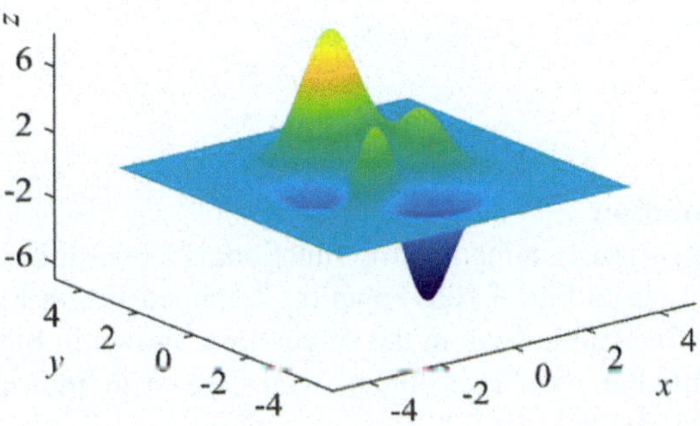

(c) Function 3: Peaks function

Table 3.3 Metrics matrix **A** of all initial samples for Function 1

Samples	e^{Loo}	*NonL*	*Dens*
$x_1 = 0.5$	0.3331	0	0.3181
$x_2 = 2$	2.1861	0.7566	0.2635
$x_3 = 3.4$	0.3958	1.1511	0.1455
$x_4 = 3.6$	1.3078	1.6744	0.1001
$x_5 = 4.5$	13.5567	0	0.1728

Table 3.4 Normalized metrics matrix **P** of all initial samples for Function 1

Samples	e^{Loo}	*NonL*	*Dens*
$x_1 = 0.5$	0.0187	0	0.3181
$x_2 = 2$	0.1230	0.2112	0.2635
$x_3 = 3.4$	0.0223	0.3214	0.1455
$x_4 = 3.6$	0.0736	0.4674	0.1001
$x_5 = 4.5$	0.7525	0	0.1728

Table 3.5 Comprehensive evaluation score of all initial samples for Function 1

s_1	s_2	s_3	s_4	s_5
0.0261	0.2156	0.2330	0.3269	0.6923

Solution

The initial matrix **A**, containing the 3 metrics for each sample, is established, as shown in Table 3.3, and the normalized matrix **P** is also shown in Table 3.4.

Example 3.5 Try to identify the most sensitive sample and its Voronoi cell, using the method proposed in Sect. 3.4.3.

Solution

The comprehensive evaluation score $S = \{s_1, s_2, \ldots, s_5\}$ of each sample are calculated, shown in Table 3.5.

Exercises

3.1 Briefly explain the differences between the several approaches to surrogate model construction presented in this chapter and illustrate them using specific one-dimensional cases.

3.2 Summarize the steps in selecting samples for adaptive sampling in the surrogate model.

3.3 Briefly describe the difference between global exploration and local exploitation in the adaptive sampling process.

3.4 Using the method presented in Sect. 3.4, calculate the newly added samples for function 1 in Example 3.3.

3.5 According to the process of solving the above Example 3.3, add new points iteratively until convergence, and draw the samples and the model's fitting curve. When the root mean square error (RMSE) is less than 0.01, the adaptive sampling is considered to be converged.

References

1. Bezerra, M. A., Santelli, R. E., Oliveira, E. P., Villar, L. S., & Escaleira, L. A. (2008). Response surface methodology (RSM) as a tool for optimization in analytical chemistry. *Talanta, 76*(5), 965–977.
2. Buhmann, M. D. (2000). Radial basis functions. *Acta Numerica, 9*, 1–38.
3. Hearst, M. A., Dumais, S. T., Osuna, E., Platt, J., & Scholkopf, B. (1998). Support vector machines. *IEEE Intelligent Systems Their Applications, 13*(4), 18–28.
4. Krige, D. G. (1951). A statistical approach to some basic mine valuation problems on the Witwatersrand. *Journal of the Southern African Institute of Mining Metallurgy, 52*(6), 119–139.
5. Du, X., He, P., & Martins, J. R. R. A. (2021). Rapid airfoil design optimization via neural networks-based parameterization and surrogate modeling. *Aerospace Science and Technology, 113*, 106701.
6. Li, L., Jiang, Z., Ong, M. C., & Hu, W. (2019). Design optimization of mooring system: An application to a vessel-shaped offshore fish farm. *Engineering Structures, 197*, 109363.
7. Zhang, W., Liang, Z., Wu, W., Ling, G., & Ma, R. (2021). Design and optimization of a hybrid battery thermal management system for electric vehicle based on surrogate model. *International Journal of Heat and Mass Transfer, 174*, 121318.
8. Jiang, Z., Hu, W., Dong, W., Gao, Z., & Ren, Z. (2017). Structural reliability analysis of wind turbines: A review. *Energies, 10*(12), 2099.
9. Su, G., Peng, L., & Hu, L. (2017). A Gaussian process-based dynamic surrogate model for complex engineering structural reliability analysis. *Structural Safety, 68*, 97–109.
10. Hu, W., Zhang, T., Deng, X., Liu, Z., & Tan, J. (2021). Digital twin: A state-of-the-art review of its enabling technologies, applications and challenges. *Journal of Intelligent Manufacturing Special Equipment, 2*(1), 1–34.
11. Thelen, A., Zhang, X., Fink, O., Lu, Y., Ghosh, S., Youn, B. D., Todd, M. D., Mahadevan, S., Hu, C., & Hu, Z. (2022). A comprehensive review of digital twin—Part 1: Modeling and twinning enabling technologies. *Structural and Multidisciplinary Optimization, 65*(12), 354.
12. Zhang, S., Liang, P., Pang, Y., Li, J., & Song, X. (2022). Multi-fidelity surrogate model ensemble based on feasible intervals. *Structural and Multidisciplinary Optimization, 65*(8), 212.
13. Shang, X., Su, L., Fang, H., Zeng, B., & Zhang, Z. (2023). An efficient multi-fidelity Kriging surrogate model-based method for global sensitivity analysis. *Reliability Engineering & System Safety, 229*, 108858.
14. Owen, A. B. (1992). Orthogonal arrays for computer experiments, integration and visualization. *Statistica Sinica*, 439–452.
15. Loh, W.-L. (1996). On Latin hypercube sampling. *The Annals of Statistics, 24*(5), 2058–2080.
16. Kleijnen, J. P. (2009). Kriging metamodeling in simulation: A review. *European Journal of Operational Research, 192*(3), 707–716.
17. Seeger, M. (2004). Gaussian processes for machine learning. *International Journal of Neural Systems, 14*(02), 69–106.
18. Husslage, B. G., Rennen, G., Van Dam, E. R., & Den Hertog, D. (2011). Space-filling Latin hypercube designs for computer experiments. *Optimization and Engineering, 12*(4), 611–630.
19. Van Dam, E. R., Husslage, B., Den Hertog, D., & Melissen, H. (2007). Maximin Latin hypercube designs in two dimensions. *Operations Research, 55*(1), 158–169.
20. Jones, D. R., Schonlau, M., & Welch, W. J. (1998). Efficient global optimization of expensive black-box functions. *Journal of Global Optimization, 13*(4), 455–492.
21. Crombecq, K., Laermans, E., & Dhaene, T. (2011). Efficient space-filling and non-collapsing sequential design strategies for simulation-based modeling. *European Journal of Operational Research, 214*(3), 683–696.
22. Jin, R., Chen, W., & Sudjianto, A. (2002). *On sequential sampling for global metamodeling in engineering design*. In Proceedings of the international design engineering technical conferences and computers and information in engineering conference, pp. 539–548.

23. Fuhg, J. N., Fau, A., & Nackenhorst, U. (2021). State-of-the-art and comparative review of adaptive sampling methods for kriging. *Archives of Computational Methods in Engineering, 28*, 2689–2747.
24. Garud, S. S., Karimi, I. A., & Kraft, M. (2017). Smart sampling algorithm for surrogate model development. *Computers & Chemical Engineering, 96*, 103–114.
25. Jiang, P., Zhang, Y., Zhou, Q., Shao, X., Hu, J., & Shu, L. (2018). An adaptive sampling strategy for Kriging metamodel based on Delaunay triangulation and TOPSIS. *Applied Intelligence, 48*(6), 1644–1656.
26. Xu, S., Liu, H., Wang, X., & Jiang, X. (2014). A robust error-pursuing sequential sampling approach for global metamodeling based on voronoi diagram and cross validation. *Journal of Mechanical Design, 136*(7), 071009.

Chapter 4
Model Verification & Validation

Nomenclature

$\mathbf{1}_A$	Indicator of event A
B_L	Lower boundary of true CDF
B_U	Upper boundary of true CDF
D	Area metric
d_L	Lower boundary of model validation interval
d_L^+	Area of lower boundary function over the standard CDF
d_L^-	Area of lower boundary function below the standard CDF
d_U	Upper boundary of model validation interval
d_U^+	Area of upper boundary function over the standard CDF
d_U^-	Area of upper boundary function below the standard CDF
$F^L(u)$	The possible CDF for lowest validation metric
$\widehat{\boldsymbol{F}}_X(\boldsymbol{x})$	Empirical CDF of random variable X
$\widehat{F}_U^T(u)$	True cumulative distribution function of U
n_d	Number of dimensions in model validation problem
n_s	Number of samples from simulation model
Q	Quantity of interest
$\overline{Q}_i$	Mean value of the i-th quantity of interest
$\boldsymbol{Q}_{\text{sim}}$	Matrix of simulation data
$\boldsymbol{q}$	Sample simulation data
$\boldsymbol{U}$	Uniform random variable from probability integral transformation
X	Random variable
$y(\boldsymbol{q}, \boldsymbol{\mu})$	Mahalanobis distance
$\boldsymbol{y}^e$	Experimental sample data
α	Confidence interval level
$\boldsymbol{\mu}$	Mean vector of simulation samples
$\boldsymbol{\Sigma}$	Covariance matrix

W. Hu, *Design Optimization Under Uncertainty*,
https://doi.org/10.1007/978-3-031-49208-2_4

4.1 Introduction

With the development of computer techniques, a variety of complex simulation models have been constructed and utilized to predict the future performance of a system instead of physical experiments to save time and economic cost. However, due to factors such as model inadequacy, numerical approximation, uncertainty, etc. [1] the output of a simulation model does not perfectly match the reality. Hence, the discrepancy between the simulation model and physical reality needs to be quantitatively assessed. The technology addressing such issue is called *Model Verification & Validation (V&V)*.

Verification and validation are independent procedures where *verification* is the process of determining that a simulation or computational model implementation matches the conceptual model and its solution, while *validation* is the process of assessing the difference between the computational model and data measured in practice [2]. In engineering, **a simplified diagram of modeling process** involving model V&V is shown in Fig. 4.1 [3]. The *quantity of interest (QoI)* is the target of the problem. In other words, it is usually the output of the models, such as strain, fatigue, remaining useful life (RUL), etc. The hidden mechanism behind the target problem is abstracted and turned into a *mathematical model* with mathematical or logical representation, e.g., partial differential equation (PDE), diagram of complex systems. Such abstraction process is referred as analysis and modeling, which requires expert knowledge in the domain [3, 4]. The mathematical model needs to be implemented as a *computational model* through programming and model construction. The consistency between the computational and mathematical model is evaluated through the model verification process. The output of the computational model is compared with the measured data of reality with the same input condition to examine the credibility of the computational model, namely the model validation process.

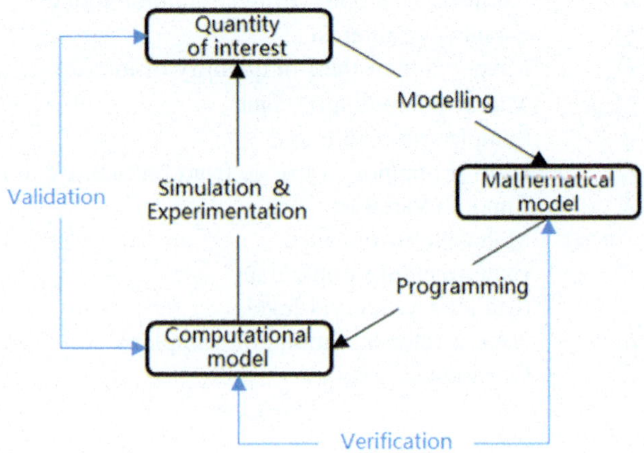

Fig. 4.1 A simplified diagram of modeling process [5]

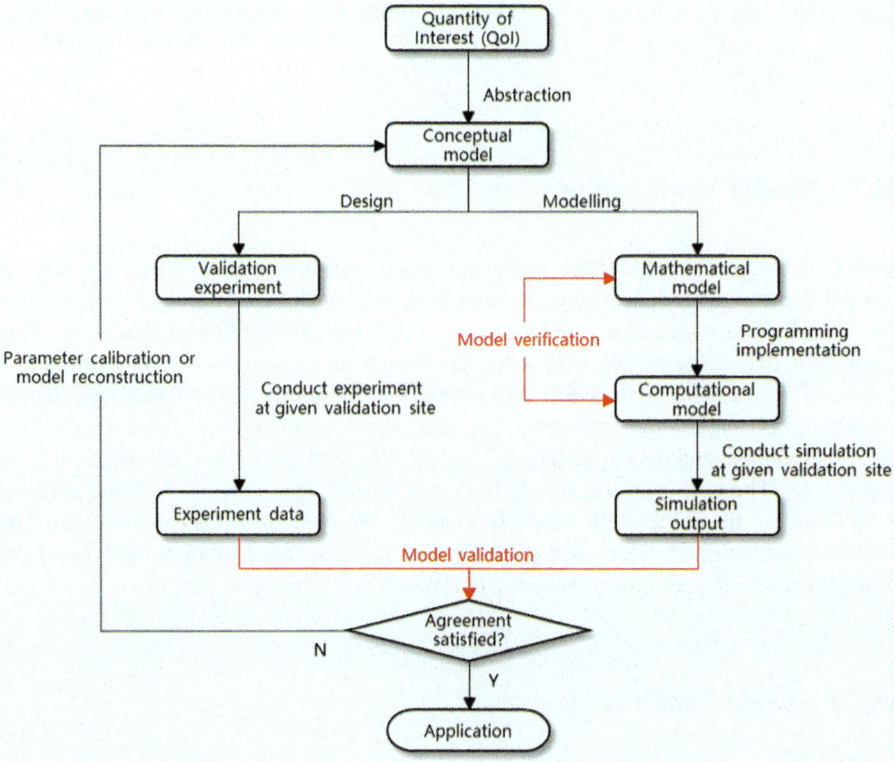

Fig. 4.2 Practical modeling process [5]

In practical engineering applications, Fig. 4.1 can be expanded to a more detailed version of Fig. 4.2 [5]. The mathematical model in Fig. 4.1 is expended as *conceptual model* and *mathematical model*. Once the quantity of interest is identified, the conceptual model should be determined including the input data, the physical mechanism, the involving parameters, the failure criteria, and the validation metric and required degree of agreement between experiment data and simulation outcome. After the confirmation of conceptual model, the design of mathematical model and experiment can be carried out simultaneously. The mathematical model includes the equation expression of physical mechanism, the initial and boundary conditions, and the description of uncertainty. The computational model is responsible for digitalized expression and mechanical but fast calculation of mathematical model with the characteristic of numerical discretization, uncertainty propagation, and iterations. The computational model should be verified against the mathematical model to ensure the correctness of programming and model construction. The *validation experiment* is designed and conducted to collect physical data at given validation sites. The measured data of validation experiment is compared with the output of simulation model to determine the agreement. If the agreement is acceptable, then the model can be utilized in engineering application, otherwise either the model should be calibrated or revised, or the validation experiment should be re-designed [5].

4.2 Model Verification

After the mathematical model is constructed and programmed into the computational model, these two models need to be compared and checked out together to make sure the implementation contains no errors, e.g., fatal bugs, result inconsistency, etc. This process is defined as *model verification*. It consists of two aspects: (1) The simulation code should be able to run stably and robustly without errors or breakdowns, which is secured through *code verification* [5]. This involves the typical software verification techniques such as expert check, logical diagram, interactive debugger, and so on [6]. (2) The output of the simulation code should be consistent with the conceptual model at the same input condition, namely the *calculation verification* [5]. The model verification is often used to make sure that the mathematical model and the computational model produces consistent result with the same inputs.

4.2.1 Code Verification

As the first step of model verification, the **robustness** of the input data should be checked. Any illegal inputs, such as inappropriate matrix size for multiplication, zero data that are to be divided. Also, **degeneracy testing** should be under consideration. In most cases, the input values should be located in an available range due to the data storage limit for different types of variables (such as integer, double, floats, etc.). When the input data are near the limit of available range, the simulation model may perform weirdly because some of the intermediate variables might exceed the limit of its available range.

The simulation model needs to ensure that there is no bugs and fatal errors. By adding **checkpoints** to crucial steps, the process flow of the simulation model can be checked and the hidden problems may be found out with an interactive debugger. Another effective way is by **structural walk-through**. The code developer is suggested to explain the code step-by-step to other people, or just to a rubber duck. By such explanation it becomes quite natural to discover the problems in the code. For the simulation model with multiple logical brunches, a logical flow diagram is also useful to compare and check whether all the logical conditions are implemented correctly.

4.2.2 Calculation Verification

After the code verification to ensure that the simulation model is able to perform from the beginning to the end, the output of the simulation model should be compared against the conceptual model to ensure credibility. Usually these simulation models contain uncertainty and the prediction result can fluctuate with repeated runs. In the computational model, uncertainties are generated and determined through **random seeds**, which can be fixed for reproducibility and to exclude the influence of randomness to the simulation output with given input. The **continuity** of the simulation model should also be tested, i.e., a slight change in input value should not result in significant difference at output. To secure the reliability of the simulation model in the verification process, the model should be checked through multiple numerical examples.

When the code does not produce the desired result, there might be some intermediate variable that goes against the expectation. Currently most of the interactive debuggers have the function of tracing the variables. By storing the intermediate result into variables, they can be compared against the corresponding steps in the conceptual or mathematical model to find out the problem. Recently, some programming platforms provide with tracing function in a graphical manner to become clearer and more intuitive. For example, Tensorboard is a useful graphical tool to overview the neural network architecture constructed with Tensorflow [7].

4.3 Model Validation

Model validation is the process to quantitively evaluate the disagreement between the simulation model and experimental data, which is a fully data-driven technique and has been widely used such as in [8–11]. Model validation is a crucial part in the modeling process since an inaccurate model may lead to a fault design. For example, a reliability analysis model might identify an unreliable design point as a reliable one due to the wrong calculation of strain, fatigue, or other performance functions. In order that a model validation metric evaluates the degree of consistency between simulation and experimental outputs, some desired features are summarized as follows [12]:

1. A validation metric should objectively assess the quantitative divergency between simulation model and physical experiment. It does not directly answer whether the model is acceptable or not. Instead, it gives a numerical result, usually ranging from 0 to 1 where 0 indicates a perfect match and 1 indicates a huge disagreement. It is the analyst and the engineer's responsibility to decide whether the candidate model produces credible result [2].
2. The uncertainty in both physical experiment and simulation data should be considered. The result of model validation changes every time as the validation process is repeated even at given validation sites. This uncertainty can be traced

back to the source data of both simulation model and physical experiment, e.g., measurement error in the physical experiment, model inadequacy, and simplification assumptions in the simulation model. Hence, a model validation metric is expected to output a range of disagreement between simulation model and experiment data. The position of the range shows the overall disagreement while the width of the range reflects the degree of uncertainty.

3. The consistency of model simulation outputs and physical experimental results should be evaluated at multiple validation sites. When a simulation model is in application, the input condition is changing all the time. For models, especially those numerical approximation models which are accurate near the training points and relatively rough at regions with sparse samples, the accuracy might be completely distinct at different validation sites. Thus, the global credibility of simulation model should be assessed by a single validation process to evaluate the whole performance of simulation model.

With the widespread usage of simulation model, model validation has also attracted considerable attention. Four types of validation metrics, i.e., classical hypothesis testing [13], Bayes factor [14, 15], frequentist's metric [3], and area metric [16–18], have become the current mainstream. Researches have been conducted to compare these branches in numerical tests, and the result shows that all these types of validation metrics have their shortcomings to the desired features mentioned above. The classical hypothesis test only gives the result of whether to accept the simulation model or not rather than quantitively assesses the consistency of the model. Bayes factor depends much on the prior knowledge of analyst. Frequentist's method is mainly dependent on the specific statistical features, such as means, standard deviation, or maximum value, to compare the similarity of simulation model and experimental data distribution. In contrast, area metric describes the overall difference of these two distributions by calculating the area difference of cumulative distribution function (CDF) curve of simulation model and experimental data. Although the uncertainty in the model validation process cannot be expressed, modifications can be applied to the area metric to easily solve this issue. Hence, only area metric-based model validation is introduced in this chapter.

4.3.1 Area Metric

Before diving deep into the area metric-based model validation, the concept of empirical cumulative distribution function (ECDF) needs to be elaborated in order to understand the principle of area metric. ECDF approximates the cumulative distribution function (CDF) based on the collected discretized data. Suppose n samples of a random variable X are acquired through measurement, denoted as $[X_1, X_2, \ldots, X_n]$, the ECDF of X, denoted as $\widehat{F}_X(x)$, is expressed as [19]:

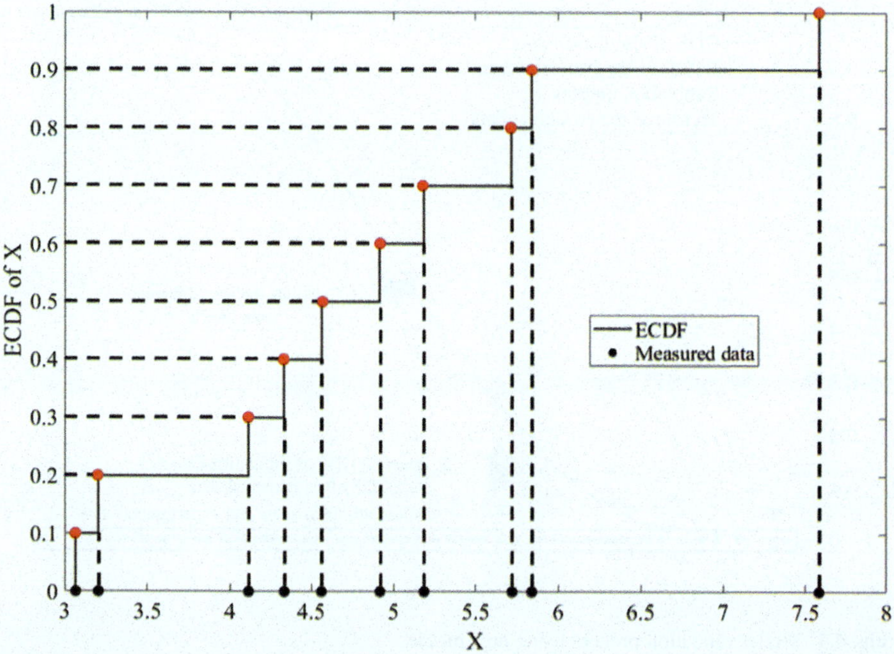

Fig. 4.3 An illustration example of ECDF

$$\widehat{F}_X(x) = \frac{1}{n}\sum_{i=1}^{n} \mathbf{1}_{X_i \leq x} \tag{4.1}$$

where $\mathbf{1}_A$ is the indicator of event A with value 1 if A is true and 0 when A is false. Figure 4.3 illustrates an ECDF of the variable X with $n = 10$ sample data. With the increase of X, each time it passes a sample point, the ECDF value increases by $1/n$ from 0 to 1, generating such a stepped curve. This is useful when expressing the distribution of validated experimental data since the sample size often is limited.

The original area metric [16] is based on the following assumption: the validation experiment requires much time and economic cost and only a small sample size of data can be obtained at a validation site through physical experiments, while it is relatively much easier for the simulation model to generate a larger number of data. Consider the simplest model validation case: only one QoI y needs to be validated at one single validation site. In this situation, the CDFs of simulation data and experimental data are utilized since they contain rich statistical information. Based on the previous assumption, the CDF of simulation data can be interpolated since there are adequate data, while for experimental data only the ECDF can be acquired due to the limited experimental data. The area metric is defined as the area difference between the CDF of simulation data and the ECDF of experimental data:

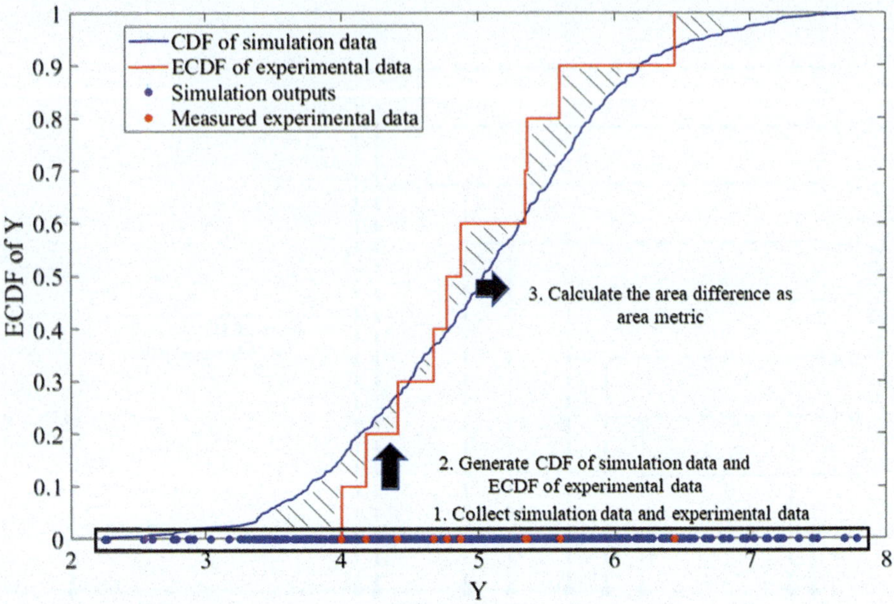

Fig. 4.4 Model validation process using area metric

$$d = \int_{-\infty}^{+\infty} \left| \widehat{F}_Y(y) - F_Y(y) \right| dy \tag{4.2}$$

where d denotes the area metric, $\widehat{F}_Y(\cdot)$ is the ECDF of experimental data, and $F_Y(\cdot)$ is the CDF of simulation data.

As shown in Fig. 4.4, the process of area metric can be summarized as the following steps:

1. Obtain simulation data and experimental data at the given validation site considering uncertainties.
2. Calculate the ECDF of experimental data $\widehat{F}_Y(\cdot)$ and the CDF of simulation data $F_Y(\cdot)$.
3. Compare the area difference expressed in Eq. (4.2) to calculate the area metric value.

However, obvious issues are exposed in the original area metric:

- The credibility of model can be evaluated *at only one validation site* in one validation process.
- The QoI is restricted to 1, i.e., only one-dimensional output simulation model can be validated. When the model outputs multiple QoIs, each QoI can only be assessed individually and an overall consistency cannot be calculated.

Table 4.1 Experimental data in Example 4.1

No.	Value
1	4.291
2	4.397
3	4.571
4	5.364
5	5.712

- The result of area metric is a determined value, which will fluctuate among repeated validation processes, causing an over- or underestimate of the model accuracy.

Thus, modifications need to be applied on the original area metric, which will be elaborated in the following sections.

Example 4.1 Consider a simulation model with one-dimensional output Y. The output of the simulation model follows a uniform distribution and its CDF is given by:

$$F_Y(y) = \begin{cases} 0 & y < 4 \\ \frac{y-4}{2} & 4 \leq y \leq 6 \\ 1 & y \geq 6 \end{cases}$$

Five experimental data are collected and shown in the following table. Calculate the area metric with the given data (Table 4.1).

Solution

Using Eq. (4.1), the ECDF of experimental data can be calculated as:

$$\widehat{F}_Y(y) = \begin{cases} 0 & y < 4.291 \\ 0.2 & 4.291 \leq y < 4.397 \\ 0.4 & 4.397 \leq y < 4.571 \\ 0.6 & 4.571 \leq y < 5.364 \\ 0.8 & 5.364 \leq y < 5.712 \\ 1 & y \geq 5.712 \end{cases}$$

The ECDF of experimental data known and CDF of simulation data are now obtained and can be plotted as the following graph.

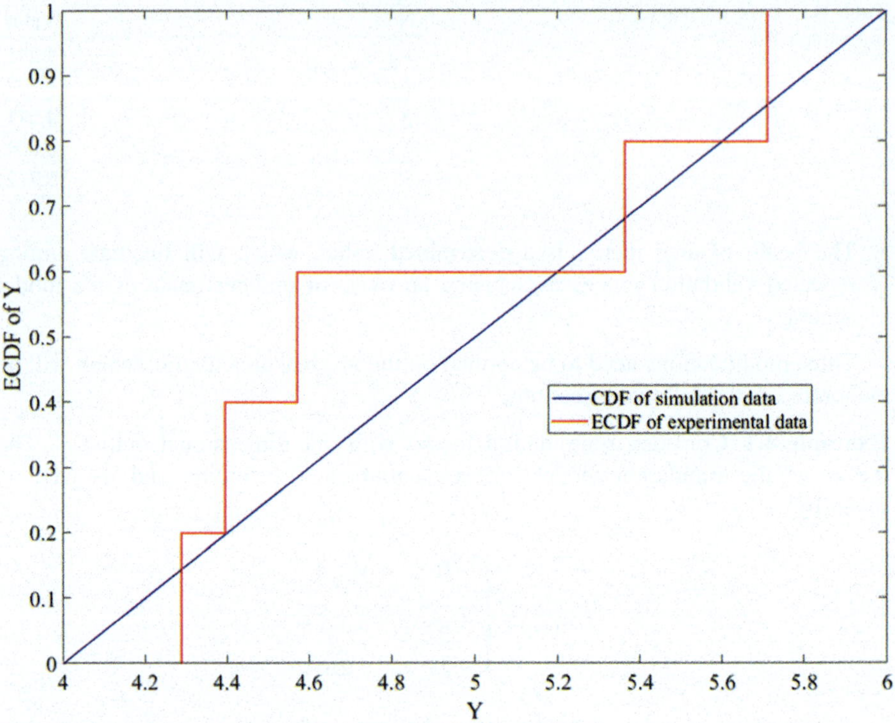

Fig. 4.5 Illustration of $\widehat{F}_Y$ and F_Y in Example 4.1

Thus, the area metric can be obtained as:

$$d = \int_{-\infty}^{+\infty} \left| \widehat{F}_Y(y) - F_Y(y) \right| dy$$

$$= \int_{-\infty}^{4} |0-0| dy + \int_{4}^{4.291} \left| \frac{y-4}{2} \right| dy + \int_{4.291}^{4.397} \left| \frac{y-4}{2} - 0.2 \right| dy + \int_{4.397}^{4.571} \left| \frac{y-4}{2} - 0.4 \right| dy$$

$$+ \int_{4.571}^{5.364} \left| \frac{y-4}{2} - 0.6 \right| dy + \int_{5.364}^{5.712} \left| \frac{y-4}{2} - 0.8 \right| dy + \int_{5.712}^{6} \left| \frac{y-4}{2} - 1 \right| dy + \int_{6}^{+\infty} |1-1| dy$$

With the illustration of Fig. 4.5, the absolute value of each integrand can be easily expanded as:

$$d=\int_{4}^{4.291}\frac{y-4}{2}dy+\int_{4.291}^{4.397}0.2-\frac{y-4}{2}dy+\int_{4.397}^{4.571}0.4-\frac{y-4}{2}dy+\int_{4.571}^{5.2}0.6-\frac{y-4}{2}dy$$
$$+\int_{5.2}^{5.364}\frac{y-4}{2}-0.6dy+\int_{5.364}^{5.6}0.8-\frac{y-4}{2}dy+\int_{5.6}^{5.712}\frac{y-4}{2}-0.8dy+\int_{5.712}^{6}1-\frac{y-4}{2}dy$$

By merging the same integrand, the calculation can be simplified as:

$$d=\int_{4}^{4.291}\frac{y-4}{2}dy+\int_{4.291}^{4.397}0.2dy+\int_{4.397}^{4.571}0.4dy+\int_{4.571}^{5.2}0.6dy-\int_{4.291}^{5.2}\frac{y-4}{2}dy$$
$$-\int_{5.2}^{5.364}0.6dy+\int_{5.364}^{5.6}0.8dy-\int_{5.6}^{5.712}0.8dy+\int_{5.712}^{6}1dy$$
$$+\int_{5.2}^{5.364}\frac{y-4}{2}dy-\int_{5.364}^{5.6}\frac{y-4}{2}dy+\int_{5.6}^{5.712}\frac{y-4}{2}dy-\int_{5.712}^{6}\frac{y-4}{2}dy$$
$$=\left(\frac{1}{4}y^2-2y\right)\Big|_{4}^{4.291}+0.2y\Big|_{4.291}^{4.397}+0.4y\Big|_{4.397}^{4.571}+0.6y\Big|_{4.571}^{5.2}-\left(\frac{1}{4}y^2-2y\right)\Big|_{4.291}^{5.2}$$
$$-0.6y\Big|_{5.2}^{5.364}+0.8y\Big|_{5.364}^{5.6}-0.8y\Big|_{5.6}^{5.712}+y\Big|_{5.712}^{6}$$
$$+\left(\frac{1}{4}y^2-2y\right)\Big|_{5.2}^{5.364}-\left(\frac{1}{4}y^2-2y\right)\Big|_{5.364}^{5.6}+\left(\frac{1}{4}y^2-2y\right)\Big|_{5.6}^{5.712}-\left(\frac{1}{4}y^2-2y\right)\Big|_{5.712}^{6}$$
$$=0.195$$

4.3.2 Evaluating at Multiple Validation Sites

It is already solved to pool the experimental data into an ECDF to compare against the simulation data distribution at one validation site. However, it is not reasonable to directly pool both simulation and experimental data at different validation sites. For example, when various stress data are collected with different input load conditions, it doesn't make sense to pool these data along the axis of input load since these data probably come from totally different distributions. One can certainly calculate a metric value for each validation site, but this is not desired and the accuracy of the whole model is not intuitive and sensible in this way. Thus, a method to unify these data from different distributions needs to be proposed.

One way to address this problem is the usage of probability integral transformation (PIT), which states that for a continuous random variable Y (which is the output of the model or QoI in this chapter) with CDF $F_Y(y)$, a new random variable denoted

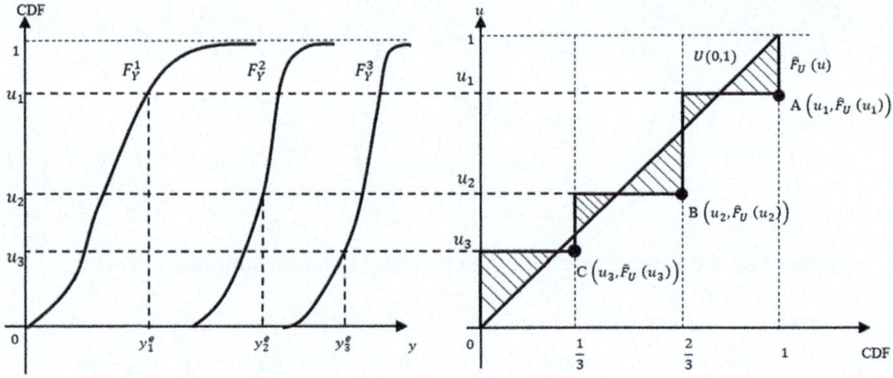

Fig. 4.6 Data transformation using PIT and process of u-pooling [21]

as $U = F_Y(Y)$ has a standard uniform distribution. The proof is simple and given as follows [20]:

$$\begin{aligned} F_U(u) &= P(U \le u) \\ &= P(F_Y(Y) \le u) \\ &= P\left(Y \le F_Y^{-1}(u)\right) \\ &= F_Y\left(F_Y^{-1}(u)\right) \\ &= u \end{aligned} \tag{4.3}$$

The discretized and sparse experimental data can be translated into a u-domain through PIT. Fig. 4.6 illustrates an example with 3 validation sites, each containing a CDF of simulation data $F_Y^1(\cdot)$, $F_Y^2(\cdot)$, and $F_Y^3(\cdot)$, respectively. The number in the superscript corresponds to the index of validation site. At each validation site, one experimental sample, namely y_1^e, y_2^e, and y_3^e, respectively, is collected and transformed into u-domain with $u_i = F_Y^i\left(y_i^e\right)$. An ECDF of these u values can be generated to calculate the overall accuracy of the model.

At single validation site, only one distribution of simulation data can be formulated. Obviously, if the experimental data come from the same distribution as the simulation data, then through PIT these experimental data are distributed to a uniform distribution over [0,1], as shown in Fig. 4.6. This situation can be certainly expanded to the multiple validation sites case with multiple distributions of simulation data at each validation site, provided that these experimental data match with their respective simulation distribution [16]. Conversely, if the experimental data are not perfectly consistent with the simulation data distribution, the distribution of experimental data will deviate from the uniform distribution to various degrees, depending on the disagreement between the model and physical reality. From this perspective, the distribution of u-values, i.e., the ECDF of u, contains features of the overall consistency between the experimental data and their corresponding simulation data distribution as shown in the right part of Fig. 4.6. Area metric can then be

Table 4.2 Experimental data in Example 4.2

Validation site	y-value
$x_1 = 0.2$	0.536
	0.672
$x_2 = 0.4$	1.240
	1.432
$x_3 = 0.6$	1.640
	1.792

Table 4.3 Transformed experimental data and corresponding CDF value

Validation site	y-value	z-value	CDF
$x_1 = 0.2$	0.536	0.096	0.538
	0.672	0.318	0.625
$x_2 = 0.4$	1.240	0.563	0.713
	1.432	0.916	0.820
$x_3 = 0.6$	1.640	0.653	0.743
	1.792	1.002	0.842

applied to the u-values to compare against the standard uniform distribution. This method was named as "u-pooling" [16] since the experimental data from various validation sites are transformed into a unified variable named U.

Example 4.2 Suppose that the relationship between the one-dimensional output y and input x is to be explored. A simulation model is constructed whose output y follows the normal distribution of $N(2.4 \sin x, (0.8 \cos x)^2)$. Three validation sites, namely $x_1 = 0.2$, $x_2 = 0.4$, and $x_3 = 0.6$, are selected to evaluate the accuracy of the model. At each validation site, two experimental data are collected respectively and recorded in the following table. Calculate the area metric of this model with the given data (Table 4.2).

Solution

Since output of the simulation data follows normal distribution, the CDF of simulation data with given output y can be easily obtained by transforming the y into standardized normal variable z as the following equation shows and look it up in the "Z-table."

$$z = \frac{y - \mu_y}{\sigma_y} = \frac{y - 2.4 \sin x}{0.8 \cos x}$$

where μ_y and σ_y are the mean and standard deviation of y, respectively. The transformed z values of experimental data, and their corresponding CDF value can be obtained and shown in Table 4.3. The relationship between experimental data and their CDF values are shown in Fig. 4.7.

The last column in Table 4.3 records all the u-values obtained from the experimental data. The ECDF of u can be easily acquired using Eq. (4.1) and the area metric between it and $U(0, 1)$ can be calculated as 0.239. The calculation of area

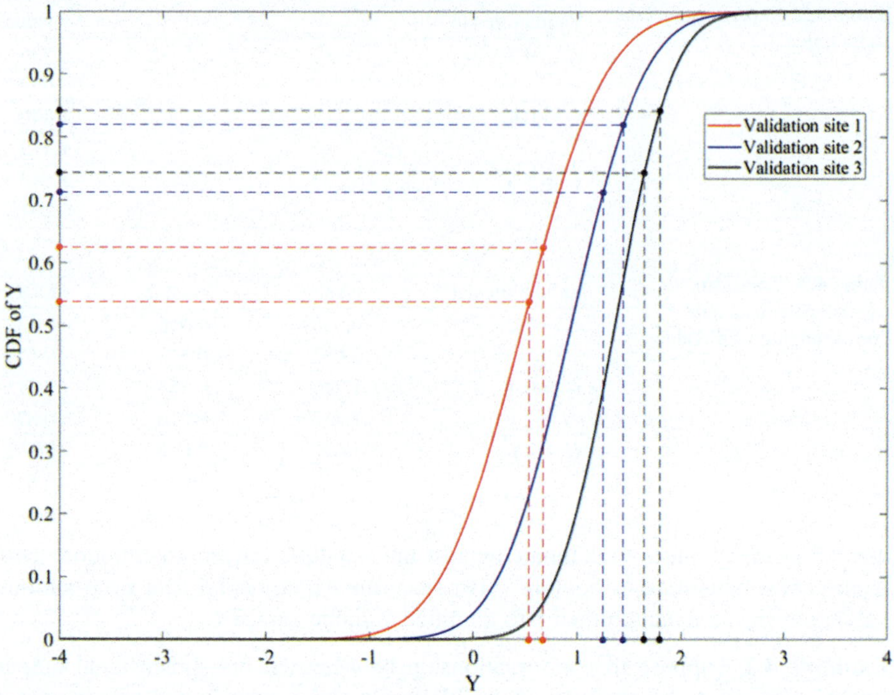

Fig. 4.7 Illustration of Example 4.2

metric is beyond the scope of this example. The readers may refer to Sect. 4.3.1 for detailed process.

4.3.3 Validating with Multiple Correlated Outputs

So far, we have only discussed model validation problem where only one QoI is under consideration. However, in the nowadays engineering applications, a simulation model usually has multiple outputs. For example, strain and stress are usually calculated together in a single finite element model (FEM) analysis. Similar with the previous section, a model validation process can be conducted for each QoI separately, but it is more desired to give an overall result with respect to all the outputs. Since model validation problem with one QoI has already been solved, it is natural to come up with the idea to merge all the QoIs into a one-dimensional "output variable," which is essentially the process of dimension reduction. Theoretically most dimension reduction methods in statistics and machine learning can be applied here and practically some statistical distances such as Mahalanobis distance (MD) [22], Bhattacharyya distance [23] and other methods such as multi-variate CDF [24], and principle component analysis (PCA) [18] have been applied. Among them, MD is the most effective and easy-to-implement way in model validation considering multi-variate outputs.

Suppose that we are facing a problem with multiple, say totally n_d, QoIs denoted as $\left[Q_1, Q_2, \ldots, Q_{n_d}\right]$. A simulation model has been repeated for a number of times at one validation site, generating n_s data denoted as:

$$\mathbf{Q}_{sim} = \begin{bmatrix} Q_{1,1} & Q_{1,2} & \cdots & Q_{1,n_d} \\ Q_{2,1} & Q_{2,2} & \cdots & Q_{2,n_d} \\ \cdots & \cdots & \cdots & \cdots \\ Q_{n_s,1} & Q_{n_s,2} & \cdots & Q_{n_s,n_d} \end{bmatrix} \tag{4.4}$$

where each row represents a simulation sample and each column stands for a QoI. A mean vector $\boldsymbol{\mu} = \left[\overline{Q}_1, \overline{Q}_2, \ldots, \overline{Q}_{n_d}\right]$ with $\overline{Q}_i$ representing the average value of the i-th QoI in simulation data, and a covariance matrix $\boldsymbol{\Sigma}$ can be calculated as follows:

$$\boldsymbol{\Sigma} = \begin{bmatrix} \sigma_{1,1} & \sigma_{1,2} & \cdots & \sigma_{1,n_d} \\ \sigma_{2,1} & \sigma_{2,2} & \cdots & \sigma_{2,n_d} \\ \cdots & \cdots & \cdots & \cdots \\ \sigma_{n_d,1} & \sigma_{n_d,2} & \cdots & \sigma_{n_d,n_d} \end{bmatrix} \tag{4.5}$$

where $\sigma_{i,\,j}$ is the covariance of i-th QoI and j-th QoI and expressed as:

$$\sigma_{i,j} = E\left[(Q_i - E(Q_i))\left(Q_j - E\left(Q_j\right)\right)\right] \tag{4.6}$$

For any data $\boldsymbol{q} = \left[q_1, q_2, \ldots, q_{n_d}\right]$, the MD $y(\boldsymbol{q}, \boldsymbol{\mu})$ calculates how far $\boldsymbol{q}$ is away from the simulation data distribution $\boldsymbol{Q}_{sim}$:

$$y(\mathbf{q}, \boldsymbol{\mu}) = \sqrt{(\mathbf{q} - \boldsymbol{\mu})^T \boldsymbol{\Sigma}^{-1} (\mathbf{q} - \boldsymbol{\mu})} \tag{4.7}$$

Note that on the right-hand side of Eq. (4.7), the data $\boldsymbol{q}$ experiences a "normalization" process. Particularly, if $\boldsymbol{q}$ is one-dimensional, it is subtracted by its mean and divided by its standard deviation, which is a typical standardization process in statistics. In this way, the multi-variate output $\left[Q_1, Q_2, \ldots, Q_{n_d}\right]$ is reduced to a single variable $y(\boldsymbol{q}, \boldsymbol{\mu})$, which can be regarded as the one-dimensional model output but still containing statistical information of the original data.

The effect of such "MD transformation" is explained by an illustrative example shown in Fig. 4.8, where $n_d = 2$. In Fig. 4.8(a), the simulation data are represented as red squares with its mean vector shown as the green square. The yellow and blue points stand for two experimental data sample. The distances from the two experimental data sample to the distribution center are the same. But due to the correlation between Q_1 and Q_2 demonstrated in the simulation data, it is natural to feel that the yellow sample should be closer to the distribution of simulation data than the blue

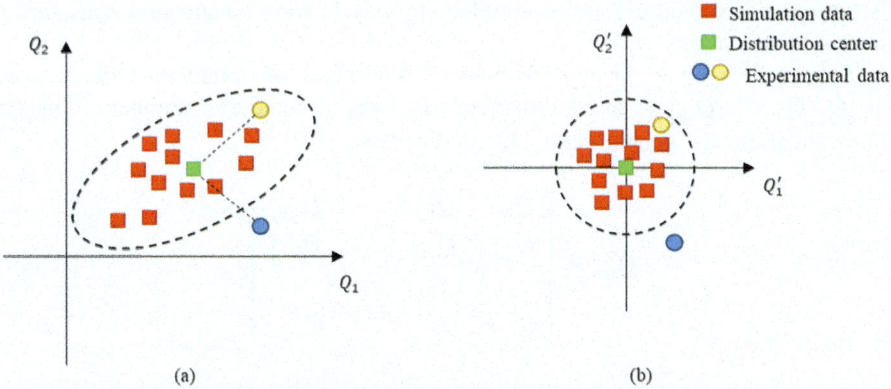

Fig. 4.8 Illustration of MD: (**a**) Original data and (**b**) Data after MD transformation

one in Fig. 4.8(a). Hence, Euclidean distance alone is not a reliable method in this case. However, by applying such "MD transformation" as shown in the right-hand side of Eq. (4.7), Q_1 and Q_2 are transformed into Q_1' and Q_2'. The MD of the two experimental data can be visualized by the Euclidean distance from the data point to the distribution center in Fig. 4.8(b). Now the yellow point is obviously closer to the simulation data distribution than the blue one. It is also worth mentioning that, the dashed circle in Fig. 4.8(b), corresponding to the dashed ellipse in Fig. 4.8(a), indicates positions with the same MD to the simulation data distribution. The points with the same MD form an ellipse rather than a circle in the original data space because MD takes the correlation between multi-variables into consideration. Thus, MD shows its superiority over Euclidean distance in the situation of model validation with multiple outputs.

Till now, a basic model validation framework for multiple QoIs at various validation sites can be shown in Fig. 4.9 and summarized as the following steps:

1. Identify all the validation sites;
2. At each validation sites, obtain the simulation and experimental data with respect to the QoIs;
3. Use dimension reduction method to process the original data;
4. Calculate the CDF of dimension-reduced simulation data;
5. Calculate the u-values of experimental data as illustrated in Fig. 4.6;
6. Once the u-values at all validation sites are obtained, the ECDF of u can be generated using Eq. (4.1);
7. Calculate the area difference between the ECDF of u and the CDF of standard uniform distribution using Eq. (4.2) as the final result of area metric.

It is worth mentioning that, the smaller the calculated area metric is, the closer the simulation data and the experimental data are, and the higher fidelity the validated model is, and vice versa.

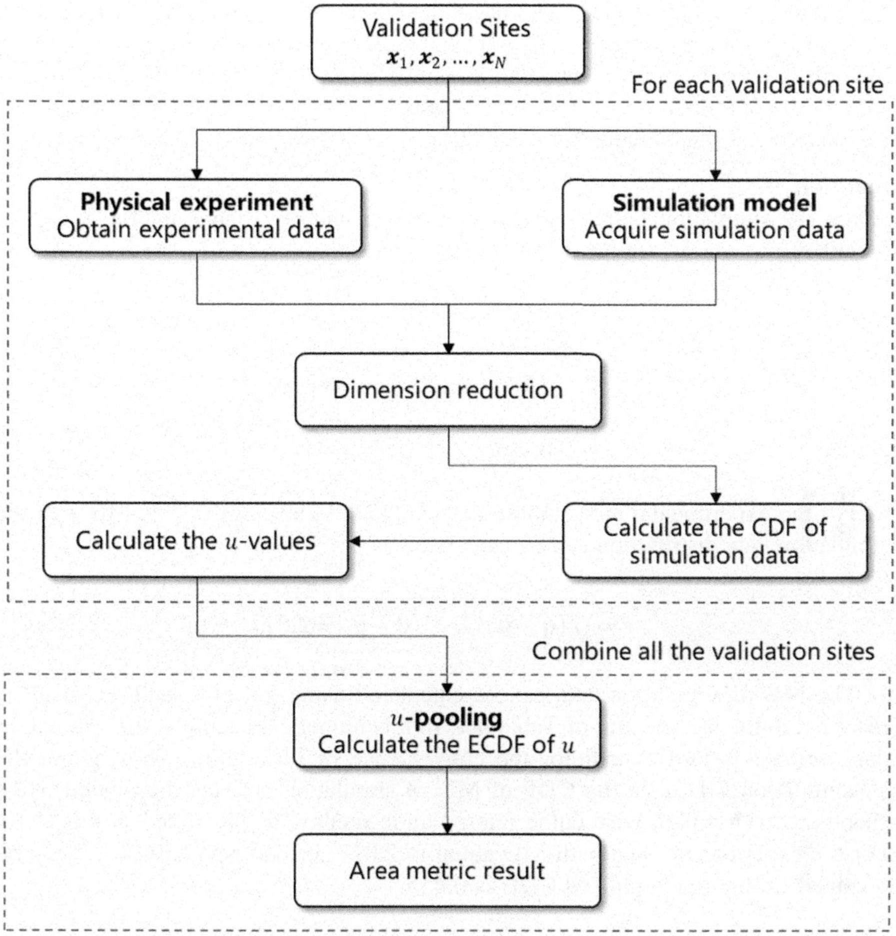

Fig. 4.9 Global model validation process for various QoIs

Table 4.4 Simulation output in Example 4.3

No.	q_1	q_2	q_3	MD to simulation distribution
1	3.326	3.063	3.262	3.404
2	3.362	3.388	3.014	3.541
3	3.051	3.383	3.340	4.893
4	3.365	3.194	3.374	3.809
5	3.253	3.320	3.271	0.611
6	3.039	3.057	3.303	4.416
7	3.111	3.169	3.297	1.245
8	3.219	3.366	3.157	1.057
9	3.383	3.317	3.262	1.845
10	3.386	3.384	3.068	2.180

Example 4.3 A simulation model has three outputs q_1, q_2, and q_3. By running the simulation model at a single validation site for 10 times, the output data are recorded in Table 4.4. For an experimental data sample $\boldsymbol{q} = [3.245 \quad 3.262 \quad 3.367]$, calculate its MD to the simulation data distribution and its corresponding CDF of MD (The assistance of computer code for calculation is recommended.).

Solution
Given the simulation output data, their mean $\boldsymbol{\mu}$ and covariance matrix $\boldsymbol{\Sigma}$ can be obtained as:

$$\boldsymbol{\mu} = [3.250 \quad 3.264 \quad 3.235]$$

$$\boldsymbol{\Sigma} = \begin{bmatrix} 0.019 & 0.005 & -0.007 \\ 0.005 & 0.018 & -0.008 \\ -0.007 & -0.008 & 0.014 \end{bmatrix}$$

For the experimental data sample $\mathbf{q} = [3.245 \quad 3.262 \quad 3.367]$, its MD y to the simulation data distribution can be calculated as

$$y = \sqrt{(\mathbf{q} - \boldsymbol{\mu})^T \boldsymbol{\Sigma}^{-1} (\mathbf{q} - \boldsymbol{\mu})} = 2.017$$

The MD of simulation data can be calculated using Eq. (4.7) and the result is recorded in the last column of Table 4.4. In this numerical example, the simulation data number is fairly small for the convenience of illustration. Thus, given the 10 simulation MD data, the CDF of MD of simulation data can be plotted using shape-preserving piecewise cubic interpolation method as illustrated in Fig. 4.10. For the experimental data with MD equal to 2.017, its corresponding CDF can be obtained from the interpolated CDF as 0.455.

4.3.4 *Interval Metric*

Various uncertainties mainly come from the model parameter, boundary condition, numerical approximation, simplification assumption, etc. which result in not only aleatory uncertainty but also epistemic uncertainty [1]. The area metric, combined with its enhancement in Sects. 4.3.2 and 4.3.3, cannot well address these uncertainties. To tackle with these uncertainties, a common idea is to use an interval to express the range of possible model validation metric value. A method is introduced here [25], which utilizes the tool of Dvoretzky-Kiefer-Wolfowitz (D-K-W) inequality to build two boundary CDF based on the ECDF of u-values in experimental data.

Once all the u-values are calculated out and the ECDF (denoted as $\widehat{F}_U(u)$) is generated according to Step (5) in Sect. 4.3.3, the D-K-W inequality states that, the

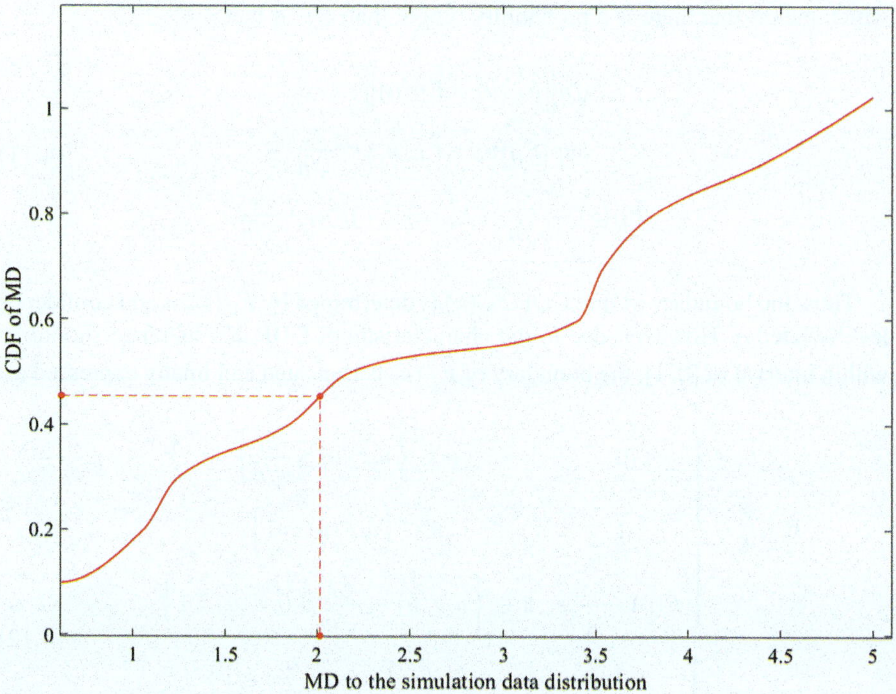

Fig. 4.10 Interpolated CDF of simulation data MD in Example 4.3

true CDF of u (denoted as $\widehat{F}_U^T(u)$ where the superscript T stands for "true") satisfies the following inequation:

$$\Pr\left(\sqrt{n}\left\|\widehat{F}_U^T(u) - \widehat{F}_U(u)\right\|_\infty > z\right) \leq 2e^{-2z^2} \tag{4.8}$$

where n is the total number of u-values, z is an arbitrary real number, and the $\left\|\widehat{F}_U^T(u) - \widehat{F}_U(u)\right\|_\infty$ is the supremum of $\widehat{F}_U^T(u) - \widehat{F}_U(u)$. The possible range of the true CDF $\widehat{F}_U^T(u)$ is determined at a confidence level $1 - \alpha$, with which a z_0 can be defined as

$$z_0 = \sqrt{-\frac{1}{2}\ln\left(\frac{\alpha}{2}\right)} \tag{4.9}$$

By substituting (4.9) into (4.8), we have

$$\Pr\left(\sqrt{n}\left\|\widehat{F}_U^T(u) - \widehat{F}_U(u)\right\|_\infty > z_0\right) \leq \alpha \tag{4.10}$$

which means that there is a probability bigger than $1 - \alpha$ that

$$
\begin{aligned}
\sqrt{n}\left\|\widehat{F}_U^T(u) - \widehat{F}_U(u)\right\|_\infty &> z_0 \\
\sup\left|\widehat{F}_U^T(u) - \widehat{F}_U(u)\right| &< \frac{z_0}{\sqrt{n}} \\
\widehat{F}_U(u) - \frac{z_0}{\sqrt{n}} < \widehat{F}_U^T(u) &< \widehat{F}_U(u) + \frac{z_0}{\sqrt{n}}
\end{aligned}
\tag{4.11}
$$

Thus, the boundary of true CDF $\widehat{F}_U^T(u)$ is determined by $\widehat{F}_U(u)$, n and confidence level-related α. However, due to the characteristic of CDF that its range should be within interval of [0, 1], the boundary of $\widehat{F}_U^T(u)$ is truncated and finally expressed as:

$$
\begin{aligned}
B_L(u) &= \begin{cases} 0 & u \in \left\{u \middle| \widehat{F}(u) \le \frac{z_0}{\sqrt{n}}\right\} \\ 1 & u \in \left\{u \middle| \widehat{F}(u) = 1\right\} \\ \widehat{F}(u) - \frac{z_0}{\sqrt{n}} & u \in \left\{u \middle| \widehat{F}(u) - \frac{z_0}{\sqrt{n}} > 0, \widehat{F}(u) \ne 1\right\} \end{cases} \\
B_U(u) &= \begin{cases} 1 & u \in \left\{u \middle| \widehat{F}(u) + \frac{z_0}{\sqrt{n_u}} \ge 1\right\} \\ 0 & u \in \left\{u \middle| \widehat{F}(u) = 0\right\} \\ \widehat{F}(u) + \frac{z_0}{\sqrt{n}} & u \in \left\{u \middle| \widehat{F}(u) + \frac{z_0}{\sqrt{n}} < 1, \widehat{F}(u) \ne 0\right\} \end{cases}
\end{aligned}
\tag{4.12}
$$

where B_L and B_U stand for the lower and upper boundary of true CDF. An illustration example is shown in Fig. 4.11a, where the black dash-dotted line shows the ECDF of u, and the red line and green line represents the upper and lower boundary, respectively.

The boundary function here gives a possible range, within which the unknown true CDF can be any monotonic curve at the given confidence level. Thus, the next step is to find out the candidate curves that induce the largest and smallest model validation metric values, which determines the final interval result. Since the u-pooling method translates all the experimental data into u-values ranging in [0,1], the CDF of standard uniform distribution, $U(0, 1)$, must cross the two boundary functions. Obviously, the **smallest interval value** corresponds to the possible CDF shown as the yellow curve in Fig. 4.11(b). The principle to construct such CDF is to make sure that it is as close to the CDF of standard uniform distribution as possible. When the CDF of standard uniform distribution, i.e., the black straight line, does not cross the area between the two boundary functions, the candidate CDF is the same as the boundary function closer to the black line. When the black line

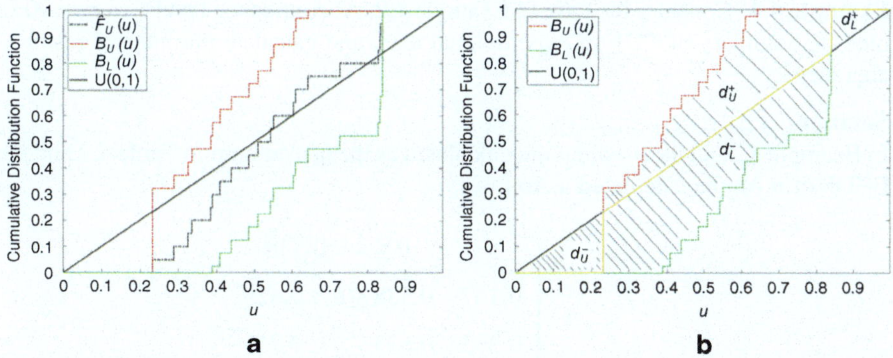

Fig. 4.11 Illustration of (**a**) boundary of experimental data CDF and (**b**) interval calculation of validation metric value

crosses the area between the boundary functions, it is able to be the same as the black line to minimize the area difference between them. With this intuitive interpretation, the construction of the CDF corresponding to the **smallest interval value**, denoted as $F^L(u)$, can be expressed mathematically as follows:

$$F^L(u) = \begin{cases} B_U(u) & B_U(u) \leq u \\ u & B_L(u) \leq u \leq B_U(u) \\ B_L(u) & u \leq B_L(u) \end{cases} \tag{4.13}$$

The situation of the **largest interval value** is relatively more complicated. Since the candidate CDF in this case should deviate from the CDF of standard uniform distribution as much as possible, it can be regarded as the either boundary function for simplicity. It cannot be directly distinguished through naked eye that which boundary function induces a larger metric value, so it is better to calculate the area metric value for both of them and take the larger one as the upper boundary of interval result.

To calculate the interval of area metric, four different areas can be divided as shown in Fig. 4.11(b), namely d_U^+, d_U^-, d_L^+, and d_L^-. Here, the subscripts "U" and "L" indicate the upper and lower boundary function, and the superscripts "+" and "−" tell whether the boundary function is over or below the CDF of standard uniform distribution. For example, d_U^+ represents the area where the *upper* boundary function is *over* the CDF of standard uniform distribution. Then the interval result $[d_L, d_U]$ can be expressed as:

$$\begin{aligned} d_L &= d_U^- + d_L^+ \\ d_U &= d_U^- + d_L^+ + \max\left(d_U^+ - d_L^+, d_L^- - d_U^-\right) \end{aligned} \tag{4.14}$$

Example 4.4 Consider the case in Example 4.2. At confidence level $1 - \alpha = 90\%$, plot the boundary of CDF of experimental data, and calculate the interval result of area metric.

Solution

In Example 4.2, all the u-values are calculated as the last column in Table 4.3 and the ECDF of u can be calculated as follows:

$$F_U(u) = \begin{cases} 0 & 0 \leq u < 0.538 \\ 0.167 & 0.538 \leq u < 0.625 \\ 0.333 & 0.625 \leq u < 0.713 \\ 0.5 & 0.713 \leq u < 0.743 \\ 0.667 & 0.743 \leq u < 0.820 \\ 0.833 & 0.820 \leq u < 0.842 \\ 1 & 0.842 \leq u \leq 1 \end{cases}$$

The boundary function, in most of the u-domain, is the combination with the ECDF of u with an offset of $\pm z_0/\sqrt{n}$. The confidence level is 90% so $\alpha = 0.1$. There are totally 6 experimental data so $n = 6$. Thus, the offset is calculated as

$$\frac{z_0}{\sqrt{n}} = \sqrt{-\frac{1}{2n}\ln\left(\frac{\alpha}{2}\right)} = \sqrt{-\frac{1}{2\times 6}\ln\left(\frac{0.1}{2}\right)} = 0.5$$

Using Eq. (4.12), the upper and lower boundary can be calculated and illustrated in Fig. 4.12. The boundary functions are expressed as follows:

$$B_L(u) = \begin{cases} 0 & 0 \leq u < 0.743 \\ 0.167 & 0.743 \leq u < 0.820 \\ 0.334 & 0.820 \leq u < 0.842 \\ 1 & 0.842 \leq u \leq 1 \end{cases}$$

$$B_U(u) = \begin{cases} 0 & 0 \leq u < 0.538 \\ 0.666 & 0.538 \leq u < 0.625 \\ 0.833 & 0.625 \leq u < 0.713 \\ 1 & 0.713 \leq u \leq 1 \end{cases}$$

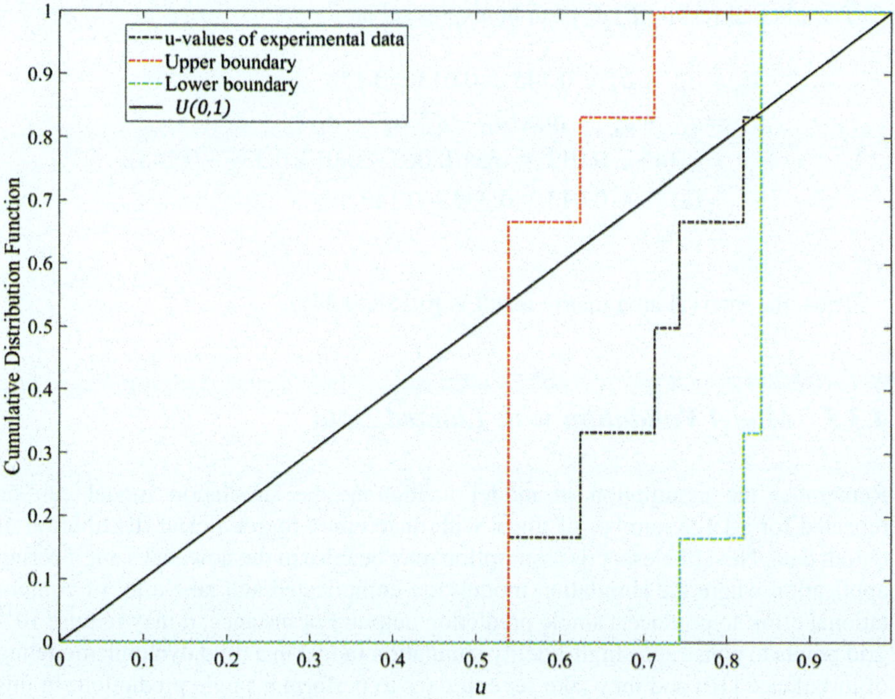

Fig. 4.12 Illustration of upper and lower boundary in Example 4.4

Since the boundary function are stepped curves, the areas d_U^+, d_U^-, d_L^+, and d_L^- can be easily calculated with simple geometry:

$$
\begin{aligned}
d_U^+ &= 0.666 \times (0.625 - 0.538) + 0.833 \times (0.713 - 0.625) + 1 \times (1 - 0.713) \\
&\quad - \frac{1}{2} \times (0.538 + 1) \times (1 - 0.538) \\
&= 0.063 \\
d_U^- &= \frac{1}{2} \times 0.538^2 = 0.145 \\
d_L^+ &= \frac{1}{2} \times (1 - 0.842)^2 = 0.013 \\
d_L^- &= \frac{1}{2} \times 0.842^2 - 0.334 \times (0.842 - 0.820) - 0.167 \times (0.820 - 0.743) = 0.334
\end{aligned}
$$

The interval result $[d_L, d_U]$ can be expressed as:

$$\begin{aligned} d_L &= d_U^- + d_L^+ = 0.145 + 0.013 = 0.158 \\ d_U &= d_U^- + d_L^+ + \max\left(d_U^+ - d_L^+, d_L^- - d_U^-\right) \\ &= 0.145 + 0.013 + \max(0.063 - 0.013, 0.334 - 0.145) \\ &= 0.145 + 0.013 + 0.334 - 0.145 \\ &= 0.347 \end{aligned}$$

Thus, the interval area metric result is [0.158, 0.347].

4.3.5 *Model Validation with Limited Data*

Remember the assumption in model validation: the simulation model can be repeated for a large number of times with uncertainty to get a clear distribution of output data. Nevertheless, this assumption may be false in the nowadays engineering application where the simulation models are complicated and take a lot of computational effort to produce a single prediction output. For instance, it may require 10^{15} grid points to construct a high-fidelity simulation model in a fluid dynamic modeling of turbulence [16] and may take several days to perform a single prediction. In this case, the underlying statistical feature of simulation data becomes fuzzy and unclear. Thus, a new model validation metric needs to be introduced using limited experimental and simulation data.

1. **Enhanced Mahalanobis Distance.**

The traditional MD relieves the burden of high-dimensional computation in model validation. But, when the sample is insufficient, it may extract wrong statistical information. Consider the following example: Model 1 and Model 2 are two simulation models with a two QoIs (y_1, y_2), which are generated as follows:

$$\begin{aligned} y_1(x,\theta) &= \sin(2x - 0.5\pi)\theta + \varepsilon_1 \\ y_2(x,\theta) &= \cos(0.25\pi x)\theta + 0.2x + \varepsilon_2 \end{aligned} \tag{4.15}$$

where y_1 and y_2 are the outputs, x is the deterministic controllable input variable and is set to be 2, and $\theta = 1.5$ is the model parameter. ε_1 and ε_2 represent the measurement errors which are correlated and follow the Gaussian distribution $N(0, 0.2^2)$ with correlation coefficient $\rho^{(1)}_{\varepsilon_1,\varepsilon_2} = 0.8$ in Model 1 and $\rho^{(2)}_{\varepsilon_1,\varepsilon_2} = -0.8$ in Model 2. In a small sample case, say, we have 50 simulation data generated using Eq. (4.15). The illustration of these generated data is shown in Fig. 4.13(a). It can be roughly inferred that despite the similarity of mean and standard deviation in both models, their covariance between the two QoIs are opposite with each other. However, the CDF of MD in both models shows no obvious difference, as shown in Fig. 4.13(c).

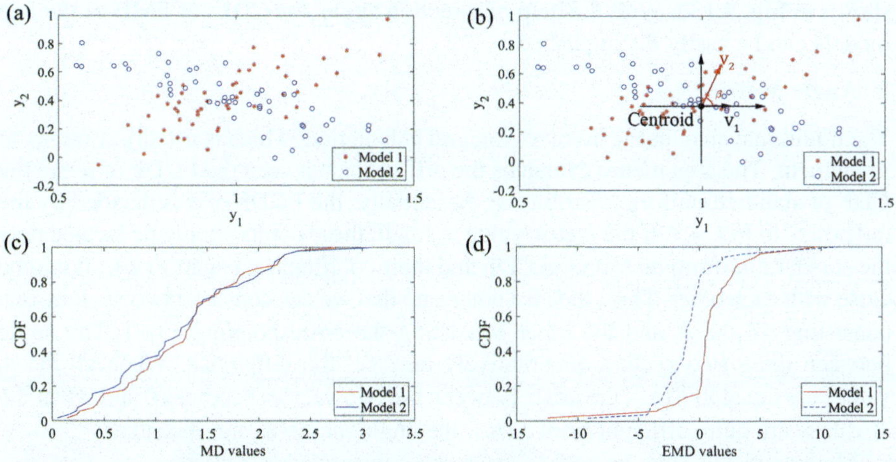

Fig. 4.13 Illustration of EMD: (**a**) Simulation output data and (**c**) CDF of MD in both models. (**b**) Illustration of EMD and (**d**) its CDF of both models [21]

In order to address this problem, some extra information needs to be appended with MD. An intuitive method is called the "direct angle" and is shown in Fig. 4.13 (b). The idea is to determine a reference direction denoted as $\boldsymbol{v}_1$, and a vector from the centroid of the distribution, $\boldsymbol{\mu}$ to the sample random point to be calculated, denoted as $\boldsymbol{v}_2$. For a sample $\boldsymbol{q} = \begin{bmatrix} q_1 & \dots & q_{n_d} \end{bmatrix}$, $\boldsymbol{v}_1$ and $\boldsymbol{v}_2$ can be expressed as follows:

$$\begin{aligned} \mathbf{v}_1 &= \begin{bmatrix} 1 & 0 & \dots & 0 \end{bmatrix} \\ \mathbf{v}_2 &= (\mathbf{q} - \boldsymbol{\mu})^T \end{aligned} \tag{4.16}$$

A direction angle β between $\boldsymbol{v}_1$ and $\boldsymbol{v}_2$ can be calculated as follows:

$$\beta = \arccos \frac{|\mathbf{v}_1 \cdot \mathbf{v}_2|}{|\mathbf{v}_1| \cdot |\mathbf{v}_2|} \tag{4.17}$$

By multiplying the cotangent of direction angle β with MD, the simulation data can be reduced to a new one-dimensional variable named Enhanced Mahalanobis Distance (EMD) y' expressed as [21]:

$$y' = y(q, \mu) \cdot \cot \beta \tag{4.18}$$

The effect of EMD can be shown in Fig. 4.13(d). Specially, for the majority data sample in Model 1, their $\cot \beta$ values are positive and for Model 2 they are negative. Thus, the CDF of EMD of Model 1 locates at the right side of that of Model 2 as shown in Fig. 4.13d. With the help of direction angle, the CDFs of EMD of the two models can be easily distinguished.

2. **Angle Metric.**

The traditional area metric involves integral calculation, which is usually complex to implement. The area metric evaluates the difference between the ECDF of u and the CDF of standard uniform distribution. Essentially, the ECDF of u is decided by the red points in Fig. 4.3. If the metric value is small, these points should be located near the standard uniform distribution CDF and their coordinates, u_i and $\widehat{F}(u_i)$, should be close with each other. Thus, the idea comes up that we can construct two vectors, one consisting all the u_i and the other consisting the corresponding $\widehat{F}(u_i)$. The angle between these two vectors quantitatively assesses the difference. When all the u_i values are equal to $\widehat{F}(u_i)$, the angle between the two vectors becomes 0 and when the u_i values are quite different from $\widehat{F}(u_i)$, the angle increases significantly.

Mathematically, two vectors $\boldsymbol{t}_1$ and $\boldsymbol{t}_2$ are constructed as follows:

$$\begin{aligned} \mathbf{t}_1 &= \begin{bmatrix} u_1 & u_2 & \ldots & u_n \end{bmatrix} \\ \mathbf{t}_2 &= \begin{bmatrix} \widehat{F}(u_1) & \widehat{F}(u_2) & \ldots & \widehat{F}(u_n) \end{bmatrix} \end{aligned} \tag{4.19}$$

and the angle between them can be obtained as follows:

$$d_a = \frac{|\mathbf{t}_1 \cdot \mathbf{t}_2|}{|\mathbf{t}_1| \cdot |\mathbf{t}_2|} \tag{4.20}$$

In order to produce an interval metric result, the method in Sect. 4.3.4 can be used to construct the corresponding candidate CDFs, and by applying Eq. (4.20), the upper and lower limit of the interval can be obtained and regarded as the final result. Details of the proposed interval EMD-pooling angle metric are elaborated in the paper [21].

Exercises

1. Suppose we are simulating a process of ultrasonic welding as shown in Fig. 4.14. The two workpieces are clamped and when the ultrasonic wave is applied, they undertake the frictional and plastic deformation due to the heat generated at the contact interface so that they are joint together.

   Now we are researching about the temperature at the given measurement point marked in Fig. 4.14 at a certain time $t = 0.03$ s. The simulation data are provided in Table 4.5 and the experimental data are provided in Table 4.6.

   (1) Plot the ECDF of measured temperature at $t = 0.03$ s.
   (2) Calculate the area metric using the simulation and measured data given.

2. Recall that in PIT, we only introduce situation with only one-dimensional variable. The reader is suggested to examine whether this principle stands with two-dimensional variables. In other words, given arbitrary two correlated random

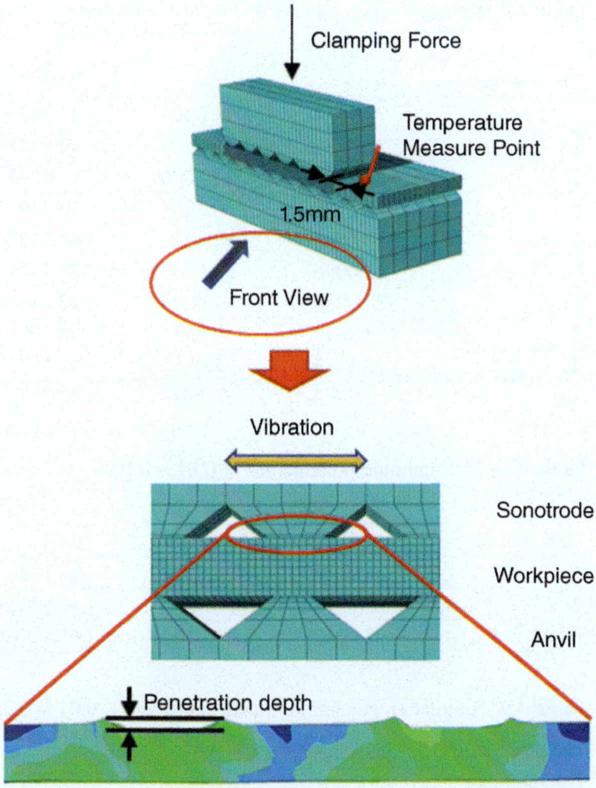

Fig. 4.14 Illustration of ultrasonic welding

Table 4.5 Simulation temperature at $t = 0.03$ s

Number	Temperature (°C)
1	41.114
2	40.824
3	41.427
4	40.559
5	40.720
6	40.742
7	40.834
8	40.798
9	41.158
10	40.717

Table 4.6 Measured temperature at $t = 0.03$ s

Number	Temperature (°C)
1	35.15
2	43.11
3	40.12

Table 4.7 Simulation temperature at $t = 0.01 \sim 0.05$ s

	Time (s)				
Data number	0.01	0.02	0.03	0.04	0.05
1	22.9454	31.0404	41.1136	50.96	60.3901
2	22.9971	30.9324	40.8243	50.1711	59.009
3	23.0042	31.2586	41.4274	51.2563	60.3456
4	22.9689	30.7804	40.5591	50.0103	58.9507
5	22.9581	30.867	40.7197	50.3604	59.4919
6	23.0096	30.9369	40.7421	50.4347	59.8674
7	22.9989	31.0268	40.8343	50.191	59.1746
8	23.0319	30.953	40.7981	50.5705	59.6997
9	22.9388	31.0388	41.1577	51.0518	60.4342
10	22.9412	30.865	40.717	50.2938	59.498

Table 4.8 Measured temperature at $t = 0.01 \sim 0.05$ s

	Time (s)				
Data number	0.01	0.02	0.03	0.04	0.05
1	22.9454	31.0404	41.1136	50.96	60.3901
2	22.9971	30.9324	40.8243	50.1711	59.009
3	23.0042	31.2586	41.4274	51.2563	60.3456

Table 4.9 Simulation penetration depth (mm) at $t = 0.01 \sim 0.05$ s

	Time (s)				
Data number	0.01	0.02	0.03	0.04	0.05
1	4.58E-03	7.12E-03	1.14E-02	1.45E-02	1.68E-02
2	3.04E-03	7.10E-03	1.13E-02	1.50E-02	1.90E-02
3	4.97E-03	7.78E-03	1.30E-02	1.82E-02	2.25E-02
4	4.06E-03	6.12E-03	9.54E-03	1.31E-02	1.55E-02
5	3.74E-03	6.12E-03	9.41E-03	1.18E-02	1.35E-02
6	4.09E-03	6.76E-03	9.80E-03	1.42E-02	1.63E-02
7	4.36E-03	7.41E-03	1.14E-02	1.47E-02	1.73E-02
8	4.02E-03	5.90E-03	8.75E-03	1.04E-02	1.27E-02
9	4.36E-03	6.87E-03	1.07E-02	1.29E-02	1.46E-02
10	4.29E-03	6.61E-03	9.81E-03	1.54E-02	1.87E-02

continuous variables Y_1 and Y_2 with joint CDF $F(y_1, y_2)$, examine whether a new random variable $U = F(Y_1, Y_2)$ is distributed with standard uniform distribution.

3. In Problem 1, this time we consider about the simulation accuracy at time 0.01 ~ 0.05 s. The simulation data and the measured data are summarized in Tables 4.7 and 4.8, respectively. Calculate the area metric using u-pooling to evaluate the overall accuracy of the simulation model.
4. In Problem 1, this time not only the temperature at given point but also the penetration depth (marked in Fig. 4.14) is considered. The additional simulated and measured penetration depth are shown in Tables 4.9 and 4.10, respectively.

Table 4.10 Measured penetration depth (mm) at $t = 0.01 \sim 0.05$ s

	Time (s)				
Data number	0.01	0.02	0.03	0.04	0.05
1	0.0177	0.0255	0.0295	0.0341	0.0379
2	0.0225	0.0288	0.032	0.034	0.0369
3	0.0242	0.0299	0.0328	0.0348	0.0371

Calculate the area metric using u-pooling and Mahalanobis distance with the given data.

References

1. Roy, C. J., & Oberkampf, W. L. (2011). A comprehensive framework for verification, validation, and uncertainty quantification in scientific computing. *Computer Methods in Applied Mechanics and Engineering, 200*, 2131–2144.
2. Oberkampf, W. L., & Barone, M. F. (2006). Measures of agreement between computation and experiment: validation metrics. *Journal of Computational Physics, 217*, 5–36.
3. Oberkampf, W. L., & Trucano, T. G. (2002). Verification and validation in computational fluid dynamics. *Progress in Aerospace Sciences, 38*, 209–272.
4. Sargent, R. G. (1991). *Simulation model verification and validation*. Institute of Electrical and Electronics Engineers (IEEE).
5. Thacker, B. H., Doebling, S. W., Hemez, F. M., Anderson, M. C., Pepin, J. E., & Rodriguez, E. A. (2004). *Concepts of model verification and validation*, p. 41.
6. Wikipedia. (2022). *Verification and validation of computer simulation models*.
7. Wongsuphasawat, K., Smilkov, D., Wexler, J., Wilson, J., Mane, D., Fritz, D., et al. (2017). Visualizing dataflow graphs of deep learning models in tensorflow. *IEEE Transactions on Visualization and Computer Graphics, 24*, 1–12.
8. Zhang, Q., Hu, W., Liu, Z., & Tan, J. (2020). TBM performance prediction with Bayesian optimization and automated machine learning. *Tunnelling and Underground Space Technology, 103*, 103493.
9. Hu, W., Chen, W., Wang, X., Jiang, Z., Wang, Y., Verma, A. S., et al. (2021). A computational framework for coating fatigue analysis of wind turbine blades due to rain erosion. *Renewable Energy, 170*, 236–250.
10. Letson, F., Barthelmie, R. J., Hu, W., & Pryor, S. C. (2019). Characterizing wind gusts in complex terrain. *Atmospheric Chemistry and Physics, 19*, 3797–3819.
11. Letson, F., Barthelmie, R., Hu, W., Brown, L., & Pryor, S. (2019). Wind gust quantification using seismic measurements. *Natural Hazards, 99*, 355–377.
12. Luk, O. O., Lakhlili, J., Hoenen, O., Toussaint, U. V., Scott, B. D., & Coster, D. P. (2021). Towards validated multiscale simulations for fusion. *Philosophical Transactions of the Royal Society A, 379*, 20200074.
13. Sornette, D., Davis, A., Ide, K., Vixie, K., Pisarenko, V., & Kamm, J. (2007). Algorithm for model validation: Theory and applications. *Proceedings of the National Academy of Sciences, 104*, 6562–6567.
14. Mahadevan, S., & Rebba, R. (2005). Validation of reliability computational models using Bayes networks. *Reliability Engineering & System Safety, 87*, 223–232.
15. Chen, W., Xiong, Y., Tsui, K.-L., & Wang, S. (2008). A design-driven validation approach using Bayesian prediction models. *Journal of Mechanical Design, 130*, 021101.

16. Ferson, S., Oberkampf, W. L., & Ginzburg, L. (2008). Model validation and predictive capability for the thermal challenge problem. *Computer Methods in Applied Mechanics and Engineering, 197*, 2408–2430.
17. Yoo, Y., Jung, U.-J., Han, Y. H., & Lee, J. (2021). Data augmentation-based prediction of system level performance under model and parameter uncertainties: role of designable generative adversarial networks (DGAN). *Reliability Engineering & System Safety, 206*, 107316.
18. Li, L., & Lu, Z. (2018). A new method for model validation with multivariate output. *Reliability Engineering & System Safety, 169*, 579–592.
19. Hogg, R. V., & Craig, A. T. (1995). *Introduction to mathematical statistics* (5th ed.). Englewood Hills.
20. Genest, C., & Rivest, L.-P. (2001). On the multivariate probability integral transformation. *Statistics & Probability Letters, 53*, 391–399.
21. Hu, W., Zhang, T., Liu, Z., Cheng, J., Lou, Y., Deng, X., et al. (2022). A new validation metric for models with correlated responses using limited experimental and simulation data. *Journal of Mechanical Design, 144*, 111702.
22. Zhao, L., Lu, Z., Yun, W., & Wang, W. (2017). Validation metric based on Mahalanobis distance for models with multiple correlated responses. *Reliability Engineering & System Safety, 159*, 80–89.
23. Hu, J., Jiang, P., Zhou, Q., McKeand, A., & Choi, S.-K. J. (2020). Model validation methods for multiple correlated responses via covariance-overlap based distance. *Journal of Mechanical Design, 142*(4).
24. Li, W., Chen, W., Jiang, Z., Lu, Z., & Liu, Y. (2014). New validation metrics for models with multiple correlated responses. *Reliability Engineering & System Safety, 127*, 1–11.
25. Wang, N., Yao, W., Zhao, Y., Chen, X., Zhang, X., & Li, L. (2018). A new interval area metric for model validation with limited experimental data. *Journal of Mechanical Design, 140*, 061403.

Chapter 5
Time-Independent Reliability Analysis

Nomenclature

ANOVA	Analysis of variance
BDR	Bivariate dimension reduction
D	Second-order Hessian matrix
DoE	Design of experiment
DR	Dimension reduction
EFF	Expected feasibility function
EGRA	Efficient global reliability analysis
FORM	First-order reliability methods
$f_X(x)$	Joint PDF
$G(\cdot)$	Performance function
$\widehat{G}$	Constructed Kriging model
HDMR	High-dimensional model representation
$h_X(x)$	Importance sampling density function
$I_{\Omega_F}(x)$	Indicator function
κ_i	Principal curvatures
LHS	Latin hypercube sampling
MCS	Monte Carlo simulation
MPP	Most probable point
N	Number of random variable dimension
n_{MC}	Number of MCS points
$n_{G \leq 0}$	Number of sample points with negative sign of the performance function
PHM	Prognostics and health management
P_F	Probability of failure
R	Reliability
R	Orthogonal rotation matrix
RBDO	Reliability-based design optimization
SORM	Second-order reliability methods

W. Hu, *Design Optimization Under Uncertainty*,
https://doi.org/10.1007/978-3-031-49208-2_5

T(x)	Transformation
T	Time factor
UDR	Univariate dimension reduction
$\boldsymbol{U}(\boldsymbol{x})$	Active learning function
u^*	MPP on the limit state surface
$\mathbf{X}$	Random input variable
$\boldsymbol{Y}^*$	MPP in Y-space
$\boldsymbol{\beta}$	Reliability index
$\boldsymbol{\mu}$	Mean
$\boldsymbol{\sigma}$	Standard deviation
$\boldsymbol{\Phi}$	Cumulative probability distribution of standard normal distribution
$\boldsymbol{\phi}$	Standard normal distribution density function
ε_{MCS}	Coefficient of variation

5.1 Basic Concept of Reliability

5.1.1 Introduction

Engineering structures and systems are vulnerable to a variety of stressors that threaten their function and integrity. Failures of engineering systems cause significant economic and societal losses. In engineering practice, there are various sources of uncertainty (e.g., material properties, loads, and geometric tolerances). These uncertainties often have varying degrees of impact on the performance of different aspects of the structure. Traditional deterministic analysis can no longer meet the need for accurate reliability evaluation. This is because traditional methods simplify the problem by treating uncertainties as deterministic parameters and relying on empirical safety factors to account for these uncertainties. Safety factors are derived from past experience but are not absolute guarantees of safety or satisfactory performance. Engineering design is usually a trade-off between maximizing safety levels and minimizing cost. Deterministic safety factors do not provide enough information to enable optimal use of available resources to maximize security. Instead, reliability analysis (RA) can consider the uncertainty in the structure or system and its stressors, and analyze the performance of the structure and system in terms of the probability that the structure and system will fail to meet the specified objectives. Therefore, the importance of reliability in engineering has been widely recognized in the past few decades. A brief discussion about the basic concept and importance of reliability has been conducted in Chap. 1, which will be reviewed in a more illustrative manner in this chapter. The main contribution of this chapter is to demonstrate the reliability calculation processes based on the most probable point (MPP) methods and the sampling methods.

5.1.2 Concept of Reliability

Recall that reliability R is defined in Chap. 1 as:

$$R = 1 - P_F \tag{5.1}$$

where P_F is probability of failure defined in Chap. 1 as:

$$P_F = \int \cdots \int_{G(\cdot)<0} f_{\mathbf{X}}(\mathbf{x}) d\mathbf{x} \tag{5.2}$$

where $\mathbf{X} = (X_1, X_2, \ldots, X_N)$ is an N-dimensional random input variable, $f_X(\boldsymbol{x})$ is the joint PDF of $\mathbf{X}$, and $G(\cdot)$ is the performance function of the system. In reliability analysis, equating the system performance function $G(\cdot)$ to zero, i.e., $G(\cdot) = 0$, gives us the so-called limit state function which separates the safe region $G(\cdot) > 0$ from the failure region $G(\cdot) < 0$.

The concept of limit state in a two-dimensional case is illustrated in Fig. 5.1.

In summary, the reliability can be sorted into two types [1]:

1. *Time-independent reliability*: referring to the probability that a system's performance (e.g., fatigue, corrosion, and fracture) meets its marginal value (or requirement) under uncertainty at very beginning of operation ($t = 0$);
2. *Time-dependent reliability*: referring to the probability that the system meets this requirement for a stated period of time over the system's expected lifespan ($t > 0$).

It should be noted that the concept of *time-independent reliability* is mainly used for designing an engineered system in a way that ensures a high *built-in* reliability; whereas, the concept of *time-dependent reliability* is often employed to design an

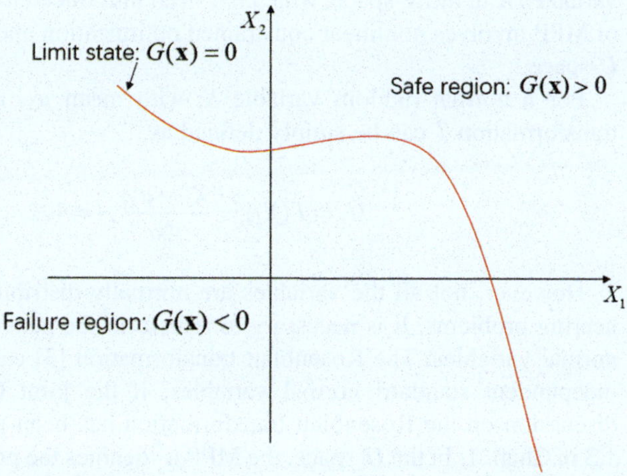

Fig. 5.1 Limit state in a two-dimensional case

engineered system and/or its affiliated prognostics and health management (PHM) system to attain a high *operational* reliability. The purpose of this chapter is to present advanced methods and tools that can be used to quantify time-independent reliability. The time-dependent reliability analysis will be introduced in detail in Chap. 6. In general, the reliability analysis methods can be categorized into two types: (i) most probable point (MPP)-based methods and (ii) sampling methods. Different methods of approximating limit state functions form the basis of different reliability analysis algorithms. First- and second-order reliability methods (FORM/SORM) [2] are standard methods of reliability analysis. They are based on linear (first-order) and quadratic (second-order) approximations of the limit state $G(x) = 0$ tangent to the MPP. Besides these MPP-based methods, there are several sampling methods (e.g., Monte Carlo simulation (MCS) and importance sampling) for reliability analysis.

From Eq. (5.1) it is obvious that, in order to calculate reliability of a system, the crucial step is to obtain the probability of failure P_F. When the performance function of system can be explicitly expressed, a theoretical method can be applied to directly solve Eq. (5.2), which is elaborated in Sect. 5.2. However, most engineering applications are too complicated that the performance functions are usually unknown. In this case, a sampling method is needed which is discussed in Sect. 5.3.

5.2 MPP-Based Methods for Reliability Analysis

5.2.1 First-Order Reliability Method

First-order reliability method is based on linear approximation of the limit state surface $(x) = 0$ at the closest point of the surface to the origin in the $\boldsymbol{U}$-space. The point is regarded as most probable point (MPP). The $\boldsymbol{U}$-space is composed of independent standard normal variables $\boldsymbol{U}$ that are transformed from the random variables $\boldsymbol{X}$ in the $\boldsymbol{X}$-space, which has been introduced in Chap. 1. The determination of MPP involves nonlinear constrained optimization and is usually performed in the $\boldsymbol{U}$-space.

For a normal random variable X_i with mean μ_{Xi} and standard deviation σ_{Xi}, transformation T can be simply defined as

$$U_i = T(X_i) = \frac{X_i - \mu_{Xi}}{\sigma_{Xi}}, i = 1, 2, \ldots, N \tag{5.3}$$

However, not all the variables are normally distributed, as is common in engineering problems. It is necessary to transform the general variables into equivalent normal variables. The Rosenblatt transformation [3] can be used to obtain a set of independent standard normal variables, if the joint CDF is available. Detailed discussion on the Rosenblatt transformation has been provided at the end of Sect. 1.3 in Chap. 1. In the $\boldsymbol{U}$-space, the MPP $\mathbf{u}^*$ denotes the point on the limit state surface which has the minimum distance to the origin. The distance, denoted as β, is called the reliability index, expressed as

$$\beta = \left\| \mathbf{u}^* \right\| = \left[\sum_{i=1}^{N} \left(u_i^* \right)^2 \right]^{1/2} \tag{5.4}$$

And the first-order estimation of the probability of failure can be computed as

$$P_F = \Phi(-\beta) \tag{5.5}$$

where Φ is the cumulative probability distribution of standard normal distribution.

Example 5.1 Consider the following performance function in ***U***-space

$$G(u_1, u_2) = 2u_1 - u_2 + 6$$

Calculate the reliability index β, and the probability of failure by FORM.

Solution

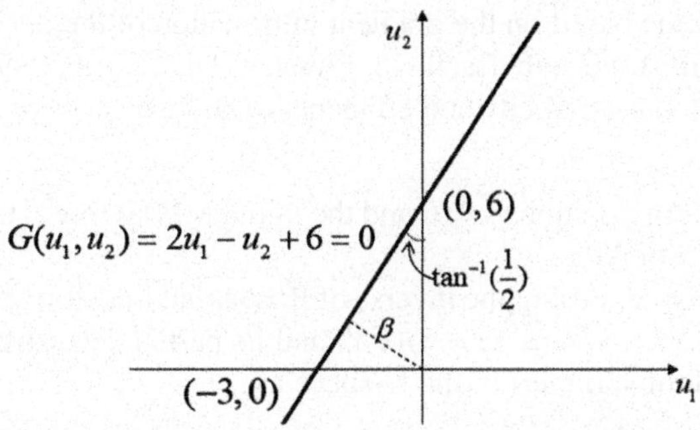

As illustrated in the figure above, the solution can be conducted in the following steps:

Step 1: Find the MPP point. Calculate the shortest distance from the origin to the LSF based on the figure above.

$$\beta = 6\sin\left(\tan^{-1}\left(\frac{1}{2}\right)\right) = \frac{6\sqrt{5}}{5} \approx 2.68$$

Step 2: According to Eq. (5.5), calculate the probability of failure.

$$P_F = \Phi(-\beta) = \Phi(-2.68) = 0.0037$$

The reliability index β is an efficient way for reliability analysis. However, this method only gives an exact reliability estimation if the limit state surface is a linear form, as is the case in the above Example 5.1. The algorithms implementing FORM involves several steps.

1. First, the $\boldsymbol{X}$-space of uncertain parameters $\boldsymbol{X}$ is transformed into a new d-dimensional $\boldsymbol{U}$-space, consisting of independent standard normal variables, $\boldsymbol{U}$. The original limit state, $G(\boldsymbol{x}) = 0$, is mapped into the new limit state $g(\boldsymbol{u}) = 0$;
2. The MPP $\mathbf{u}^*$ is determined by using an appropriate nonlinear optimization algorithm;
3. The limit state, $G(\boldsymbol{u}) = 0$, is approximated by a surface tangent to it at the MPP.

The remaining task is to search for the MPP. Generally, this task can be formulated as an optimization problem with a deterministic constraint in the $\boldsymbol{U}$-space, expressed as

$$\begin{aligned} &\text{Minimize } \left\|\mathbf{u}\right\| \\ &\text{Subject to } G(\mathbf{u}) = 0 \end{aligned} \tag{5.6}$$

where the optimal result is the MPP $\mathbf{u}^*$. The MPP search requires an iterative optimization scheme based on the gradient information of the performance function $G(\boldsymbol{u})$. The Hasofer-Lind and Rackwitz-Fiessler (HL-RF) method [4] is the most widely used due to its simplicity and efficiency. The main steps of HL-RF are shown as following.

1. Set the number of iterations $k = 0$ and the initial MPP $\boldsymbol{u} = \boldsymbol{u}^{(0)}$ that corresponds to the mean value of $\boldsymbol{X}$.
2. Transform $\boldsymbol{u}^{(k)}$ to $\boldsymbol{x}^{(k)}$ using the inverse of Rosenblatt transformation. Compute the performance function $G(\boldsymbol{u}^{(k)}) = G(\boldsymbol{x}^{(k)})$ and its partial derivatives with respect to the input random variables in the U-space as

$$\nabla_{\mathrm{U}} G\left(\mathbf{u}^{(k)}\right) = \left[\frac{\partial G}{\partial U_1}, \frac{\partial G}{\partial U_1}, \ldots, \frac{\partial G}{\partial U_1}\right]_{\mathrm{U}=\mathbf{u}^{(k)}} \tag{5.7}$$

3. Update the search point at current iteration as

$$\begin{aligned} \mathbf{u}^{(k+1)} &= \left(\mathbf{u}^{(k)} \cdot \mathbf{n}^{(k)} - \frac{G\left(\mathbf{u}^{(k)}\right)}{\left\|\nabla_{\mathrm{U}} G(\mathbf{u}^{(k)})\right\|}\right) \mathbf{n}^{(k)} \\ &= \left[\mathbf{u}^{(k)} \cdot \nabla_{\mathrm{U}} G\left(\mathbf{u}^{(k)}\right) - G\left(\mathbf{u}^{(k)}\right)\right] \frac{\nabla_{\mathrm{U}} G(\mathbf{u}^{(k)})}{\|\nabla_{\mathrm{U}} G(\mathbf{u}^{(k)})\|^2} \end{aligned} \tag{5.8}$$

where $\boldsymbol{n}^{(k)}$ is the normalized steepest ascent direction of $G(\boldsymbol{U})$ at $\boldsymbol{u}^{(k)}$, expressed as:

$$\mathbf{n}^{(k)} = \frac{\nabla_{\mathbf{U}} G(\mathbf{u}^{(k)})}{\left\| \nabla_{\mathbf{U}} G(\mathbf{u}^{(k)}) \right\|} \tag{5.9}$$

4. Go to Steps 2 and 3 until the convergence of $\boldsymbol{u}$.

Example 5.2 Consider the following performance function

$$G(X_1, X_2) = 1 - \frac{20}{X_1 + X_2 + 5}$$

where X_1 and X_2 each follows a normal distribution with the mean 4 and the standard deviation 1. Find the MPP and compute the probability of failure by FORM.

Solution:

The HL-RF method is used here, and the first iteration is shown as follows:

Step 1: Set the number of iterations $k = 0$ and the initial point $\boldsymbol{u}^{(0)} = (0, 0)$;
Step 2: Transform $\boldsymbol{u}^{(0)}$ to $\boldsymbol{x}^{(0)}$ with the inverse of Eq. (5.2) to obtain:

$$\begin{aligned}
x_1^{(0)} &= \mu_{X1} + U_1^{(0)} \sigma_{X1} = 4 + 0 \times 1 = 4 \\
x_2^{(0)} &= \mu_{X2} + U_2^{(0)} \sigma_{X2} = 4 + 0 \times 1 = 4 \\
\mathbf{x}^{(0)} &= \left(x_1^{(0)}, x_2^{(0)} \right) = (4, 4)
\end{aligned}$$

Compute the performance function $G(\boldsymbol{x}^{(0)})$ as

$$G(X_1, X_2) = 1 - \frac{20}{4 + 4 + 5} \approx -0.5385$$

and the partial derivatives as

$$\frac{\partial G}{\partial U_1} = \frac{\partial G}{\partial X_1} \frac{\partial X_1}{\partial U_1} = \frac{20}{(X_1 + X_2 + 5)^2} \sigma_{X1} = \frac{20}{(4 + 4 + 5)^2} \times 1 \approx 0.1183$$

$$\frac{\partial G}{\partial U_2} = \frac{\partial G}{\partial X_2} \frac{\partial X_2}{\partial U_2} = \frac{20}{(X_1 + X_2 + 5)^2} \sigma_{X2} = \frac{20}{(4 + 4 + 5)^2} \times 1 \approx 0.1183$$

Step 3: Update the search point at the current iteration as

Table 5.1 Iteration result in Example 5.2

Iteration	U_1	U_2	$G(U)$	$\frac{\partial G}{\partial U_1}$	$\frac{\partial G}{\partial U_2}$
0	0	0	0.5385	0.1183	0.1183
1	2.2760	2.2760	0.1395	0.0649	0.0649
2	3.3505	3.3505	0.0152	0.0515	0.0515
3	3.4981	3.4981	0.0002	0.0500	0.0500
4	3.5	3.5	0		

$$\mathbf{u}^{(1)} = \left[\mathbf{u}^{(0)} \cdot \nabla_U G\left(\mathbf{u}^{(0)}\right) - G\left(\mathbf{u}^{(0)}\right)\right] \frac{\nabla_U G(\mathbf{u}^{(0)})}{\|\nabla_U G(\mathbf{u}^{(0)})\|^2}$$
$$= [(0,0) \cdot (0.1183, 0.1183) - (-0.5385)] \frac{(0.1183, 0.1183)}{0.1183^2 + 0.1183^2}$$
$$= (2.2760, 2.2760)$$

$$\mathbf{x}^{(1)} = T\left(\mathbf{u}^{(1)}\right) = \sigma \mathbf{u}^{(1)} + \mu = (6.2760, 6.2760)$$

The subsequent iteration process is analogous to the first iteration and the result is listed in Table 5.1.

Therefore, the MPP $\mathbf{u}^*$ is (3.5, 3.5). The reliability index and the probability of failure are calculated.

$$\beta = \left\|\mathbf{u}^*\right\| = \left[\sum_{i=1}^{N} \left(\mathbf{u}_i^*\right)^2\right]^{1/2} = \|(3.5, 3.5)\| \approx 4.9497$$

$$P_F = \Phi(-\beta) = \Phi(-4.9497) = 3.7164 \times 10^{-7}$$

5.2.2 *Second-Order Reliability Method*

As can be seen in Fig. 5.2, FORM constructs a first-order or linear approximation to the limit state surface at MPP. If $G(u)$ is nonlinear, the resulting error will be large. In such cases, SORM, which considers the curvature information in the MPP, is more credible. The comparison between FORM and SORM is shown in Fig. 5.2. The accuracy of reliability estimation can be improved by nonlinear approximation of the limit state surface using SORM. This increase in accuracy is achieved by utilizing more information, especially about the second derivatives of the input random variables.

There are many SORM forms (e.g., Breitung [5], Hohenbichler and Rackwitz [6], and Tvede [7]). This chapter details one of these methods, Breitung's asymptotic

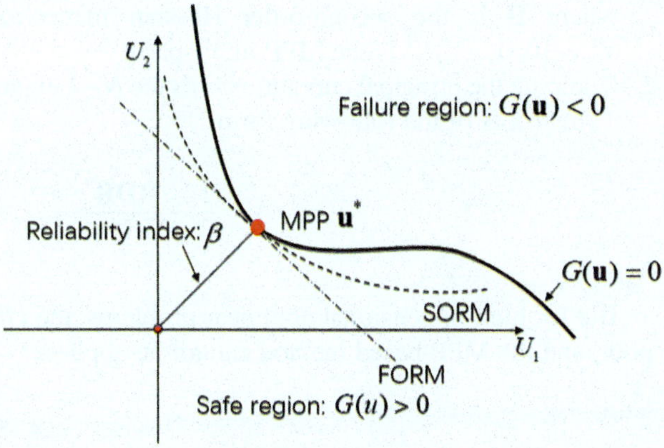

Fig. 5.2 FORM/SORM approximations of a performance function at U-space

solution. The Breitung's SORM approximation of the probability of failure can be expressed in an explicit form as

$$P_F = \Phi(-\beta)\prod_{i=1}^{N-1}(1+\beta\kappa_i)^{-\frac{1}{2}} \tag{5.10}$$

where κ_i, $i = 1, 2, \ldots, N$, are the principal curvatures of $G(u)$ at MPP. Therefore, κ_i need to be calculated, which can be completed in two steps:

1. Rotate the standard normal variables U_i (in U-space) to a new set of standard normal variables Y_i (in Y-space), where the last variable Y_N is in the same direction as the unit gradient vector of $G(u)$ in MPP. To do this, we generate an orthogonal rotation matrix $\mathbf{R}$, which can be derived from a simple matrix $\mathbf{R}_0$ expressed as

$$\mathbf{R}_0 = \begin{bmatrix} 1 & 0 & \cdots & 0 \\ 0 & 1 & \cdots & 0 \\ \vdots & \vdots & \ddots & \vdots \\ \frac{\partial G(\mathbf{u}^*)/\partial u_1}{|\nabla G(\mathbf{u}^*)|} & \frac{\partial G(\mathbf{u}^*)/\partial u_2}{|\nabla G(\mathbf{u}^*)|} & \cdots & \frac{\partial G(\mathbf{u}^*)/\partial u_N}{|\nabla G(\mathbf{u}^*)|} \end{bmatrix} \tag{5.11}$$

where the last row consists of the components of the unit gradient vector of the limit state function at the MPP. Next, an orthogonal matrix $\mathbf{R}$ can be obtained by orthogonalizing $\mathbf{R}_0$ using the Gram-Schmidt algorithm. In the Y-space, the second-order approximation of the limit state function at the MPP can be expressed as

$$G(\mathbf{Y}) = -Y_N + \beta + \frac{1}{2}(\mathbf{Y}-\mathbf{Y}^*)^{\mathrm{T}}\mathbf{R}\mathbf{D}\mathbf{R}^{\mathrm{T}}(\mathbf{Y}-\mathbf{Y}^*) \tag{5.12}$$

where **D** is the second-order Hessian matrix of the size N by N; and $Y^* = [0, 0, \ldots, \beta]^T$ is the MPP in Y-space.

2. Compute the principal curvatures κ_i as the N - 1 eigenvalues of an $(N-1) \times (N-1)$ matrix A of the following form

$$\mathbf{A} = \frac{1}{2} \frac{\mathbf{RDR}^{\mathrm{T}}}{|\nabla \mathrm{G}(\mathbf{u}^*)|} \tag{5.13}$$

But for high-dimensional nonlinear problems, the effect of FORM and SORM is poor, and the MPP-based method cannot be applied.

5.3 Sampling Methods for Reliability Analysis

During the last several decades, sampling methods have played an important role in advancing research in reliability analysis. These methods generally involve generation of random samples of input random variables, deterministic evaluations of the performance function at these random samples, and post-processing to extract the probabilistic characteristics (e.g., statistical moments, reliability, and PDF) of the performance function. In this section, direct MCS, the simplest yet very useful sampling method, is briefly introduced, followed by a smart MCS method that borrows ideas from MPP-based methods, namely the importance sampling method.

5.3.1 Monte Carlo Simulation

The MCS captures the frequency of the target event by means of "experimentation". It is utilized to estimate the probability of the target event without knowing the nature of the target event, which is very suitable for the black-box model. The basic idea behind MCS is to approximate the underlying distribution and associated probabilistic characteristics (e.g., mean, variance, and higher-order moments) of a random function by computing the value of the function from a certain amount of simulated random samples. In addition to the direct Monte Carlo method, many Monte Carlo methods based on surrogate models have been proposed [8, 9].

1. **Direct MCS**

Define the performance function $G(\boldsymbol{x}) < 0$ as failure event. The sign of the performance function can be used to judge whether the constraint fails (positive or negative). In addition, the boundary state between the positive and negative domains is defined as the limit state (i.e., $G(\boldsymbol{x}) = 0$). Hence, as illustrated in Fig. 5.3, by spreading a large number n_{MC} of random sample points in the design space, the performance function could be evaluated. After counting the number of sample

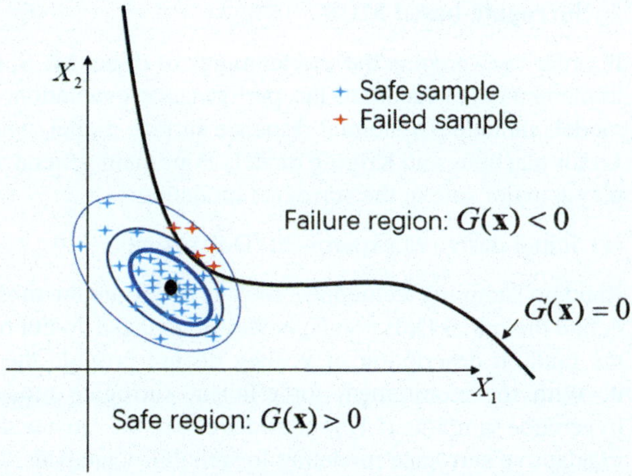

Fig. 5.3 Concept of reliability analysis using direct MCS

points with negative sign of the performance function $n_{G \le 0}$, the failure probability of the constraint can be approximated by the ratio of $n_{G \le 0}$ to n_{MC} [10].

$$P_F = \frac{n_{G \le 0}}{n_{MC}} \tag{5.14}$$

Besides, the failure domain set Ω_F is introduced as the collection of $\boldsymbol{x}$ which yields $G(\boldsymbol{x}) < 0$. The sample points in the failure domain set are the failure points. An indicator function $I_{\Omega_F}(x)$ is used to determine whether the sample points are in the failure domain:

$$I_{\Omega_F}(x) = \begin{cases} 1, x \in \Omega_F \\ 0, \text{otherwise} \end{cases} \tag{5.15}$$

Direct MCS becomes more precise as the number of sample points increases. This is also its biggest advantage. In theory, as long as there are enough sample points, the accuracy of the obtained results can be guaranteed. Its shortcomings are also particularly obvious. For the problem of low probability of failure, the required sample size is very large, which brings huge calculation and time costs. The number of random sample points needed is estimated by giving a coefficient of variation (COV) ε_{MCS} for quantifying the probability of failure:

$$\varepsilon_{MCS} = \sqrt{\frac{(1 - P_F)}{n_{MC} \times P_F}} \times 100\% \tag{5.16}$$

For example, there is a reliability problem with the $P_F = 0.5\%$. If we want to make the coefficient of variation (COV) ε_{MCS} less than 0.05, the MCS sample n_{MC} should be more than 79600.

2. **Surrogate-based MCS**

In order to overcome the shortcomings of direct MCS, surrogate models have been implemented to facilitate the performance evaluation. Commonly used surrogate models include polynomial response surface model, neural network model, support vector machine, and Kriging model. Four main general aspects can be highlighted to play a major role in the surrogate modeling:

(1) Initial design of experiment (DoE) points

Random sampling techniques, such as MCS, are the most fundamental techniques to define the initial DoE points; however, as these do not obey any criterion other than the random description of $\boldsymbol{x}$, they do not provide the most efficient approach to it. With the requirement for efficient surrogate modeling techniques, the Latin hypercube sampling (LHS) becames one of the most widely implemented techniques in adaptive surrogate modeling for reliability analysis. In order to meet the demand for some complex engineering problems, [11, 12] recently proposed the usage of uniform DoE points, and [13] utilized the Sobol sequences for the selection of initial DoE points.

(2) DoE enrichment and stopping criterion

Despite the importance of the initial DoE, the possibility to adaptively enrich the initial DoE is one of the main features of adaptive surrogate modeling. It includes improvements to establish $G(\boldsymbol{x})$ substitution capabilities in order to select other DoE points that are expected to improve this approximation. This happens repeatedly until the stop criterion is satisfied. Learning functions are convenient mathematical functions that weight alternative model properties to find the best candidates for improved DoE. They evaluate a set of candidate criteria, which are essentially based on consideration of uncertainty in model approximation to failure regions, and select the new most promising DoE-enriching criteria.

Learning functions are the present state-of-art technique for DoE enrichment. Echard et al. [14] proposed the adaptive Kriging model method (AK-MCS). The Adaptive Kriging Monte Carlo method consists of an active learning reliability method combining Kriging and Monte Carlo simulation. Kriging active learning methods for reliability analysis have actually been proposed in previous works [15]. Two learning functions are used here, namely the expected feasibility function (EFF) and the active learning function $U(\boldsymbol{x})$.

EFF comes from the global optimization (efficient global reliability analysis, EGRA) method [15]. It is able to provide at a certain sample point the degree to which the response value of its performance equation satisfies the equality constraint $G(\boldsymbol{x}) = a$, which is in a self-defined area $a \pm \varepsilon$. EEF is defined as shown in Eq. (5.17).

$$EFF(x) = \left(\widehat{G}(x) - a\right)\left[2\Phi\left(\frac{a - \widehat{G}(x)}{\sigma_{\widehat{G}}(x)}\right) - \Phi\left(\frac{(a-\varepsilon) - \widehat{G}(x)}{\sigma_{\widehat{G}}(x)}\right) - \Phi\left(\frac{(a+\varepsilon) - \widehat{G}(x)}{\sigma_{\widehat{G}}(x)}\right)\right]$$
$$- \sigma_{\widehat{G}}(x)\left[2\phi\left(\frac{a - \widehat{G}(x)}{\sigma_G(x)}\right) - \phi\left(\frac{(a-\varepsilon) - \widehat{G}(x)}{\sigma_G(x)}\right) - \phi\left(\frac{(a+\varepsilon) - \widehat{G}(x)}{\sigma_G(x)}\right)\right]$$
$$+ \left[\Phi\left(\frac{(a+\varepsilon) - \widehat{G}(x)}{\sigma_{\widehat{G}}(x)}\right) - \Phi\left(\frac{(a-\varepsilon) - \widehat{G}(x)}{\sigma_{\widehat{G}}(x)}\right)\right] \tag{5.17}$$

where $\widehat{G}$ is the constructed Kriging model, Φ is the standard normal cumulative distribution function, and ϕ is the standard normal distribution density function. For reliability analysis, the threshold of a is 0. In EGRA, the value of ε in the expected feasibility function is $2\sigma^2_{\widehat{G}}$.

The active learning function U($\boldsymbol{x}$) is a learning function based on a different concept from EFF based on the statement that only the sign of the response value of the sample point is concerned when the Monte Carlo method is used to calculate the failure probability in the reliability analysis. In fact, in AK-MCS, the exact estimation of the sign of the response value only needs to be performed among the Monte Carlo population candidate sample points, and according to the distribution of random variables, the limit state can be roughly approximated in the case of low probability density distribution. Those points with a high potential risk of crossing G($\boldsymbol{x}$) = 0 must be added to the training set for model building and their response values evaluated. In fact, uncertainty at these points can cause their predicted values to change from positive to negative (or from negative to positive). This results in a change in the probability of failure. Potential "danger" points can exhibit three characteristics: being close to a limit state, having a high degree of uncertainty (high Kriging variance), or both. To identify them, a learning function U($\boldsymbol{x}$) is proposed, specifically as shown in Eq. (5.18) [14]:

$$|\widehat{G}(x)| - \mathrm{U}(x)\sigma_{\widehat{G}}(x) = 0 \tag{5.18}$$

It represents the distance of the Kriging standard deviation between the predicted and estimated limit states. It represents a measure of reliability, expressing the probability that the symbol estimate at a certain point is wrong. The smaller the U($\boldsymbol{x}$) is, the more likely it is that the dot symbol is incorrectly estimated, and the more it needs to be added to our training set.

The above content is the earlier research results of the active learning algorithm of the Kriging model. In recent years, based on AK-MCS, EFF, and active learning function U($\boldsymbol{x}$), scholars have proposed many improvement methods to improve their sampling efficiency. The adaptive Kriging-oriented importance sampling method proposed by Zhang et al. [16] and the error rate-based adaptive Kriging reliability analysis method proposed by Wang et al. [17] can effectively improve the sampling efficiency. To guide the choice of samples, Jiang et al. [18] proposed a failure-pursuing sampling framework with sensitive Voronoi cell.

The probability of failure can be calculated by MCS after constructing a response surrogate model $\widehat{G}(\boldsymbol{X})$.

$$P_F = \frac{\sum_{j=1}^{n_{MC}} I\left(\widehat{G}(\mathbf{X})\right)}{n_{MC}} \tag{5.19}$$

(3) DoE size

In the definition of the initial and posterior DoE, there is interest in considering the number of variables that are strictly necessary to define an accurate surrogate model. High-dimensional spaces demand additional effort in the analysis. Sensitivity analyses are an effective method to reduce the DoE to the variables of interest. Adaptive reduction of the DoE random variables, such as applied in the PCE-RBDO of [19], is an efficient method to address dimensionality in complex problems. Recent research works [20–24] have shown that dimensionality dependence is still relevant in reliability surrogate modeling implementations.

(4) Surrogate model parameters

As seen in Chap. 3, all surrogate models have a priori assumptions and parameters to estimate that are expected to have a large impact on the performance of the surrogate for $G(\boldsymbol{x})$. This fact has given rise to a range of techniques for improving the reliability of parameter estimates of surrogate models. Nonetheless, the relevance of [25] emphasizing model assumptions and parameter estimates in the application of alternative models in engineering remains largely underestimated and ignored.

5.3.2 *Importance Sampling*

To alleviate this computational burden and reduce the variance of reliability estimates, researchers have developed several modified MCS methods, the most popular of which is the importance sampling method [26]. The basic idea of importance sampling is to assign more sample points to regions that have a greater influence on the probability of failure, i.e., the neighborhoods of $G(\boldsymbol{x}) = 0$. The evaluation of sampling points far away from $G(\boldsymbol{x}) = 0$ becomes less significant and can be reduced to improve efficiency. Therefore, one of the most important elements in importance sampling is choosing an appropriate sampling distribution that encourages random samples to be placed in these regions. The $\boldsymbol{U}$-space in Fig. 5.4 shows an illustrative comparison between sampling distributions centered at origin and at the MPP in the standard normal space.

The efficiency of importance sampling is improved by introducing an importance sampling density function $h_X(x)$ that favors failure regions. Xiao et al. [27] proposed a hierarchical importance sampling method based on importance sampling. Even without an optimal important sampling density, further improvements in sampling

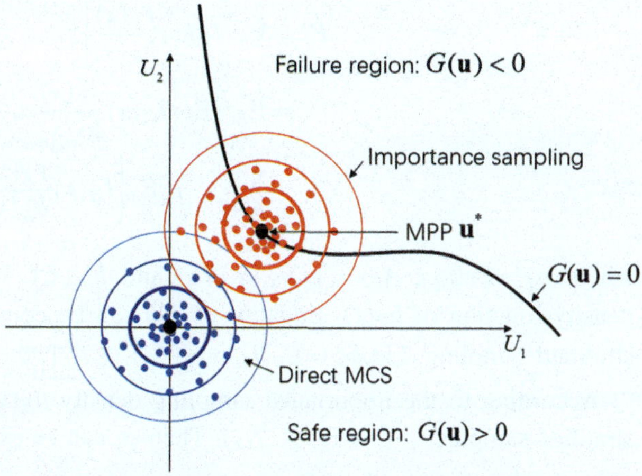

Fig. 5.4 Comparison between direct MCS and importance sampling

efficiency can be achieved. Under the same sample size, stratified importance sampling can obtain the estimation of failure probability with lower variance. If the importance sampling density $h_X(x)$ can be obtained, the probability of failure can be expressed as:

$$
\begin{aligned}
P_F &= \int \cdots \int_{\Omega} I_{\Omega}(x) f_X(x) dx \\
&= \int \cdots \int_{\Omega} \left[I_{\Omega}(x) \frac{f_X(x)}{h_X(x)} \right] h_X(x) dx \\
&= \mathrm{E}_h \left[I_{\Omega}(x) \frac{f_X(x)}{h_X(x)} \right] \\
&= \frac{1}{n_{MC}} \sum_{j=1}^{n_{MC}} I_{\Omega}(x_j) \frac{f_X(x_j)}{h_X(x_j)}
\end{aligned}
\tag{5.20}
$$

It can be expected that in the new sampling distribution, the probability of the index value of random sampling points is greater, that is, the probability of falling into the failure area is greater. However, the optimal importance sampling density is not implementable since it involves the failure probability which is unknown beforehand. To approximate the optimal importance sampling density, several adaptive methods have been proposed [28, 29].

Xiao et al. [27] proposed the stratified importance sampling method to improve the efficiency of importance sampling. Suppose the range of values of the i-th variable x_i $[b_1, b_2]$ is divided into m subintervals, i.e., $A_k = [a_{k-1}, a_k]$, $k = 1, 2, \ldots, m$, $a_0 = b_1$, $a_m = b_2$. The failure probability in Eq. (5.19) can be represented as

$$\begin{aligned} P_F &= \mathrm{E}_h\left[I_\Omega(x)\frac{f_X(x)}{h_X(x)}\right] \\ &= \mathrm{E}_{A_k}\left[\mathrm{E}_h\left[I_\Omega(x)\frac{f_X(x)}{h_X(x)}|x_i \in A_k\right]\right] \\ &= \sum\nolimits_{k=1}^{m} p_k \mathrm{E}_h\left[I_\Omega(x)\frac{f_X(x)}{h_X(x)}|x_i \in A_k\right] \end{aligned} \tag{5.21}$$

where $p_k = Pr(x_i \in A_k) = \int_{A_k} h_{X_i}(x_i)dx_i$ and $h_{X_i}(x_i)$ is the marginal probability density function of $h_X(\boldsymbol{x})$. Equation (5.21) can be considered as a special case of stratified sampling. Let $\varphi_k = E_h\left[I_\Omega(x)\frac{f_X(x)}{h_X(x)}|x_i \in A_k\right]$, then $P_F = \sum_{k=1}^{m} p_k \varphi_k$.

According to the importance sampling density $h_X(\boldsymbol{x})$, generate N_k independent stratified samples $x_k^{(j)}\, (j = 1, \ldots, N_k)$. Then φ_k can be estimated as

$$\widehat{\varphi}_k = \frac{1}{N_k}\sum_{j=1}^{N_k}\left[I_\Omega\left(x_k^{(j)}\right)\frac{f_X\left(x_k^{(j)}\right)}{h_X\left(x_k^{(j)}\right)}\right] \tag{5.22}$$

And the failure probability can be estimated as

$$\widehat{P}_F = \sum_{k=1}^{m} p_k \widehat{\varphi}_k \tag{5.23}$$

Since all the samples are independently and identically distributed, the expectation and variance of $\widehat{\varphi}_k$ can be easily obtained as

$$\mathrm{E}(\widehat{\varphi}_k) = \mathrm{E}_h\left[I_\Omega(x)\frac{f_X(x)}{h_X(x)}|x_i \in A_k\right] \tag{5.24}$$

$$\mathrm{V}(\widehat{\varphi}_k) = \frac{1}{N_k}\mathrm{V}_h\left[I_\Omega(x)\frac{f_X(x)}{h_X(x)}|x_i \in A_k\right] \tag{5.25}$$

Then the expectation of $\widehat{P}_F$ can be represented as

$$\begin{aligned} \mathrm{E}\left(\widehat{P}_F\right) &= \mathrm{E}\left(\sum\nolimits_{k=1}^{m} p_k \widehat{\varphi}_k\right) \\ &= \sum\nolimits_{k=1}^{m} p_k \mathrm{E}(\widehat{\varphi}_k) \\ &= \sum\nolimits_{k=1}^{m} p_k \mathrm{E}_h\left[I_\Omega(x)\frac{f_X(x)}{h_X(x)}|x_i \in A_k\right] \\ &= P_F \end{aligned} \tag{5.26}$$

This shows that $\widehat{P}_F$ is an unbiased estimation of P_F. The variance of $\widehat{P}_F$ can be written as

$$\begin{aligned} \mathrm{V}\left(\widehat{P}_F\right) &= \sum_{k=1}^{m} p_k^2 \mathrm{V}(\widehat{\varphi}_k) \\ &= \sum_{k=1}^{m} p_k^2 \frac{1}{N_k} \mathrm{V}_h \left[I_\Omega(x) \frac{f_X(x)}{h_X(x)} | x_i \in A_k \right] \end{aligned} \tag{5.27}$$

In order to facilitate a proof of improvement, take the case that N_k is proportional to p_k [30], i.e., $N_k = p_k N$. Then, Eq. (5.27) can be further written as

$$\begin{aligned} \mathrm{V}\left(\widehat{P}_F\right) &= \sum_{k=1}^{m} p_k \frac{N_k}{N} \frac{1}{N_k} \mathrm{V}_h \left[I_\Omega(x) \frac{f_X(x)}{h_X(x)} | x_i \in A_k \right] \\ &= \sum_{k=1}^{m} p_k \frac{1}{N} \mathrm{V}_h \left[I_\Omega(x) \frac{f_X(x)}{h_X(x)} | x_i \in A_k \right] \\ &= \frac{1}{N} \sum_{k=1}^{m} p_k \mathrm{V}_h \left[I_\Omega(x) \frac{f_X(x)}{h_X(x)} | x_i \in A_k \right] \\ &= \frac{1}{N} \mathrm{E}_{A_k} \left[\mathrm{V}_h \left[I_\Omega(x) \frac{f_X(x)}{h_X(x)} | x_i \in A_k \right] \right] \end{aligned} \tag{5.28}$$

According to the law of total variance in subintervals [31], as shown in Eqs. (5.29), (5.28) can also be written as

$$\mathrm{V}(Y) = \mathrm{E}_{A_k}[\mathrm{V}(Y|X \in A_k)] + \mathrm{V}_{A_k}[\mathrm{E}(Y|X \in A_k)] \tag{5.29}$$

$$\mathrm{V}\left(\widehat{P}_F\right) = \frac{1}{N} \left[\mathrm{V}_h \left[I_\Omega(x) \frac{f_X(x)}{h_X(x)} \right] - \mathrm{V}_{A_k} \left[\mathrm{E}_h \left[I_\Omega(x) \frac{f_X(x)}{h_X(x)} | x_i \in A_k \right] \right] \right] \tag{5.30}$$

Since $V_{A_k}\left[E_h\left[I_\Omega(x)\frac{f_X(x)}{h_X(x)}|x_i \in A_k\right]\right]$ is nonnegative, stratified IS can get an estimate of failure probability with lower variance for the same sample size. Additionally, to obtain even more variance reduction, the input variable with the highest value of $V_{A_k}\left[E_h\left[I_\Omega(x)\frac{f_X(x)}{h_X(x)}|x_i \in A_k\right]\right]$ should be chosen as the variable for stratification. Actually, $V_{A_k}\left[E_h\left[I_\Omega(x)\frac{f_X(x)}{h_X(x)}|x_i \in A_k\right]\right]$ is an estimate of the variance-based sensitivity measure [32].

5.3.3 *Other Methods*

1. **Dimension Reduction Method**

The search for efficient computational procedures to handle high-dimensional problems has led to the development of dimension reduction (DR) methods

[33, 34]. Outside the field of engineering design, DR methods are widely known as high-dimensional model representation (HDMR) methods, which were originally developed for efficient multivariate model representation in chemical system modeling [35, 36]. This method approximates a multidimensional response function using a set of component functions, where the number of random variables is increased from a constant function to a multidimensional function. For system responses with negligible higher-order variable interactions, HDMR methods can efficiently and accurately formulate this response function using lower-order component functions (usually second-order or bivariate are sufficient). In fact, the response of most real physical systems is only significantly affected by low-order interactions of the input random variables. Depending on the way in which the component functions are determined, HDMR methods can be categorized into two types: ANOVA-HDMR and Cut-HDMR [35]. ANOVA-HDMR exactly follows the analysis of variance (ANOVA) procedure, and is useful for measuring the contributions of the variance of each component function to the output variance [37]. However, the multidimensional integrations involved in ANOVA-HDMR make this expansion computationally unattractive. On the other hand, Cut-HDMR expansion exactly represents the response function in the hyperplane that passes through a reference point in the input random space. This expansion does not require multidimensional integrations and is computationally much more efficient than ANOVA-HDMR. It is fair to say that the DR method is essentially Cut-HDMR designated for the purpose of reliability analysis. Specialized versions of this method include the univariate dimension reduction (UDR) method that simplifies one multidimensional response function to multiple one-dimensional component functions [33, 35] and the bivariate dimension reduction (BDR) method that simplifies one multidimensional response function to multiple one- and two-dimensional integrations [34, 36].

2. **System Reliability Analysis Method**

In actual engineering, a system may have multiple failure modes, so it is necessary to study system reliability analysis and propose a new active learning strategy for multiple failure modes. Fauriat et al. [39] proposed a system reliability analysis method, which mainly defined the expression of the limit state function for the distribution of series system, parallel system, and composite system for system reliability analysis. On this basis, Yun et al. [39] proposed a further improved system reliability analysis method, and proposed an improved learning function U($\boldsymbol{x}$) for system reliability analysis under multiple failure modes.

The key issue to analyze the system reliability (only the series system and the parallel system are considered here) is to estimate Eqs. (5.31) and (5.32). MCS methods for estimating Eqs. (5.31) and (5.32) are

$$P_f^{\text{series}} = \Pr\left(\bigcup_{i=1}^{n} g_i(x) \leq 0\right) \tag{5.31}$$

$$P_f^{\text{parallel}} = \Pr\left(\bigcap_{i=1}^{n} g_i(x) \leq 0\right) \tag{5.32}$$

Exercises

5.1 Consider a reliability problem for a mechanical component involving two random variables X_1 and X_2. The variables are independent with their marginal probability density functions defined by

$$f_{X_1}(x_1) = 2 - 2x_1$$
$$f_{X_2}(x_2) = 2 - 2x_2$$

And the limit state function for this component is represented by

$$G(X_1, X_2) = 1.8 - X_1 - X_2$$

Calculate the probability of failure by FORM and SORM.

5.2 Consider a reliability problem for a mechanical system with four constrains involving two basic random variables X_1 and X_2. Both of the random variables obey the standard normal distribution. The limit state function of this problem can be expressed as

$$G(x_1, x_2) = \min \begin{cases} 3 + 0.1(x_1 - x_2)^2 - \dfrac{(x_1 + x_2)}{\sqrt{2}} \\ 3 + 0.1(x_1 - x_2)^2 + \dfrac{(x_1 + x_2)}{\sqrt{2}} \\ (x_1 - x_2) + \dfrac{6}{\sqrt{2}} \\ -(x_1 - x_2) + \dfrac{6}{\sqrt{2}} \end{cases}$$

Calculate the probability of failure by MCS method. If possible, calculate the probability of failure by other simulation methods (i.e., the surrogate-based method or the importance sampling method).

References

1. Hu, C., Youn, B. D., & Wang, P. (2019). *Engineering design under uncertainty and health prognostics*. Springer.
2. Hasofer, A. M., & Lind, N. C. (1974). Exact and invariant second-moment code format. *Journal of the Engineering Mechanics Division, 100*(1), 111–121.
3. Rosenblatt, M. (1952). Remarks on a multivariate transformation. *The Annals of Mathematical Statistics, **23***(3), 470–472.
4. Hohenbichler, M., & Rackwitz, R. (1981). Non-normal dependent vectors in structural safety. *Journal of the Engineering Mechanics Division, 107*(6), 1227–1238.

5. Breitung, K. (1984). Asymptotic approximations for multinormal integrals. *Journal of Engineering Mechanics, 110*(3), 357–366.
6. Hohenbichler, M., & Rackwitz, R. (1988). Improvement of second-order reliability estimates by importance sampling. *Journal of Engineering Mechanics, 114*(12), 2195–2199.
7. Tvedt, L. (1983). *Two second-order approximations to the failure probability*. Veritas report RDIV/20-004083.
8. Peng, X., et al. (2022). Construction of adaptive Kriging metamodel for failure probability estimation considering the uncertainties of distribution parameters. *Probabilistic Engineering Mechanics, 70*, 103353.
9. Peng, X., et al. (2022). Estimation of small failure probability based on adaptive subset simulation and deep neural network. *Journal of Mechanical Design, 144*(10), 101704 (1–13).
10. Cruse, T. A. (1997). *Reliability-based mechanical design* (Vol. 108). CRC Press.
11. Jiang, Y., et al. (2015). An efficient method for generation of uniform support vector and its application in structural failure function fitting. *Structural Safety, **54***, 1–9.
12. Zhao, W., Fan, F., & Wang, W. (2017). Non-linear partial least squares response surface method for structural reliability analysis. *Reliability Engineering System Safety, 161*, 69–77.
13. Zhang, J., et al. (2019). Probability and interval hybrid reliability analysis based on adaptive local approximation of projection outlines using support vector machine. *Computer-Aided Civil Infrastructure Engineering, 34*(11), 991–1009.
14. Echard, B., Gayton, N., & Lemaire, M. (2011). AK-MCS: An active learning reliability method combining Kriging and Monte Carlo simulation. *Structural Safety, **33***(2), 145–154.
15. Bichon, B. J., et al. (2008). Efficient global reliability analysis for nonlinear implicit performance functions. *AIAA Journal, 46*(10), 2459–2468.
16. Zhang, X., Wang, L., & Sørensen, J. D. (2020). AKOIS: An adaptive Kriging oriented importance sampling method for structural system reliability analysis. *Structural Safety, 82*, 101876.
17. Wang, Z., & Shafieezadeh, A. (2019). REAK: Reliability analysis through Error rate-based Adaptive Kriging. *Reliability Engineering and System Safety, 182*, 33–45.
18. Jiang, C., et al. (2019). A general failure-pursuing sampling framework for surrogate-based reliability analysis. *Reliability Engineering and System Safety, 183*, 47–59.
19. Kim, N., Wang, H., & Queipo, N. (2004). Adaptive reduction of design variables using global sensitivity in reliability-based optimization. In *10th AIAA/ISSMO multidisciplinary analysis and optimization conference* (p. 4515).
20. Pan, Q., & Dias, D. (2017). Sliced inverse regression-based sparse polynomial chaos expansions for reliability analysis in high dimensions. *Reliability Engineering System Safety, 167*, 484–493.
21. Xu, J., & Kong, F. (2018). A cubature collocation based sparse polynomial chaos expansion for efficient structural reliability analysis. *Structural Safety, 74*, 24–31.
22. Cheng, K., & Lu, Z. (2018). Adaptive sparse polynomial chaos expansions for global sensitivity analysis based on support vector regression. *Computers Structures, 194*, 86–96.
23. Fang, H., et al. (2019). A gradient-based uncertainty optimization framework utilizing dimensional adaptive polynomial chaos expansion. *Structural Multidisciplinary Optimization, 59*, 1199–1219.
24. Xu, J., & Wang, D. (2019). Structural reliability analysis based on polynomial chaos, Voronoi cells and dimension reduction technique. *Reliability Engineering System Safety, 185*, 329–340.
25. Abdallah, I., Lataniotis, C., & Sudret, B. (2019). Parametric hierarchical kriging for multi-fidelity aero-servo-elastic simulators—Application to extreme loads on wind turbines. *Probabilistic Engineering Mechanics, 55*, 67–77.
26. Echard, B., et al. (2013). A combined importance sampling and kriging reliability method for small failure probabilities with time-demanding numerical models. *Reliability Engineering and System Safety, 111*, 232–240.
27. Xiao, S., Oladyshkin, S., & Nowak, W. (2020). Reliability analysis with stratified importance sampling based on adaptive Kriging. *Reliability Engineering and System Safety, 197*, 106852.

28. Dubourg, V., Sudret, B., & Deheeger, F. (2013). Metamodel-based importance sampling for structural reliability analysis. *Probabilistic Engineering Mechanics, 33*, 47–57.
29. Au, S.-K., & Beck, J. L. (1999). A new adaptive importance sampling scheme for reliability calculations. *Structural Safety, **21***(2), 135–158.
30. Fishman, G. (2013). *Monte Carlo: Concepts, algorithms, and applications*. Springer.
31. Xiao, S., & Lu, Z. (2020). Structural reliability analysis with conditional importance sampling method based on the law of total expectation and variance in subintervals. *Journal of Engineering Mechanics, 146*(1), 04019111.
32. Sobol, I.M. (1993) Sensitivity Estimates for Nonlinear Mathematical Models. *Mathematical Modelling and Computational Experiments, 4*, 407–414.
33. Rahman, S., & Xu, H. (2004). A univariate dimension-reduction method for multi-dimensional integration in stochastic mechanics. *Probabilistic Engineering Mechanics, 19*(4), 393–408.
34. Xu, H., & Rahman, S. (2004). A generalized dimension-reduction method for multidimensional integration in stochastic mechanics. *International Journal for Numerical Methods in Engineering, 61*(12), 1992–2019.
35. Rabitz, H., et al. (1999). Efficient input—output model representations. *Computer Physics Communications, 117*(1-2), 11–20.
36. Li, G., Rosenthal, C., & Rabitz, H. (2001). High dimensional model representations. *The Journal of Physical Chemistry A, 105*(33), 7765–7777.
37. Sobol', I. M. (2003). Theorems and examples on high dimensional model representation. *Reliability Engineering System Safety, 79*(2), 187–193.
38. Xu, H., & Rahman, S. (2005). Decomposition methods for structural reliability analysis. *Probabilistic Engineering Mechanics, 20*(3), 239–250.
39. Fauriat, W., & Gayton, N. (2014). AK-SYS: An adaptation of the AK-MCS method for system reliability. *Reliability Engineering and System Safety, 123*, 137–144.
40. Yun, W., et al. (2019). AK-SYSi: An improved adaptive Kriging model for system reliability analysis with multiple failure modes by a refined U learning function. *Structural Multidisciplinary Optimization, 59*, 263–278.

Chapter 6
Time-Dependent Reliability Analysis

Nomenclature

AERS	Adaptive extreme response surface method
$\boldsymbol{C_Y}$	Correlation matrix
$\boldsymbol{C}^{l}_{\text{sensitive}}$	Sensitive Voronoi cells
EGO	Efficient global optimization
EOLE	Expansion optimal linear estimation
eSPT	Equivalent stochastic process transformation method
$\boldsymbol{e(t)}$	Stochastic process with zero mean
$\boldsymbol{e(x)}$	Mean squared error
$\boldsymbol{e^{i}_{LOO*}}$	Failure probability error
$\boldsymbol{\bar{e}_{LOO*}}$	Average failure probability error
FORM	First-order reliability methods
$\boldsymbol{F(t)}$	Stochastic process model
$\boldsymbol{F_{min}}$	Approximated global minimum value
$\boldsymbol{f(t)}$	Approximation of stochastic process model
$G(\mathbf{X},\mathbf{Y}(t),t)$	LSF containing time factors
LSF	Limit state function
mVar	Average variances of error
NERS	Nested extreme response surface
NTPM	Nested time-prediction model
$\boldsymbol{N^{+}(0, T_L)}$	Number of outcrossing from safe domain to failure domain within $[\boldsymbol{0}, \boldsymbol{T_L}]$
n_X	Number of random variables
n_Y	Number of stochastic processes
OSE	Orthogonal series expansion
$\boldsymbol{P}$	Sample set
$\boldsymbol{P \backslash p_i}$	Sample set without $\boldsymbol{p_i}$
$\boldsymbol{P_F(0)}$	Instantaneous probability of failure at initial time

W. Hu, *Design Optimization Under Uncertainty*,
https://doi.org/10.1007/978-3-031-49208-2_6

$\widehat{\boldsymbol{P}}_f$	Predicted failure probability evaluated by using the sample set $\boldsymbol{P}$
$\widehat{\boldsymbol{P}}_f^{P\backslash p_i}$	Predicted failure probability evaluated by using the sample set $\boldsymbol{P}\backslash \boldsymbol{p}_i$
p	Number of dominant eigenvalues
$\boldsymbol{R}_c$	Covariance matrix
SORM	Second-order reliability methods
STRA	Surrogate-based time-dependent reliability analysis
s	Number of discrete time nodes
T_L	Specified time period
TRA	Time-dependent reliability analysis
t	Time parameter
$\boldsymbol{t}_i$	Discrete time nodes
$\boldsymbol{\nu}^+(\boldsymbol{t})$	Instantaneous outcrossing rate at time t
WEFF	Weighted expected feasibility function
WSE	Wrong state estimation
WSIE	Wrong sign estimation
$\mathbf{X}$	Vector of random variables
$\mathbf{Y}(t)$	Vector of stochastic processes
$\mathbf{Z}$	Standard random variables
$\boldsymbol{\mu}_Y(\boldsymbol{t})$	Mean function
$\boldsymbol{\sigma}_Y(\boldsymbol{t})$	Standard deviation function
$\boldsymbol{\rho}_Y(\boldsymbol{t})$	Vector of the correlation function
$\boldsymbol{\rho}_Y(\boldsymbol{t}_i, \boldsymbol{t}_j)$	Autocorrelation coefficient function
$\boldsymbol{\lambda}_i$	Eigenvalues of the correlation matrix
$\boldsymbol{\Phi}_i$	Eigenvectors of the correlation matrix
$\boldsymbol{\Phi}(\cdot)$	Cumulative distribution function for the standard Gaussian distribution
$\boldsymbol{\phi}(\cdot)$	Probability density function for the standard Gaussian distribution
$\xi(\boldsymbol{x})$	Relative error
$\boldsymbol{\varepsilon}_{max}$	Maximum real-time estimation error
$\boldsymbol{\varepsilon}_{tar}$	Target prediction error

6.1 Basic Concept of Time-Dependent Reliability Analysis

6.1.1 Introduction

In recent years, with the expansion of production scale and complexity of large-scale equipment, the probability of failure is also increasing, causing a large number of economic losses and casualties. There are various time-varying variables in the actual design and manufacturing process, which are regarded as stochastic processes. For example, external loads vary in time and material properties degenerate over time. While one of the main concerns of product design is to ensure a high level of system reliability throughout the product life cycle, it is crucial to perform time-dependent reliability analysis in the design. Here, time-dependent reliability is defined as the probability that a time-dependent probability constraint is satisfied over the design lifetime. Unlike time-independent reliability analysis, time-dependent reliability analysis involves limit state functions that vary with time; therefore, time-dependent reliability analysis requires a series of instantaneous limit states over the design life of the system. Existing time-independent probabilistic analysis methods are difficult to handle the time dependence of limit state functions. To estimate the realistic reliability of complex equipment, researches on time-dependent reliability analysis (TRA) [1, 2] were the hotspot in the past few decades.

6.1.2 Mathematical Expression of Time-Dependent Reliability Analysis Problem

The theoretical basis of time-dependent reliability mainly includes the analysis of random variables and stochastic processes. The Kriging model is widely used in time-dependent reliability problems, which can provide not only a prediction of the model response but also the predicted variance.

In time-dependent reliability analysis, the limit state function (LSF) represents the working state of the system or structure. An LSF containing time factors can be expressed as $G(\boldsymbol{X}, \boldsymbol{Y}(t), t)$, where $\boldsymbol{X} = [X_1, X_2, \ldots, X_{n_X}]$ is the vector of random variables, $\boldsymbol{Y}(t) = [Y_1(t), Y_2(t), \ldots, Y_{n_Y}(t)]$ is the vector of stochastic processes, t is the time parameter, and n_X and n_Y are the number of random variables and stochastic processes, respectively. The failure event occurs when the limit state function is less than zero in arbitrary time node within a specified time period $[0, T_L]$. Hence, the time-dependent failure probability can be expressed by

$$P_F(0, T_L) = \Pr\{G(\mathbf{X}, \mathbf{Y}(t), t) < 0), \exists t \in [0, T_L]\} \tag{6.1}$$

In order to solve the problem of TRA, several time-dependent reliability analysis methods have been proposed, such as the out-crossing rate method, extreme value methods. This chapter will introduce these methods in detail.

6.2 Expansion of the Stochastic Process

As mentioned in the previous section, stochastic process is widely involved in time-dependent reliability analysis. Stochastic process models have been widely used for function approximation, where details can be find in [3, 4]. In TRA, the inputs of the limit state function may contain stochastic processes. However, stochastic processes are hard to handle in TRA. Discretization is a common strategy for dealing with stochastic processes. Discretization methods commonly used in stochastic problems mainly include Karhunen-Loeve expansion [5], the Orthogonal Series Expansion (OSE) [6], and the Expansion Optimal Linear Estimation (EOLE) [7].

For time-dependent reliability analysis problems, the EOLE is often used to expand the stochastic process $\boldsymbol{Y}(t)$ by a series of random variables $\boldsymbol{Z} = (Z_1, Z_2, \ldots, Z_p)$, which is shown as follows:

$$Y(t) = \mu_Y(t) + \sigma_Y(t) \sum_{i=1}^{p} \frac{1}{\sqrt{\lambda_i}} \boldsymbol{\Phi}_i Z_i \boldsymbol{\rho}_Y(t) \tag{6.2}$$

where p is the number of dominant eigenvalues, which should be smaller than or equal to the number of discrete time nodes s; $\mu_Y(t)$ and $\sigma_Y(t)$ are the mean function and standard deviation function of $Y(t)$, respectively; Z_i is the independent standard normal variable; $\boldsymbol{\rho}_Y(t) = [\rho_Y(t, t_1), \rho_Y(t, t_2), \ldots, \rho_Y(t, t_i), \ldots, \rho_Y(t, t_s)]^T$ is a vector of the correlation function; $t_i (i = 1, 2, \ldots, s)$ are the discrete time nodes; and λ_i and $\boldsymbol{\Phi}_i$ are the eigenvalues and eigenvectors, respectively, of the correlation matrix $\boldsymbol{C}_Y$. The correlation matrix $\boldsymbol{C}_Y$ is expressed as

$$\mathbf{C}_Y = \begin{bmatrix} \rho_Y(t_1, t_1) & \rho_Y(t_1, t_2) & \cdots & \rho_Y(t_1, t_s) \\ \rho_Y(t_2, t_1) & \rho_Y(t_2, t_2) & \cdots & \rho_Y(t_2, t_s) \\ \vdots & \vdots & \ddots & \vdots \\ \rho_Y(t_s, t_1) & \rho_Y(t_s, t_2) & \cdots & \rho_Y(t_s, t_s) \end{bmatrix}_{s \times s} \tag{6.3}$$

where $\rho_Y(t_i, t_j)$ is the autocorrelation coefficient function.

First, the time interval $[0, T]$ is divided into time nodes, and the EOLE method is used to convert the stochastic process $\boldsymbol{Y}(t)$ into random variables $\boldsymbol{Z}$. Then $[\boldsymbol{X}, \boldsymbol{Y}(t)]$ is transformed into $[\boldsymbol{X}, \boldsymbol{Z}]$, followed by the use of the sampling strategy to select next sample. Because the sample is at a specific time node, $\left[Z_{j1}^*, \ldots, Z_{jp_z}^*, t_s\right]$ is put into Eq. (6.2), $Y_j^*(t_s)$ can be obtained, and the surrogate model is constructed.

6.3 Outcrossing Rate Methods

The outcrossing rate method is developed based on the "outcrossing event" proposed by Rice [8]. As is shown in Fig. 6.1, an outcrossing event takes place when the time-variant system performance function $G(\boldsymbol{X}, \boldsymbol{Y}(t), t)$ is about to cross the $G = 0$ axis with respect to time t. In TRA, this indicates that the system is about to fail, and the first derivative of reliability with respect to time can be approximated by the outcrossing rate [9]. Therefore, the key to solve TRA problems lies in the calculation of outcrossing rate.

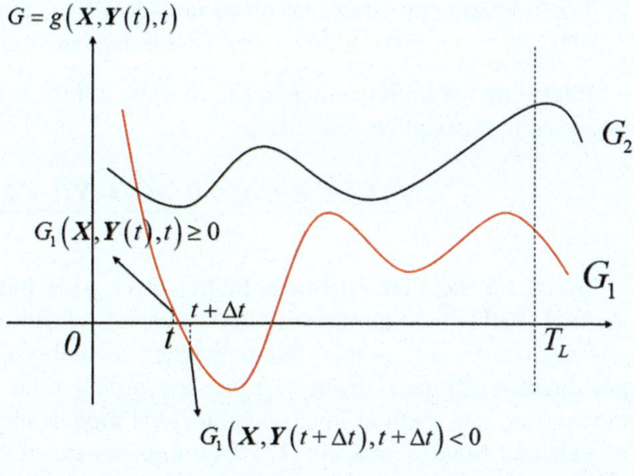

Fig. 6.1 The outcrossing event within the $[0, T_L]$

The approach based on the outcrossing rate calculates the probability of failure according to the expected average number of outcrossing events. The instantaneous outcrossing rate at time t is defined as

$$v^+(t) = \lim_{\Delta t \to 0, \Delta t > 0} \frac{P\{G(\mathbf{X}, \mathbf{Y}(t), t) > 0 \cap G(\mathbf{X}, \mathbf{Y}(t + \Delta t), t + \Delta t) \leq 0\}}{\Delta t} \tag{6.4}$$

The outcrossing events are assumed to be statistically independent and be approximated using the Poisson's distribution, which is shown in Fig. 6.1.

The probability of failure can be defined as [1]

$$\begin{aligned} P_F(0, T_L) &\approx 1 - \exp[-\mathrm{E}[N^+(0, T_L)]] \\ &= 1 - (1 - P_F(0)) \exp\left[-\frac{v^+ T_L}{1 - P_F(0)}\right] \end{aligned} \tag{6.5}$$

where $[N^+(0, T_L)]$ denote the number of outcrossing from safe domain to failure domain within $[0, T_L]$, $P_F(0)$ is the instantaneous probability of failure at initial time t.

Although outcrossing rate methods tend to provide a mathematical derivation of time-dependent reliability, it is difficult to assess outcrossing rates for general stochastic processes. Only special stochastic processes (e.g., stationary Gaussian processes) can be used to analyze outcrossing rates [10]. A large number of methods have focused on the asymptotic integration method to calculate the outcrossing rate [11, 12]. In this section, two methods for calculating the outcrossing rate, namely the PHI2 method [13] and its improved PHI2+ method [14], will be briefly discussed.

The differences between PHI2 and the traditional methods are:

1. PHI2 considers the LSF itself as the stochastic process whose outcrossing of the zero level is to characterize;

2. The stochastic processes involved in $G(\mathbf{X}, \mathbf{Y}(t), t)$ don't need to be discretized;
3. PHI2 does not make a first-order Taylor expansion of the limit state at $t + \Delta t$.

Replacing the limit passage in Eq. (6.4) by a finite difference-like operations, the outcrossing rate can be rewritten as

$$\nu_{\mathrm{PHI2}}{}^{+}(t) = \frac{P\{G(\mathbf{X}, \mathbf{Y}(t), t) > 0 \cap G(\mathbf{X}, \mathbf{Y}(t + \Delta t), t + \Delta t) \leq 0\}}{\Delta t} \tag{6.6}$$

Based on the PHI2 method, Sudret et al. [14] proposed the improved PHI2 method (PHI2+) to stabilize the PHI2 method, which considers the stationary and non-stationary case. In this time-dependent reliability approach, reliability indexes are directly estimated from first-passage probabilities [15, 16]. When using the outcrossing rate method, many hypothetical models already exist, such as Poisson models and Markov models [17]. The time-dependent failure probability and reliability indexes vary for these different hypothetical models. Although the outcrossing rate method has been widely used in time-dependent reliability problems in the past [18], its calculation accuracy is relatively low due to the assumption that all outcrossings are independent with each other and follow a Poisson distribution.

6.4 Extreme Value Methods

Instead of calculating the outcrossing rate, the extreme value methods calculate the probability that the extreme value of performance exceeding the specified performance threshold of an engineering system. These methods substitute the performance function with surrogate model, as introduced in Chap. 3. The extreme value methods mainly include two types, i.e., the extreme response surrogate-based methods and the response surrogate-based methods. The extreme response surrogate-based method aims at constructing the extreme surface with respect to time by performing global optimization at every training point, while the response surrogate-based method constructs a global surrogate model for the performance function with the random variables, stochastic processes, and time parameter as inputs.

6.4.1 Nested Extreme Response Surface Approach

This section presents the details about one of the extreme response surrogate-based methods for time-dependent reliability analysis, namely, the nested extreme response surface (NERS) approach [19]. The NERS approach is based on nested time-prediction model (NTPM). For a given system, the performance function $G(\mathbf{X}, \mathbf{Y}(t), t)$ may change over time due to time-dependent uncertainty. Figure 6.2 shows

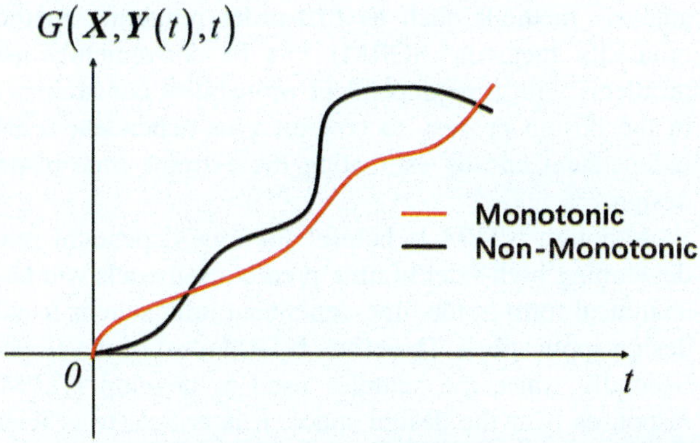

Fig. 6.2 Examples of time-dependent performance functions

the random realizations of two different types of system responses, where the red and black lines represent monotonically and non-monotonically increasing performance, respectively. If $G(\boldsymbol{X}, \boldsymbol{Y}(t), t)$ increases or decreases monotonically with time, extreme responses will usually occur at the time interval boundary, where the probability of failure will also approach its maximum value. For such time-dependent reliability constraints, reliability analysis only needs to be performed at the time interval boundary. However, the situation is more complicated when the system response $G(\boldsymbol{X}, \boldsymbol{Y}(t), t)$ is non-monotonic. In this case, the key is to carry out the reliability analysis at the instantaneous time when the extreme response of the performance function is obtained.

The time that leads to the extreme response of the system performance function varies with different design $\mathbf{X}$, and thus a response surface of the time with respect to system design variables can be determined as

$$T = f(\mathbf{X}) : \left\{ \max_t S(x, t), \forall x \in \mathbf{X} \right\} \tag{6.7}$$

For illustrative purpose, consider a time-varying LSF $G(\boldsymbol{X}, t) = 20 - X_1^2 X_2 + 5X_1 t - (X_2 + 1)t^2$ where $t \in [0, 25]$. For any given design $x : \{x_1 \in [0, 10], x_2 \in [0, 10]\}$, there exists an instantaneous time t^* that maximizes $G(\boldsymbol{x}, t)$. To obtain t^*, one may set the derivative of G with respect to t equal to zero to yield $t^* = 5x_1/(2x_2 + 2)$. The time instant t^* for extreme response of G varies with the design variables x_1 and x_2. For example, for a specific implementation of the system design, where $x_1 = 4$, $x_2 = 1$, the system approaches the maximum response at time $t = 5$; while for another specific implementation of the system design, where $x_1 = 5$, $x_2 = 0$, the system's maximum response is reached at $t = 12.5$. Here the response surface at time t is defined as an NTPM in the NERS method. The inputs of NTPM are the design variables and parameters of interest, which can be stochastic or deterministic. The output is the time at which the system response approaches an extremum. With NTPM, time-dependent reliability analysis problems can be transformed into time-independent ones, so that existing advanced reliability analysis methods such as first-order reliability methods (FORM), second-order reliability methods (SORM), can be conveniently used. In design optimization problems with time-dependent probability constraints, NTPM can be fully nested in the design process to convert time-dependent reliability analysis into a time-independent one by estimating the extreme time responses for any given system design.

Although NTPM is helpful for time-dependent reliability analysis, effectively developing high-fidelity time prediction models can be very challenging. First, the analytical form of the time-dependent limit state is usually not available in practical design applications. Therefore, NTPM must be developed based on a limited sample. Secondly, since the samples used to develop NTPM require extreme temporal responses over the design space, it is critical to efficiently extract these responses to make the design process computationally affordable. Thirdly, NTPM must be adapted to perform two different roles: predicting the extreme temporal response of reliability analysis, and including additional sample points in necessary regions during an iterative design process to improve the model itself. This section presents the NERS approach, which addresses the three challenges above.

The key to NERS approach is the efficient construction of nested time prediction models in the design space of interest, which can be used to predict when the system response approaches an extreme value. The NERS approach consists of three main techniques in three sequential steps: (1) efficient global optimization (EGO) for extreme time response identification, (2) construction of Kriging-based NTPM, and (3) adaptive time response prediction and model maturation. The first step, EGO, is used to efficiently extract a certain number of extreme time response samples for the development of the second step NTPM. Once the Kriging-based NTPM for extreme time responses is established, then use the adaptive response prediction and model maturation mechanism to ensure the accuracy and efficiency of the predictions by autonomously enrolling new sample points as needed during the analysis. The NERS method is outlined in the flowchart shown in Fig. 6.3, and the three key techniques mentioned above are explained in detail in the rest of this section.

1. Efficient Global Optimization for Extreme Time Response Identification

For the reliability analysis of time-dependent system response, it is very important to effectively calculate the extreme value response of the LSF and effectively locate the corresponding time when the system response is close to the extreme value. For a given system design, the response of LSF is time-dependent and can be a monotonic or non-monotonic one-dimensional function with respect to time. Here, the EGO technique [20] can be used to effectively locate the extreme system response and the corresponding time when it is close to the extreme response, mainly because of its ability to handle non-monotonic limit states while ensuring excellent computational efficiency. This section focuses on the application of the EGO technique to the identification of extreme time responses. Discussion of the EGO technique itself is omitted, as more details on the technique can be obtained from references [21, 22].

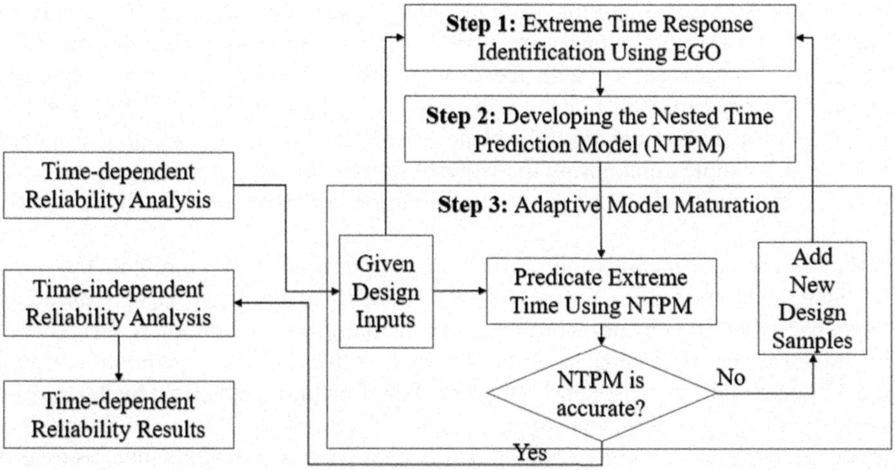

Fig. 6.3 Flowchart of NERS approach for reliability analysis

In order to find the globally optimal time leading to the extreme value response of the LSF, the EGO technique generates a one-dimensional stochastic process model from the existing sample response as a function of time. Stochastic process models have been used extensively for function approximation; more information on these models can be found in reference [3, 4]. In this study, the time-dependent response of the limit state function at a particular design point is represented by a one-dimensional stochastic process model with a constant global mean in EGO as

$$F(t) = \mu + e(t) \tag{6.8}$$

where μ is the global model representing the function mean, and $e(t)$ is a stochastic process with zero mean and variance σ_e^2. The covariance between $e(t)$ at two different points t_i and t_j is defined as $Cov(e(t_i), e(t_j)) = \sigma_e^2 R_c(t_i, t_j)$, in which the correlation function is given by

$$R_c(t_i, t_j) = \exp\left(-a|t_i - t_j|^b\right) \tag{6.9}$$

where a and b are unknown model hyperparameters. Based on a set of initial samples, $F(t_1), F(t_2), \ldots, F(t_k)$, an initial stochastic process model of $F(t)$ can always be constructed by maximizing the likelihood function

$$L_F = \frac{1}{(2\pi)^{n/2}(\sigma^2)|R_c|^{1/2}} \exp\left(-\frac{(F-\mu)' R_c^{-1}(Y-\mu)}{2\sigma^2}\right) \tag{6.10}$$

where $F = (F(t_1), F(t_2), \ldots, F(t_k))$ denote the sample responses of the limit state function, and R_c is the covariance matrix which is defined by Eq. (6.9). After developing an initial one-dimensional stochastic process model, EGO iteratively updates the model by continuously searching for the most useful sample points to ensure maximum accuracy until the convergence criterion is met. To update the stochastic process model, EGO employs the expected improvement [23] metric to quantify the potential contribution of new sample points to the existing response surface; the sample point giving the largest expected improvement value is selected in the next iteration. Next, the expected improvement metrics will be briefly described.

Let us consider a continuous function $F(t)$ over time t that represents the time-dependent LSF for a given system design point in the design space. Here, we employ the expected improvement metric to determine the global minimum of $F(t)$. Due to the finite sample size of $F(t)$, the initial stochastic process model may introduce large model uncertainties, so the approximation of the function represented by $f(t)$ may be significantly biased compared to the actual function $F(t)$. Due to the uncertainty involved in this model, in EGO, the function approximation of $f(t)$ at time t is treated as a normal random variable whose mean and standard deviation are computed by an approximated response $F_r(t)$, and its standard error $e(t)$ is determined from the stochastic process model. With these notations, the improvement at time t can be defined as

$$I(t) = \max(F_{\min} - f(t), 0) \tag{6.11}$$

where F_{min} indicates the approximated global minimum value at the current EGO iteration. By integrating the expectation of the right part of Eq. (6.11), the expected improvement at any given time t can be presented as [20]

$$\begin{aligned} E[I(t)] &= E[\max(F_{\min} - f, 0)] \\ &= (F_{\min} - F_r(t))\Phi\left(\frac{F_{\min} - F_r(t)}{e(t)}\right) + e(t)\phi\left(\frac{F_{\min} - F_r(t)}{e(t)}\right) \end{aligned} \tag{6.12}$$

where $\Phi(\cdot)$ and $\phi(\cdot)$ are the cumulative distribution function and the probability density function for the standard Gaussian distribution, respectively. The larger the expected improvement at time t, the more likely it is to achieve a better approximation to the global minimum. Therefore, a new sample should be evaluated at a specific time t_i where the maximum expected improvement is obtained to update the stochastic process model. Using the updated model, a new global minimum approximation to $F(t)$ can be obtained. The same process can be iteratively repeated by evaluating new samples at time t_i, which provides the value of the maximum expected improvement and updating the stochastic process model for the new global minimum approximation until the maximum expected improvement is sufficiently small, and less than a critical value I_c (in [19], $I_c = |F_{min}|\%$, which is 1% of the absolute value of the current best global minimum approximation).

2. Kriging-Based Nested Time Prediction Model

This section describes the process of developing a nested temporal prediction model using Kriging. After repeating the EGO process for different system sample points in the design space, a set of data can be obtained, including the initial sample point $\boldsymbol{x}_0$ in the design space, and the corresponding time t_0 when the system response is close to its extreme value. To construct NTPM using the NERS method, design points are randomly generated in the design space according to the stochastic properties of the design variables. To balance the accuracy and efficiency of NTPM, it is recommended to initially use $10 * (n_X - 1)$ samples to build a Kriging-based NTPM for n_X-dimensional problems ($n_X > 1$). The accuracy and efficiency of NTPM are controlled by the Adaptive Response Prediction and Model Maturity (ARPMM) mechanism, which is detailed in the next subsection. The goal here is to develop a predictive model to estimate the times leading to extreme performance of the limit state function for any given system design in the design space. For this purpose, a Kriging model was constructed based on the sample dataset obtained during the EGO process. It is worth noting that different response surface approaches, such as simple linear regression models or artificial neural networks [24–26], can be applied here for the development of NTPM. In this study, Kriging was used because it can model nonlinear relationships between extreme time responses of system design variables well. The detail of Kriging has been shown in Chap. 3.

With the Kriging time prediction model, the extreme time response for any given new point $\boldsymbol{x}'$ can be estimated as

$$\mathbf{t}(\mathbf{x}') = \mu + \mathbf{r}^T(\mathbf{x}')R_c^{-1}(t - A\mu) \tag{6.13}$$

where $\boldsymbol{r}(\boldsymbol{x}')$ is the correlation vector between $\boldsymbol{x}'$ and the sampled points x_1~x_n, in which the i-th element of $\boldsymbol{r}$ is given by $r_i(\boldsymbol{x}') = R_c(\boldsymbol{x}', x_i)$.

3. Adaptive Response Prediction and Model Maturation

When designing with nested temporal prediction models, model prediction accuracy is critical. Therefore, during the design process, a model maturation mechanism is needed to automatically register new sample points to improve the accuracy of the nested temporal prediction model when the accuracy conditions are not met. Wang et al. [19] developed an Adaptive Response Prediction and Model Maturity (ARPMM) mechanism based on the mean squared error $e(\boldsymbol{x})$ of the current best prediction. The detail procedure can be find in [19].

Before predicting the time response at a new design point $\boldsymbol{x}$ using the latest update of NTPM, the ARPMM mechanism can be employed by first computing the current mean squared error $e(\boldsymbol{x})$ of the best prediction.

$$e(\mathbf{x}) = \sigma^2\left[1 - \mathbf{r}^T R^{-1}\mathbf{r} + \frac{\left(1 - A^T R^{-1}\mathbf{r}\right)^2}{A^T R^{-1} A}\right] \tag{6.14}$$

To reduce the numerical error, the relative MSE is suggested as a prediction performance measure for the NTPM, which is given by

$$\xi(\mathbf{x}) = \frac{e(\mathbf{x})}{\mu} \tag{6.15}$$

The predicted time response t' for a new design point $\boldsymbol{x}$ using NTPM is accepted only if the relative error $\xi(\boldsymbol{x})$ is less than a user-defined threshold ξ_t. In order to balance a smooth design process with the required prediction accuracy, the value of $\xi(\boldsymbol{x})$ is recommended to be in the range $[10^{-3}, 10^{-2}]$. Once the prediction for this particular design point $\boldsymbol{x}$ is accepted, the time response t' of extreme performance can be estimated using Eq. (6.13) and return to the time-dependent reliability analysis procedure. If the relative error is greater than a threshold, $\boldsymbol{x}$ will be registered as a new sample input, and the EGO procedure will be employed to extract the true time response for $\boldsymbol{x}$ when the limit state function approaches its limit performance. Using the new design point $\boldsymbol{x}$ and the real-time response of $\boldsymbol{x}$, NTPM will be updated. Through the developed ARPMM mechanism, NTPM can be adaptively updated during time-dependent reliability analysis to guarantee accuracy while maintaining efficiency. Note that the ARPMM mechanism automatically improves the Kriging model during the design process; in rare cases, when multiple design points close together are used to seed the Kriging model, stability caused by singular matrices may occur question. Therefore, we recommend including an extra step in the ARPMM procedure to improve the predictive accuracy of the Kriging model by checking for singularities after adding new sample points.

6.4.2 *Other Methods*

Hu et al. [27] proposed a mixed efficient global optimization for time-dependent reliability analysis. Different from the current EGO method, which draws samples of random variables and time independently, the m-EGO method draws samples for the two types of samples simultaneously. The m-EGO method employs the adaptive Kriging–Monte Carlo simulation (AK–MCS) so that high accuracy is also achieved. Then, Monte Carlo simulation (MCS) is applied to calculate the time-dependent reliability based on the surrogate model. The detail of m-EGO can be found in [27]. Nevertheless, finding its extreme time at each point will bring much computational burden.

6.5 Response Surrogate-Based Methods

Instead of constructing an extreme response surface, the response surrogate-based approach constructs a global surrogate model with random variables, random processes, and time parameters as input as a performance function. Subsequently, the extremum surfaces can be approximated based on the global surrogate model. The real concern of the global surrogate model is whether the random point will fall into the failure state in any discrete time node, rather than whether the extreme value of the random point in the time interval meets the failure condition. It is obvious that the response surrogate-based approach is more efficient than the extreme response surrogate-based method. This section will introduce several response surrogate-based methods. The confidence-based adaptive extreme response surface method (AERS) [9] has to build many Kriging models at all discretized time nodes, and the training samples of different Kriging models cannot be reused. The stopping criterion of the equivalent stochastic process transformation method (eSPT) [28] varies according to the specific problems. Li et al. [29] presents a new instantaneous response surface method t-IRS for time-dependent reliability analysis. Hu et al. [31] have proposed a surrogate-based time-dependent reliability analysis (STRA) for a digital twin (DT).

6.5.1 Confidence-Based Adaptive Extreme Response Surface Method

This method makes spectral decomposition of input stochastic processes at first, representing a stochastic process *Y*(*t*) by a set of deterministic functions with corresponding random coefficients. The EOLE method usually be adopted in this part. With the employment of spectral decomposition, input stochastic process *Y*(*t*) is represented by a formulation of p random variables $\mathbf{Z} = [Z_1, Z_2, \ldots, Z_p]$ with the corresponding eigenfunctions. The time-dependent limit state function is derived as

$$G(\mathbf{X}, \mathbf{Y}(t), t) = G_T(\mathbf{X}, \mathbf{Z}, t) \tag{6.16}$$

If a sample $[\boldsymbol{x}, z]$ is drawn from the joint distribution of random variables $[\mathbf{X}, \mathbf{Z}]$, we then obtain a deterministic realization of stochastic process. The time-dependent extreme value function is defined as

$$G_{e,T}(\mathbf{X}, \mathbf{Z}, T) = \max_{t \in [0, T]} G_T(\mathbf{X}, \mathbf{Z}, t) \tag{6.17}$$

where the extreme value $G_{e,T}(\cdot)$ is a function of random variables **X** and **Z**, as well as time parameter *T*. Note that the time-dependent extreme value is a monotonic increasing function with time for a realization of random parameters **X** and **Z**. The probability of failure for time interval $[0, T]$ is calculated by

$$\begin{aligned} P_f(0, T) &= \Pr(G_{e,T}(\mathbf{X}, \mathbf{Z}, T) > 0) \\ &= \int \cdots \int_{G_{e,T}(\mathbf{X},\mathbf{Z},T) > 0} f_{\mathbf{X}}(\mathbf{X}) f_{\mathbf{Z}}(\mathbf{Z}) d\mathbf{x} dz \end{aligned} \tag{6.18}$$

Since it is unlikely to derive the time-dependent extreme value function and evaluate the multi-dimension integration analytically in Eq. (6.18), in this method, a set of surrogate models will be constructed for predicting the time-dependent extreme value function and approximation of the time-dependent probability of failure in Eq. (6.18) using MCS. This method also adopts some advanced sampling strategies to improve the efficiency of building surrogate models, which can be found in [9].

6.5.2 Equivalent Stochastic Process Transformation Method

This approach presents an equivalent stochastic process transformation method for cost-effectively predicting reliability degradation over the lifetime of engineering systems, taking into account random variables and stochastic process parameters. A new concept called transient failure surface is introduced to cover all potential failure events that may occur within a certain time period. To obtain the instantaneous fault surface, the time-independent reliability problem is formulated by transforming the stochastic process and time parameters into random variables. A surrogate model is constructed using Kriging techniques to predict the instantaneous failure surface, and the surrogate model is efficiently updated using a maximum confidence enhanced sequential sampling scheme. Time-dependent reliability of dynamical systems is assessed by Monte Carlo simulations based on an updated high-fidelity Kriging model. The detail of this method can be found in [28].

6.5.3 Instantaneous Response Surface Method

There is a new instantaneous response surface method t-IRS for time-dependent reliability analysis. The procedure of this method can be expressed as:

1. The time interval of interest $[0, T]$ is discretized into s time nodes t_i using $\Delta t = t_i - t_{i-1} = \frac{T}{s-1}, i = 1, 2, \ldots, s$. $Y(t)$ is reconstructed and transformed into $\mathbf{Z}$ by using the EOLE approach.
2. Instantaneous response surrogate model construction and update of the instantaneous response surrogate model.
3. The time-dependent reliability is calculated by MCS.

The t-IRS method only needs to build one instantaneous response surrogate model, and thus computational efficiency can be greatly improved. And the detail of this method can be found in [29].

6.5.4 Real-Time Estimation Error-Guided Active Learning Kriging Method

Since the Kriging model can provide local uncertainty information, it is easy to obtain the probability of wrong sign estimation (WSIE) of each random candidate at discrete time nodes. When estimating time-dependent failure probabilities by Monte Carlo simulation (MCS), the Kriging model classifies all stochastic candidates as failure candidates and safe candidates. However, the classification is not entirely accurate. Based on the probability of WSIE, the probability of wrong state estimation (WSE) for failed and safe candidates can be calculated separately. In each iteration, the candidate with the highest WSE probability will be selected as a new training sample, and the corresponding time node will be determined according to the WSIE probability. Additionally, the probability of WSE is used to calculate a confidence interval for the total number of misclassified safe or failed candidates. According to the confidence interval, the real-time maximum estimation error of the predicted failure probability can be obtained. Then use the maximum error to judge whether to stop the training of the surrogate. The maximum real-time estimation error can be calculated by [30].

$$\varepsilon_{\max}^{k} = \max\left(\left|1 + \frac{\widehat{N}_f}{\widehat{N}_f + \widehat{n}_s^l - \widehat{n}_s^u}\right|, \left|1 - \frac{\widehat{N}_f}{\widehat{N}_f + \widehat{n}_s^l - \widehat{n}_s^u}\right|\right) \times 100\% \tag{6.19}$$

6.5.5 Surrogate-Based Time-Dependent Reliability Analysis Method

The authors proposed a new surrogate-based time-dependent reliability analysis method for digital twin, particularly focusing on stochastic process discretization and the adaptive sampling strategy [31].

The EOLE method requires two parameters: the number of discrete time nodes s and the number of dominated eigenvalues p. For the same p, the different s can make representation of the stochastic processes with different accuracy. In addition, the number of discrete time nodes s affects the calculation time of TRA, which affects the real-time feature of a DT. The STRA uses the variance of error to select best s. The variance of the error for the EOLE method at j-th test time node t_j is given by Eq. (6.20):

$$Var\left[Y(t_j) - \widetilde{Y}(t_j)\right] = \sigma_Y^2(t_j) - \sum_{i=1}^{p} \frac{1}{\lambda_i}\left(\Phi_i^T C_Y(t_j)\right)^2 \tag{6.20}$$

For a given s, the variances of the error at 10^6 test time nodes are calculated. For each stochastic process, the average of the variances of error at all test time nodes is set as the error of the discretization of the stochastic process, expressed by Eq. (6.21). If there are several stochastic processes in a TRA problem, the error of the discretization of all the stochastic processes is equal to summation of the average of the error of each stochastic process. For different s, the one corresponding to the minimum error of the discretization is selected for discretizing the stochastic process (es). Therefore, STRA can make a trade-off between accuracy and real-time.

$$mVar = \frac{\sum_{j=1}^{10^6} Var\left[Y(t_j) - \widetilde{Y}(t_j)\right]}{10^6} \tag{6.21}$$

To decrease the number of random variables, the dominated eigenvalues need to be selected. For a given s, there will be s eigenvalues and eigenvectors. Arrange these eigenvalues in descending order, and the first p are selected as dominated eigenvalues. The value of p is defined by Eq. (6.22):

$$\frac{\sum_{i=1}^{p} \lambda_i}{\sum_{i=1}^{s} \lambda_i} > \theta \tag{6.22}$$

Here, the θ is set as 0.99. Therefore, the stochastic process $Y(t)$ can be represented by several independent standard normal variables $Z = (Z_1, Z_2, \ldots, Z_p)$.

The STRA method aims to improve the efficiency of constructing the Kriging model by sampling in the regions with large failure probability prediction errors and shifts the adaptive sampling domain from the global space to the vicinity of the LSF. Two major issues need to be addressed: (1) properly partitioning the sampling space and (2) identifying the sensitive regions near the LSF.

1. Partitioning the Sampling Region Based on Voronoi Tessellation

As shown in Chap. 3, the Voronoi diagram is used for sampling region partition. Figure 6.4 shows a two-dimensional example partitioned by the Voronoi tessellation. The asterisks in Fig. 6.4 are the existing samples. All the candidate samples are then categorized into each Voronoi cell based on the Voronoi tessellation. The candidate samples in the entire design space are first generated by the Monte Carlo simulation (MCS) method, as demonstrated by the triangles.

2. Identification of Sensitive Voronoi Cells

For efficient TRA, each new sample is expected to contribute to the improvement of the accuracy of reliability analysis using the surrogate model. To select samples more efficiently, it is necessary to filter out the Voronoi cells with large prediction errors. the modified (LOO) cross-validation is developed using the error of the

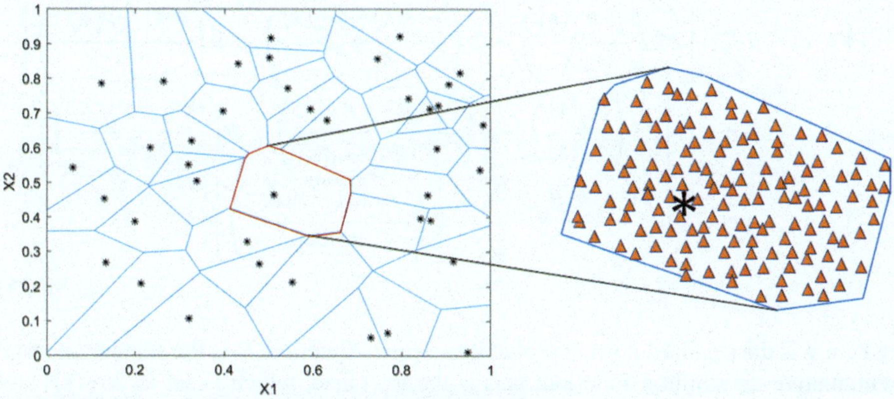

Fig. 6.4 Example of two-dimensional Voronoi tessellation

failure probability as the error-index to select the sensitive Voronoi cells. The modified LOO error index of a point p_i can be defined by

$$e^i_{LOO*} = \mid \widehat{P}_f - \widehat{P}_f^{\,P\backslash p_i} \mid \tag{6.23}$$

where $\widehat{P}_f$ is the predicted failure probability evaluated by using the sample set P and $\widehat{P}_f^{\,P\backslash p_i}$ is the predicted failure probability evaluated by using the sample set $P\backslash p_i$. The above process is repeated for each point in P, and eventually the failure probability errors of all existing samples will be obtained.

It is observed that large failure probability error e^i_{LOO*} of sample p_i indicates that the constructed Kriging model deviates from the real model largely in cell C_i. The formula for calculating the average failure probability error $\overline{e}_{LOO*}$ is as follows:

$$\overline{e}_{LOO*} = \frac{\sum_{i=1}^{m} e^i_{LOO*}}{m} \tag{6.24}$$

where m is the number of Voronoi cells. The sensitive Voronoi cells, expressed by $C^l_{\text{sensitive}}, l = 1, 2, \ldots, q$, are defined as the cells whose error of failure probability e^i_{LOO*} is greater than the average failure probability error $\overline{e}_{LOO*}$. The new points will be sampled in these sensitive cells.

After the sensitive Voronoi cells are selected, the next step is to select new training samples among all the candidate MCS samples in the sensitive cells to update the Kriging model. One of the most popular learning functions used to identify new training samples of the adaptive Kriging model is the expected feasibility function (EFF) proposed in the efficient global reliability analysis (EGRA) algorithm [32]. The EFF tends to select points that are close to the LSF with high uncertainty [32]. It is expressed as

$$\begin{aligned}\mathrm{EFF}(x) = \mu_{\widehat{G}}(x)\left[2\Phi\left(\frac{-\mu_{\widehat{G}}(x)}{\sigma_{\widehat{G}}(x)}\right) - \Phi\left(\frac{-\varepsilon(x)-\mu_{\widehat{G}}(x)}{\sigma_{\widehat{G}}(x)}\right) - \Phi\left(\frac{\varepsilon(x)-\mu_{\widehat{G}}(x)}{\sigma_{\widehat{G}}(x)}\right)\right] \\ - \sigma_{\widehat{G}}(x)\left[2\phi\left(\frac{\mu_{\widehat{G}}(x)}{\sigma_G(x)}\right) - \phi\left(\frac{-\varepsilon(x)-\mu_{\widehat{G}}(x)}{\sigma_G(x)}\right) - \phi\left(\frac{\varepsilon(x)-\mu_{\widehat{G}}(x)}{\sigma_G(x)}\right)\right] \\ + \varepsilon(x)\left[\Phi\left(\frac{\varepsilon(x)-\mu_{\widehat{G}}(x)}{\sigma_{\widehat{G}}(x)}\right) - \Phi\left(\frac{-\varepsilon(x)-\mu_{\widehat{G}}(x)}{\sigma_{\widehat{G}}(x)}\right)\right]\end{aligned} \tag{6.25}$$

where ϕ is the standard normal probability density function; Φ is the standard normal cumulative distribution function; $\varepsilon(x)$ is the tolerance, which is set to $2\sigma_{\widehat{G}}(x)$; and $\mu_{\widehat{G}}(x)$ and $\sigma_{\widehat{G}}(x)$ are the predicted mean and standard deviation, respectively, for sample x.

To be applied in DT, the efficiency of EFF need to be enhanced and decrease the number of DoE samples. The weighted expected feasibility function (WEFF) is proposed to improve the sampling efficiency, which combines weights based on EFF. The WEFFs balance the importance of samples and their positional relationship. For each sensitive Voronoi cell $C^l_{\text{sensitive}}, l = 1, 2, \ldots, q$, there is a corresponding weight which is expressed as

$$weight(i) = \frac{e^i_{LOO*}}{\max(e_{LOO*})} \tag{6.26}$$

where $\max(e_{LOO*})$ is the maximum failure probability error over all the identified sensitive cells.

For candidate point x, which belongs to the Voronoi cell $C^l_{\text{sensitive}}$, its weighted expected feasibility values can be defined by

$$\mathrm{WEFF}(x) = weight(l) \times \mathrm{EFF}(x) \tag{6.27}$$

By filter out the candidate sample points located in sensitive Voronoi cells $C^l_{\text{sensitive}}, l = 1, 2, \ldots, q$, the best next sample to be evaluated and added for updating the surrogate model is the sample having the maximum WEFF value. In this way, the Kriging model is iteratively updated until the following stopping criterion is satisfied:

$$\varepsilon_{\max} \le \varepsilon_{tar} \tag{6.28}$$

where ε_{max} is the predicted error, which can be calculated according to the maximum real-time estimation error calculation method [30], and ε_{tar} is the target prediction error, which is equal to 0.05 [30]. After the surrogate models $\widehat{G}(\boldsymbol{X}, \boldsymbol{Y}(t), t)$ are successfully constructed, the time-dependent probability of failure can be instantly calculated by MCS. A flowchart of the STRA method is shown in Fig. 6.5.

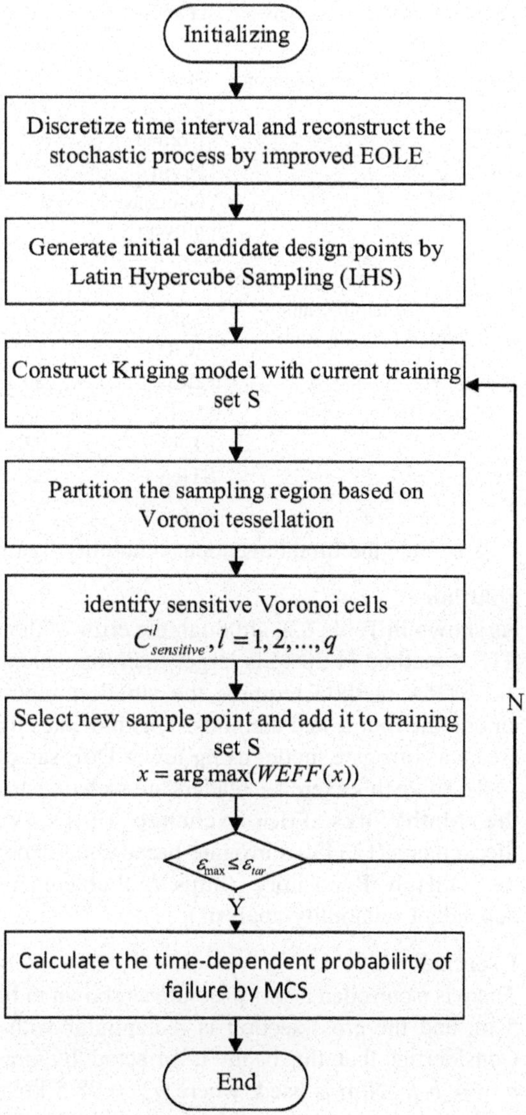

Fig. 6.5 Flowchart of the STRA method

Example 6.1 Consider the following performance function

$$G(\boldsymbol{X}, Y(t), t) = -20 + X_1^2 X_2 - 5X_1(1 + Y(t))t + (X_2 + 1)t^2$$

where t represents the time parameter, which varies from 0 to 1; $\boldsymbol{X} = [X_1, X_2]$ denotes two independent random variables; and $Y(t)$ is a Gaussian process. The statistical parameters of the random variables and the stochastic process are shown in Table 6.1.

Table 6.1 Statistical parameters of the random variables and the stochastic process used in the case study

Random variable or stochastic process	Distribution type	Mean	Standard deviation	Autocorrelation function
X_1	Normal	3.5	0.25	
X_2	Normal	3.5	0.25	
$Y(t)$	Gaussian process	0	1	$\exp(-(t_2 - t_1)^2)$

Table 6.2 Reliability analysis results of the case study

Method	NFE	P_f	Error (%)
MCS	10^6	0.307814	
t-IRS	80.3	0.308050	0.08
eSPT	48.1	0.299643	2.65
REAL	32.1	0.299636	2.66
STRA	27.4	0.304104	1.20

Calculate the time-dependent reliability by response surrogate methods.

Solution

As shown in Table 6.2, although the error of the failure probability calculated by the STRA method (1.20%) is larger than that calculated by the t-IRS method (0.08%), the STRA method requires the smallest number of function evaluations (NFE) among the four tested methods. This indicates that the STRA method can efficiently create a surrogate model using fewer DoE samples.

To solve this exercise, readers are welcome to use the codes which are available at the MathWorks File Exchange (https://ww2.mathworks.cn/matlabcentral/fileexchange/123480-surrogate-based-time-dependent-reliabillity-analysis) and at the GitHub Repository (https://github.com/WeifeiHuZJU/Surrogate-based-time-dependent-reliabillity-analysis).

Exercises

There is a corroded beam structure as shown in Fig. 6.6. The length of the beam, L, is 5 m, and the cross-section is rectangular with an initial width b_0 and height h_0. Considering that the beam is affected by gravity, a uniformly distributed load $p = \rho_{st} b_0 h_0$ N/m is used, where $\rho_{st} = 78.5$ kN/m^3 is the steel density. In addition, the beam is subjected to a concentrated force $F(t)$ at the center point.

The beam will be corroded by external effects such as raindrop erosion during its life cycle. Its cross-sectional area is assumed to linearly decrease with time, and the mechanical strength of the corroded area will be gradually lost. The remaining uncorroded cross-sectional area $A(t)$ can be expressed as

$$A(t) = b(t) \times h(t) \tag{6.29}$$

where $b(t) = b_0 - 2\kappa t$, $h(t) = h_0 - 2\kappa t$, $\kappa = 0.03$ mm/year denotes a constant corrosion rate, and t represents the time parameter, which varies within $[0, 20]$ years.

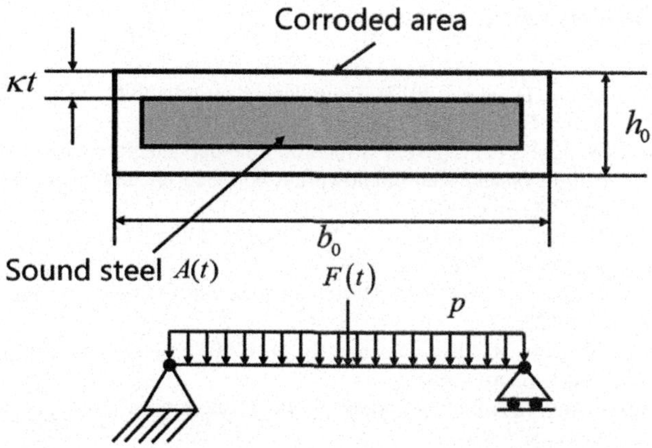

Fig. 6.6 Corroded beam structure

Table 6.3 Statistical information on the random parameters of the corroded beam structure

Parameter	Distribution	Mean	Standard deviation	Autocorrelation function
f_y	Lognormal	240 MPa	24 MPa	
b_0	Lognormal	0.2 m	0.01 m	
h_0	Lognormal	0.03 m	0.003 m	
$F(t)$	Gaussian process	3500 N	700 N	$\exp(-(t_2 - t_1)^2/\lambda^2)$

The performance function of the corroded beam can be expressed as

$$G(X, Y(t), t) = M_u(t) - M(t) = \frac{b(t)h^2 f_y}{4} - \left(\frac{F(t)L}{4} + \frac{\rho_{st} b_0 h_0 L^2}{8}\right)$$

where $M_u(t) = b(t)h^2(t)f_y/4$ is the ultimate bending moment capacity of the beam and f_y is the steel yield stress. The bending moment at the midpoint of the beam is the largest, which can be expressed as

$$M(t) = ((F(t)L/4) + (\rho_{st} b_0 h_0 L^2/8))$$

Let $\boldsymbol{X} = [f_y, b_0, h_0]$ be lognormally distributed random variables, $F(t)$ be a Gaussian process, and the correlation parameter λ in the autocorrelation function be equal to 5 years. Detailed statistical information on the random variables and the stochastic process is presented in Table 6.3.

Calculate the time-dependent reliability of this corroded beam in 5 years. Note: this exercise is one of the case studies from the reference [31]. For detailed methodology, results, and discussion, readers are suggested to check the reference [31].

References

1. Zhang, D., et al. (2017). Time-dependent reliability analysis through response surface method. *Journal of Mechanical Design, 139*(4), 041404.
2. Wang, L., et al. (2017). Structural time-dependent reliability assessment of the vibration active control system with unknown-but-bounded uncertainties. *Structural Control Health Monitoring, 24*(10), e1965.
3. Koehler, J. R., & Owen, A. B. (1996). 9 Computer experiments. In *Handbook of statistics* (Vol. 13, pp. 261–308).
4. Sacks, J., et al. (1989). Design and analysis of computer experiments. *Statistical Science, 4*(4), 409–423.
5. Ghanem, R. G., & Spanos, P. D. (2003). *Stochastic finite elements: A spectral approach*. Courier Corporation.
6. Zhang, J., & Ellingwood, B. (1994). Orthogonal series expansions of random fields in reliability analysis. *Journal of Engineering Mechanics, 120*(12), 2660–2677.
7. Li, C.-C., & Der Kiureghian, A. (1993). Optimal discretization of random fields. *Journal of Engineering Mechanics, 119*(6), 1136–1154.
8. Rice, S. O. (1944). Mathematical analysis of random noise. *The Bell System Technical Journal, 23*(3), 282–332.
9. Wang, Z., & Chen, W. (2017). Confidence-based adaptive extreme response surface for time-variant reliability analysis under random excitation. *Structural Safety, 64*, 76–86.
10. Breitung, K. (1988). Asymptotic crossing rates for stationary Gaussian vector processes. *Stochastic Processes Their Applications, 29*(2), 195–207.
11. Schrupp, K., & Rackwitz, R. (1988). Outcrossing rates of marked poisson cluster processes in structural reliability. *Applied Mathematical Modelling, 12*(5), 482–490.
12. Breitung, K. (1994). *Asymptotic approximations for the crossing rates of Poisson square waves* (pp. 75–75). Nist Special Publication Sp.
13. Andrieu-Renaud, C., Sudret, B., & Lemaire, M. (2004). The PHI2 method: A way to compute time-variant reliability. *Reliability Engineering System Safety, 84*(1), 75–86.
14. Sudret, B. (2008). Analytical derivation of the outcrossing rate in time-variant reliability problems. *Structure Infrastructure Engineering, 4*(5), 353–362.
15. Lutes, L.D. and S. Sarkani, Reliability analysis of systems subject to first-passage failure. 2009.
16. Song, J., & Der Kiureghian, A. (2006). Joint first-passage probability and reliability of systems under stochastic excitation. *Journal of Engineering Mechanics, 132*(1), 65–77.
17. Jiang, C., et al. (2014). A time-variant reliability analysis method based on stochastic process discretization. *Journal of Mechanical Design, 136*(9), 091009.
18. Hu, Z., & Du, X. (2015). First order reliability method for time-variant problems using series expansions. *Structural Multidisciplinary Optimization, 51*, 1–21.
19. Wang, Z., & Wang, P. (2012). A nested extreme response surface approach for time-dependent reliability-based design optimization. *Journal of Mechanical Design, 134*(12).
20. Jones, D. R., Schonlau, M., & Welch, W. J. (1998). Efficient global optimization of expensive black-box functions. *Journal of Global optimization, 13*(4), 455.
21. Schonlau, M., Computer experiments and global optimization. 1997.
22. Stuckman, B. E. (1988). A global search method for optimizing nonlinear systems. *IEEE Transactions on Systems, Man, Cybernetics, 18*(6), 965–977.
23. Mockus, J. (1998). The application of Bayesian methods for seeking the extremum. *Towards Global Optimization, 2*, 117.
24. Knill, D. L., et al. (1999). Response surface models combining linear and Euler aerodynamics for supersonic transport design. *Journal of Aircraft, 36*(1), 75–86.
25. Madsen, J. I., Shyy, W., & Haftka, R. T. (2000). Response surface techniques for diffuser shape optimization. *AIAA Journal, 38*(9), 1512–1518.
26. Welch, W. J., et al. (1992). Screening, predicting, and computer experiments. *Technometrics, 34*(1), 15–25.

27. Hu, Z., & Du, X. (2015). Mixed efficient global optimization for time-dependent reliability analysis. *Journal of Mechanical Design, 137*(5), 051401.
28. Wang, Z., & Chen, W. (2016). Time-variant reliability assessment through equivalent stochastic process transformation. *Reliability Engineering System Safety, 152*, 166–175.
29. Li, J., et al. (2019). Developing an instantaneous response surface method t-IRS for time-dependent reliability analysis. *Acta Mechanica Solida Sinica, 32*, 446–462.
30. Jiang, C., et al. (2020). Real-time estimation error-guided active learning Kriging method for time-dependent reliability analysis. *Applied Mathematical Modelling, 77*, 82–98.
31. Hu, W., et al. (2023). Surrogate-based time-dependent reliability analysis for a digital twin. *Journal of Mechanical Design*, 1–36.
32. Bichon, B. J., et al. (2008). Efficient global reliability analysis for nonlinear implicit performance functions. *AIAA Journal, 46*(10), 2459–2468.

Chapter 7
Reliability-Based Design Optimization

Nomenclature

c	Constant number
C	Copula function
d	Design variable
$\boldsymbol{d}^{(k)}$	Design vector of the *k*-th iteration
$\boldsymbol{d}_k$	Searching direction
$f_x(\boldsymbol{x})$	Joint probability density function (JPDF) of all random system
$f_X(x_i, \mu_i)$	Marginal PDF corresponding to the *i*-th random variable X_{i}
$F_G(\boldsymbol{g})$	Statistic description of $\boldsymbol{G}(\boldsymbol{x})$
g^*	Target probability metric
$g_j(\boldsymbol{x})$	Inequality constraints
$G(\boldsymbol{x})$	System performance function
$h_i(\boldsymbol{x})$	Equality constraints
m_1	Number of equality constraints
m_2	Number of inequality constraints
$\boldsymbol{m}(\boldsymbol{y})$	Value function
$\boldsymbol{M_f}$	Number of failed samples
nc	Number of probabilistic constraints
nd	Number of design variable
nr	Sum of random variables and parameters
P_{Fj}^{Tar}	Target probability of failure
r	Nonlinear indicator
s	Coordinate offset coefficient
u	Marginal CDFs for $\boldsymbol{X_i}$
v	Marginal CDFs for $\boldsymbol{X_j}$
X^{rp}	Random parameter of the input
$\boldsymbol{x}^{*(k-1)}$	The MPP point in the *k*-1th iteration
$\boldsymbol{x}^{(m)}$	*m*-th realization of *X*

W. Hu, *Design Optimization Under Uncertainty*,
https://doi.org/10.1007/978-3-031-49208-2_7

Z(Y)	Approximation function
β_s	Reliability index
ε_d	Constant of the design tightness criterion
Ψ	Vector of distribution parameters
μ_i	First-order score function
μ_{s_f}	Mean value of the score function
Ω_F	Failure set
θ	Correlation coefficient between X_i and X_j

7.1 Basic Concept

The computational ability of computers has greatly enhanced in last few decades, which decrease the computational time of the large-scale simulation techniques (e.g., finite element analysis (FEA), computational fluid dynamics (CFD)) for complex systems. By utilizing the large-scale simulation techniques, the performance (e.g., stress, strain, heat field, etc.) can be analyzed and further provide the possibility to find an optimal design of the complex systems by using different input of simulation. The process of obtaining optimal design is known as *design optimization*. Optimization is used to find optimal design, which is achieved by achieving a higher performance or a lower cost (i.e., the *objective function*) while satisfying the performance requirements (i.e., *constraints*) by changing the *design variable*. Typical objective functions include minimizing weights, improving performances, and increasing the production, etc. Typical constraints include satisfying maximum stress and allowable deflection/strain, design within the feasible design domain, etc. The basic paradigm in design optimization is to find a set of design variables that optimizes an objective function while satisfying the constraints, which can be seen in Eq. (7.1).

$$\begin{aligned} &\text{minimize} && \text{Cost}(\mathbf{X}) \\ & && h_i(x)=0, i=1, \ldots, m_1 \\ &\text{subjectto} && g_j(x)\leq 0, j=1, \ldots, m_2 \\ & && x\in \mathbf{X}\in \mathbf{R}^{nr} \end{aligned} \tag{7.1}$$

where **X** is the design variable, cost is one type of object function, $h_i(x)$ are equality constraints, $g_j(x)$ is the inequality constraints, nr is the sum of random variables, and parameters, m_1 and m_2 are the number of equality constraints and inequality constraints, respectively. In the process of design optimization, the design variables in the problems are assumed to be deterministic [6]. In this situation, these problems with deterministic design variable are defined as *deterministic design optimization*. Existing design optimization methods can be divided into two groups, the sensitivity-based methods, and the non-sensitivity-based methods. For the sensitivity-based optimization methods, gradient information of the objective function with respect to design variables is applied in finding an optimal design. However, the gradient information may not be available for some optimization problems with comprehensive simulation. To overcome these shortcomings, the non-sensitivity-based methods are utilized for design optimization. The advantage of non-sensitivity-based methods include: (1) it requires no gradients, which can be utilized for high-dimensional multi-physical problems and (2) it has a better global searching ability than those of sensitivity-based method. However, such methods can be computationally inefficient, as they do not use the gradient information while exploring the same design space. The computational time of non-sensitivity-based algorithm are large. Additionally, uncertainties may lead to large variations in the performance characteristics of the system. In real world, there must be various uncertainties, which need to be considered for safety and reliability of design optimization process. Otherwise, results of deterministic optimal problem may be associated with high possibility of failure without considering uncertainties.

To associate the actual engineering problems with uncertainties, the *reliability-based design optimization* (RBDO) is proposed to optimize design characterized by a low probability of failure [8]. The RBDO tries to find the optimum design by fine-tuning the design variables and satisfying probabilistic constraints at a specific level of reliability or failure probability. In the process of RBDO, the uncertain variables and the failure modes are determined initially carried by probability theory, which can be referred in Chap. 1. Then in an RBDO formulation, the critical failure modes in deterministic optimization are replaced by these failure modes accompanied with uncertain variables, which can be analyzed by reliability analysis described in Chaps. 5 and 6. In the optimization process, the RBDO is traditionally treated as a nested optimization problem (i.e., double-loop optimization problem), which contains the analysis of deterministic optimization as well as reliability analysis [4]. However, such methods may be computationally intensive when the number of design variables or random variables increases. It is even impossible to obtain an accurate result when the design parameters obey a complex joint probability density function. To tackle the mentioned issues, researchers have proposed most probable point (MPP) and sampling-based RBDO methods from the perspective of probability resolution and the surrogate model, respectively. The main goal of this chapter is to introduce the reliability-based design optimization methods including the MPP-based RBDO and sampling-based RBDO.

7.2 Problem Statement and Formulation

The mathematical formulation of general component level of RBDO problem is expressed as

$$\begin{aligned} &\text{minimize} \quad \text{Cost}(\mathbf{d}) \\ &\text{subjectto} \quad \begin{array}{l} P\big[G_j(\mathbf{X}) > 0\big] \le P_{F_j}^{Tar}, j = 1, \ldots, nc \\ \mathbf{d}^L \le \mathbf{d} \le \mathbf{d}^U, \mathbf{d} \in \mathbf{R}^{nd} \text{and} \mathbf{X} \in \mathbf{R}^{nr} \end{array} \end{aligned} \tag{7.2}$$

where $\mathbf{d} = \mu(\mathbf{X^{rv}})$, $i = 1 \sim nd$ is the design variable, which is the mean of the nd-dimensional random variable $\boldsymbol{X_{rv}} = \{X_1, X_2, \ldots, X_{nd}\}^T$; $\boldsymbol{X} = \{\boldsymbol{X_{rv}}, \boldsymbol{X_{rp}}\}^T$, where $\mathbf{X^{rp}}$ represent the random parameter of the random input $\mathbf{X}$; ${P_{Fj}}^{\text{Tar}}$ is the target probability of failure, which is achieved by reliability analysis, for the j-th constraint; nc and nd are the number of probabilistic constraints, design variables, respectively.

The system performance criteria are described by system performance functions. Consider a system performance function $G(x)$, where the system fails if $G(x) > 0$. The statistic description of $G(x)$ is characterized by its CDF $F_G(g)$ as

$$F_G(g) = P(G(x) > g) = \int_{G(x) > g} \cdots \int f_x(x) dx_1 \ldots dx_n, x^L \le x \le x^U \tag{7.3}$$

where $f_x(x)$ is the joint probability density function (JPDF) of all random system, $F_G(x)$ is the cumulative distribution function of G. The reliability analysis for both component and systems level involves calculation of the probability of failure, which has been described in Chaps. 5 and 6. The statistical sampling methods (e.g., MCS) is utilized to estimate the true responses of probability from computer simulation. However, the statistical sampling methods may be computationally inefficient and almost prohibited. To achieve a fast and accurate estimating of the probability of failure, the surrogate model is adopted in RBDO, which is introduced in Chap. 3.

7.3 Most Probable Point-Based RBDO

7.3.1 Reliability Index Approach and Performance Measure Approach

1. **Mathematical Model of RIA and PMA**

Reliability index approach (RIA) and performance measure approach (PMA) are two traditional RBDO methods based on most probable point (MPP) [13]. The RIA and PMA methods are equivalent for probabilistic constraints. However, the two methods have different perspectives on the evaluation of probabilistic constraints, and the study shows that the PMA method has better robustness and efficiency compared to the RIA method.

As mentioned in Chap. 5, the failure of probability can be expressed as

$$P_{F_j}^{Tar} = \Phi(-\beta_t) \tag{7.4}$$

Hence, the formulation of RBDO can be represented as

$$F_{G_j}(0) \geq \Phi(-\beta_t) \tag{7.5}$$

Equation (7.5) can be expressed in two ways using the following inverse transformations, respectively, as

$$\beta_s = -\Phi^{-1}(F_G(0)) \leq \beta_t \tag{7.6}$$

$$g^* = F_G^{-1}(\Phi(-\beta_t)) \leq 0 \tag{7.7}$$

where β_s is the reliability index and g^* is the target probability metric. Hence, the RIA replaces the probability constraint with Eq. (7.6), which can be expressed as

$$\begin{array}{ll} \text{minimize} & \text{Cost}(\mathbf{d}) \\ \text{subjectto} & \begin{array}{l} \beta_{s,j} \leq \beta_{t,j}, j = 1, \ldots, nc \\ \mathbf{d}^L \leq \mathbf{d} \leq \mathbf{d}^U, \mathbf{d} \in \mathbf{R}^{nd} \text{and} \mathbf{X} \in \mathbf{R}^{nr} \end{array} \end{array} \tag{7.8}$$

Besides, when the probability constraint is replaced with Eq. (7.7), the PMA is formulated, expressed as

$$\begin{array}{ll} \text{minimize} & \text{Cost}(\mathbf{d}) \\ \text{subjectto} & \begin{array}{l} g_j^* \leq 0, j = 1, \ldots, nc \\ \mathbf{d}^L \leq \mathbf{d} \leq \mathbf{d}^U, \mathbf{d} \in \mathbf{R}^{nd} \text{and} \mathbf{X} \in \mathbf{R}^{nr} \end{array} \end{array} \tag{7.9}$$

where the meaning of other letters in Eqs. (7.8) and (7.9) is the same as that in Eq. (7.1) RBDO model.

2. **Reliability Analysis Process and Inverse Analysis Process**

To use RIA, PMA for solving RBDO problems, the calculation of reliability index β_G and target probability metric g^* needs to be determined, which can be achieved by reliability analysis and inverse reliability analysis for the design point according to the derivation process β_G and g^* are calculated as follows.

$$\beta_G = -\Phi^{-1}(F_G(0)) = -\Phi^{-1}\left(\int_{G(x)>g} \cdots \int f_x(x) dx_1 \ldots dx_n\right) \tag{7.10}$$

$$g^* = F_G^{-1}(\Phi(-\beta_t)) = F_G^{-1}\left(\int_{G(x)>g} \cdots \int f_x(x) dx_1 \ldots dx_n\right) \tag{7.11}$$

Obviously, the applied both methods always need to take multiple integrals, which are difficult to calculate directly to obtain accurate results in practical engineering applications. To take these issues, some approximate probability integration methods have been developed to provide effective solutions [1, 5, 14, 17], such as the first-order reliability method (FORM) or the second-order reliability method (SORM) which are introduced in Chap. 5. Among these methods, FORM is widely accepted for RBDO applications because of its accuracy and less computational burden than SORM. Since the reliability index and performance function are obtained by two different inverse transformations according to Eqs. (7.10) and (7.11), the calculation process for RIA and PMA is called reliability analysis process and inverse reliability analysis process, respectively. Next, the process of reliability analysis and inverse reliability analysis based on FORM applied to RIA and PMA is introduced.

It has been shown that the RIA method produces singularities in its results for certain problems, while the PMA direction has better robustness with respect to the RIA method. Besides, the computational efficiency of both methods depends on the activity of the probabilistic constraints. For the optimal solution of the RBDO model, the robustness and efficiency of the algorithm are important. Hence, previous authors have proposed some numerical methods for RIA, PMA methods to improve the efficiency and stability of the probabilistic constraint evaluation of the RBDO model, which will be introduced in the next section.

7.3.2 Numerical Reliability Analysis Method Based on RIA

1. **Hasofer-Lind Rackwitz-Fiessler Method**

The Hasofer-Lind Rackwitz-Fiessler (HL-RF) method is similar to FORM, which performs a Taylor first-order expansion to linearize the constraint at design point [18] and has been introduced in Chap. 5. In most cases, the method can converge quickly. However, it has been shown [11] that under certain conditions, this method may converge slowly or even lead to divergence when the limit state function is complex and highly nonlinear. Therefore, an improved method is proposed to improve its robustness HL-RF method called improved HL-RF method, which will be introduced in the next section.

2. **Improved HL-RF Method**

Compared to the HL-RF method, the improved HL-RF algorithm introduces a value function $m(y)$ to evaluate the convergence of the sequence

$$m(y) = \frac{1}{2}\left| y - \frac{\nabla G(y) y}{|\nabla G(y)^2|^2} \nabla G(y) \right|^2 + \frac{1}{2} c G(y)^2 \tag{7.12}$$

where c is a constant number. During the optimization process, the next point is selected by the linear search along the direction.

$$d_k = \frac{1}{|\nabla G(y_k)|^2}[\nabla G(y_k)y_k - G(y_k)]\nabla G(y_k)^T - y_k \tag{7.13}$$

The improved HL-RF method, searching along the direction of d_k until the value function $m(y)$ decreases sufficiently. The d_k does not always search the direction of gradient descent of $m(y)$, which resulting in the global convergence of the improved HL-RF method is still not guaranteed. But the method improves the robustness of the original HL-RF method.

3. **Two-Point Approximation Method**

The two-point approximation method introduces adaptive intervening variables based on the HL-RF method to propose an approximation function Z(Y) to solve for the exact function value and gradient, with the following expressions.

$$Z(Y) = g(Y_k) + \frac{1}{r}\sum_{i=1}^{n} y_{i,k}^{1-r}\frac{\partial g(Y_k)}{\partial u_i} \times \left[\left(u_i + \frac{\overline{x}_i}{\sigma_i}\right)^r - \left(u_{i,k} + \frac{\overline{x}_i}{\sigma_i}s\right)^r\right] \tag{7.14}$$

where r is the nonlinear indicator, which controls the generalized function nonlinearity, s is the coordinate offset coefficient. The nonlinear indicator and the coordinate offset coefficient can be determined by the following equation.

$$g(Y_{k-1}) - \left[g(Y_k) + \frac{1}{r}\sum_{i=1}^{n} y_{i,k}^{(1-r)}\frac{\partial g(Y_k)}{\partial u_i}\right] \times \left[\left(u_i + \frac{\overline{x}_i}{\sigma_i}\right)^r - \left(u_{i,k} + \frac{\overline{x}_i}{\sigma_i}s\right)^r\right] = 0 \tag{7.15}$$

Since the obtained function values and gradients are exact, the method is suitable for highly nonlinear and implicit performance function problems that require large-scale FEM for structural analysis.

7.3.3 Numerical Reliability Analysis Method Based on PMA

1. **Advanced Mean Value (AMV) Method**

The most commonly used numerical methods for PMA are advanced mean value (AMV) [16], and conjugate mean value (CMV) [22]. The advanced mean value (AMV) method is simple and efficient for convex performance functions, but for concave performance functions will show numerical defects such as slow convergence and even divergence, while the conjugate mean value (CMV) method is always convergent for the results, but inefficient for convex functions when solving. Therefore, according to the characteristics of AMV and CMV methods applied to PMA, once the type of performance function (whether it is convex or concave) is determined, the adaptive selection of AMV and MCV methods for PMA can evaluate the probability constraint more effectively, which is called hybrid mean value (HMV) method [19, 20].

The AMV method is widely used by researchers due to its simple formula and fast solution. The AMV method starts from the mean value (MV), where the mean value is defined as

$$u_{MV}^{*} = \beta_t n(0) \tag{7.16}$$

$$n(0) = -\frac{\nabla_X G(\mu)}{\|\nabla_X G(\mu)\|} = -\frac{\nabla_U G(0)}{\|\nabla_U G(0)\|} \tag{7.17}$$

where $n(0)$ represents the direction that the performance function G is minimized, i.e., the normalized steepest descent direction. The AMV method iteratively updates the direction vector of the steepest descent method. After using the MV method to obtain the initial value u_{MV}^{*}, the AMV method can be expressed as

$$u_{\mathrm{AMV}}^{(1)} = u_{MV}^{*}, u_{\mathrm{AMV}}^{(k+1)} = \beta_t n\left(u_{\mathrm{AMV}}^{(k)}\right) \tag{7.18}$$

where

$$n\left(u_{\mathrm{AMV}}^{(k)}\right) = -\frac{\nabla_U G\left(u_{\mathrm{AMV}}^{(k)}\right)}{\left\|\nabla_U G\left(u_{\mathrm{AMV}}^{(k)}\right)\right\|} \tag{7.19}$$

Existing study shows that the AMV method exhibits fast and efficient properties for convex performance functions, but slow convergence or even divergence for concave performance functions, which can be seen in Figs. 7.1 and 7.2. The concave and convex properties of the performance function are defined based on the performance function around the MPP with respect to the U-space origin, taking a convex performance function $G_1(X)$ and a concave performance function $G_2(X)$, respectively, as examples

$$G_1(X) = -e^{(X_1-7)} - X_2 + 10, X_i \sim N(6, 0.8), i = 1, 2, \beta_t = 3$$

$$G_2(X) = \frac{\left[e^{(0.8X_1-1.2)} + e^{(0.7X_2-0.6)} - 5\right]}{10}, X_i \sim N(5, 0.8), i = 1, 2, \beta_t = 3$$

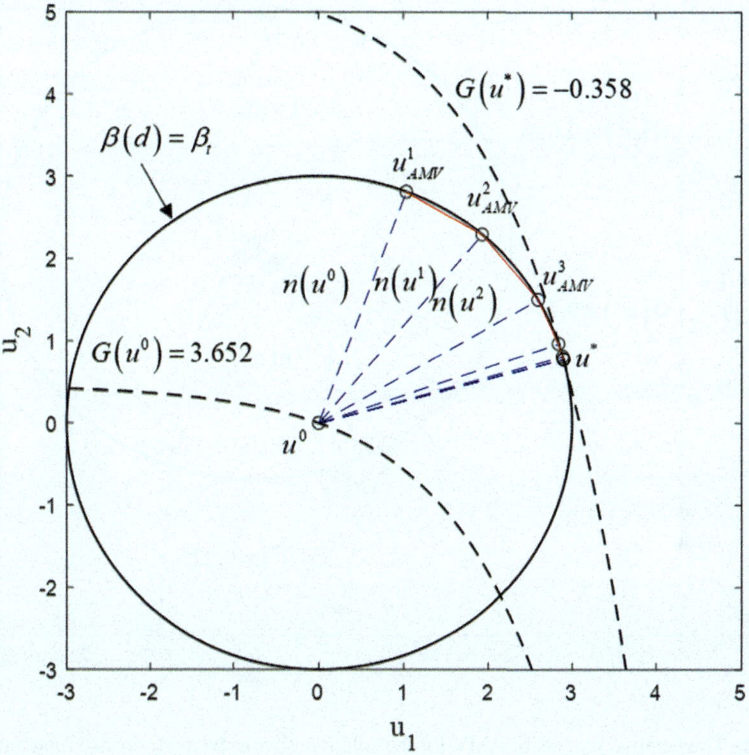

Fig. 7.1 The iterative process of AMV for the solution of convex performance function adapted from Youn et al. [21]

Because the update point of AMV method is the steepest descent direction of the performance function, it has a faster convergence speed for the gradient descent direction of the performance function, but for the concave performance function, the oscillation phenomenon as shown in Fig. 7.3.

2. **Conjugate Mean Value (CMV) Method**

In the AMV method, the lack of sufficient update information in conducting reliability analysis causes it to be unstable and inefficient. To address the shortcomings of the AMV method, the conjugate mean value (CMV) method is formed by considering the current as well as previous MPP information so that it points to the diagonal of the three consecutive steepest descent directions. The CMV method is expressed as

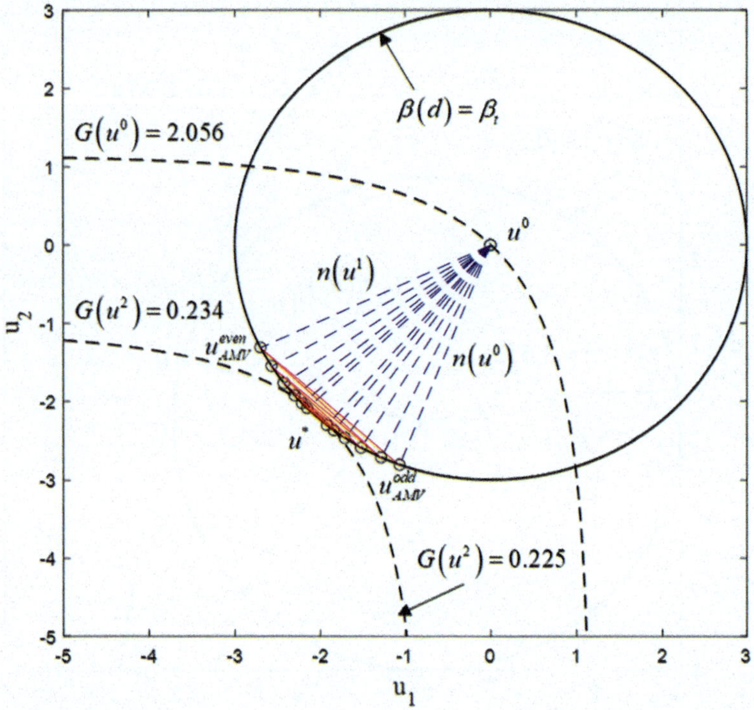

Fig. 7.2 The iterative process of AMV for the solution of concave performance function adapted from Youn et al. [21]

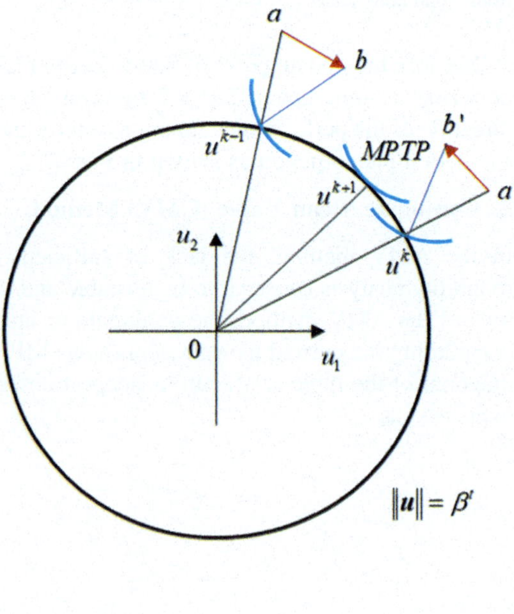

Fig. 7.3 AMV method solving result oscillation phenomenon adapted from Hao et al. [7]

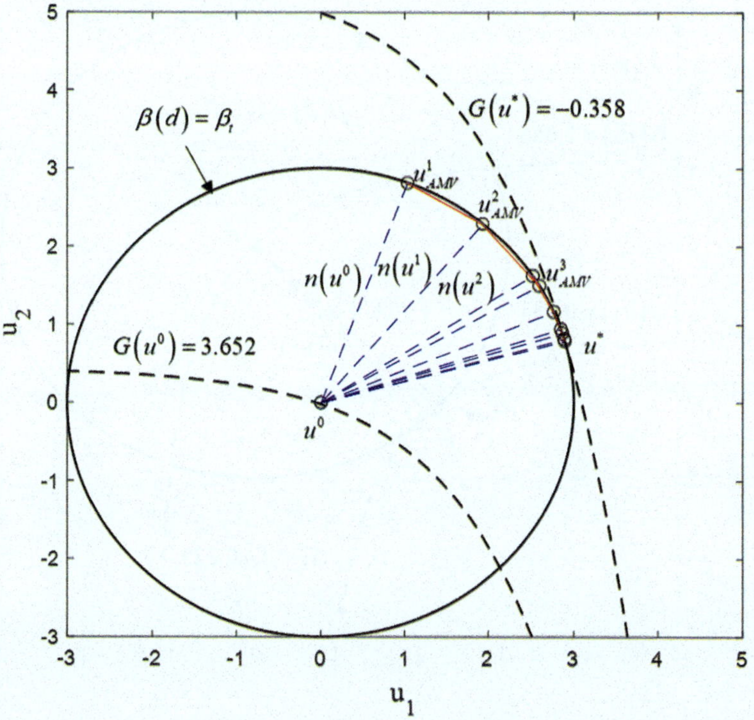

Fig. 7.4 Iterative process of CMV for solving convex performance functions adapted from Youn et al. [21]

$$
\begin{aligned}
&u_{CMV}^{(0)} = 0, u_{CMV}^{(1)} = u_{AMV}^{(1)}, u_{CMV}^{(2)} = u_{AMV}^{(2)} \\
&u_{CMV}^{(k+1)} = \beta_t \frac{n\left(u_{CMV}^{(k)}\right) + n\left(u_{CMV}^{(k-1)}\right) + n\left(u_{CMV}^{(k-2)}\right)}{\left\| n\left(u_{CMV}^{(k)}\right) + n\left(u_{CMV}^{(k-1)}\right) + n\left(u_{CMV}^{(k-2)}\right) \right\|}, k \geq 2 \\
&\text{where} n\left(u_{CMV}^{(k)}\right) = -\frac{\nabla_U G\left(u_{CMV}^{(k)}\right)}{\left\| \nabla_U G\left(u_{CMV}^{(k)}\right) \right\|}
\end{aligned}
\tag{7.20}
$$

A comparison of the AMV method with the CMV method using the same numerical cases as in the AMV method is shown in Figs. 7.4 and 7.5.

Figure 7.4 shows the MPP iteration history applied to the convex performance function using the CMV method, and Fig. 7.5 shows the MPP iteration history applied to the concave performance function using the CMV method. Clearly, for convex performance functions, the CMV method has more iterations compared to the AMV method. The numerical efficiency is not as good as AMV method for convex performance functions. While for the concave performance function, the numerical results of CMV method are convergent and stable.

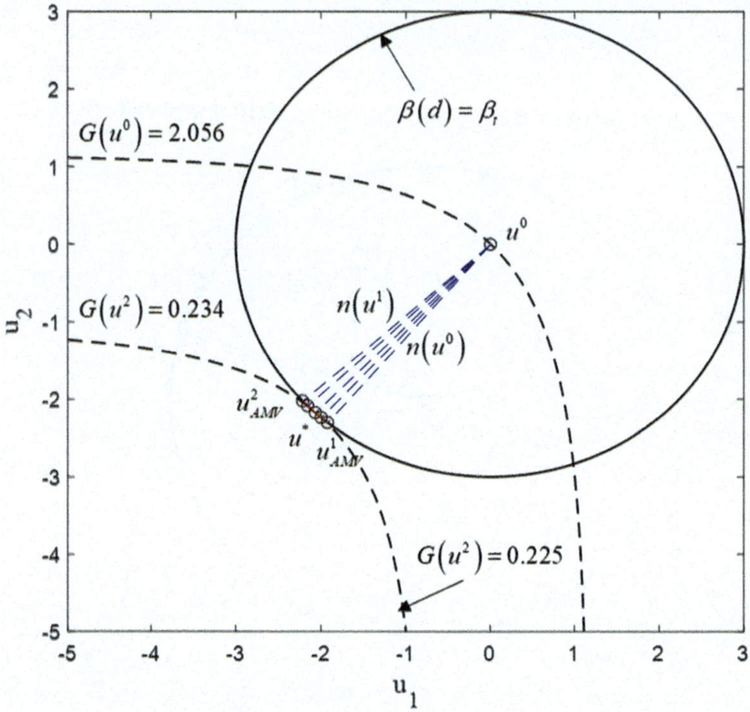

Fig. 7.5 Iterative process of CMV for solving concave performance functions adapted from Youn et al. [21]

Figure 7.6 can explain the reason why the CMV method is suitable for solving MPP for concave performance functions. In Fig. 7.6, $n(u_0)$ is the normalized steepest descent direction for the performance function at u_0. Since the CMV method considers the previous MPP information in the new search direction at the same time, even if the MPP of the current and subsequent iterations oscillates, the next update point is restricted to the iterative oscillation interval, thus gradually converging to the final MPP.

The AMV method is efficient for convex performance functions and the CMV method is stable for concave performance functions. The applicability of the PMA method can be greatly improved if the AMV method and CMV method can be adaptively selected for the information of the performance function types around the current iteration of MPPs, and thus the hybrid mean value (HMV) method is proposed, which will be introduced in the next section.

3. **Hybrid Mean Value (HMV) Method**

The core of the HMV method is to adaptively select the AMV method or the CMV method to solve the next updated MPP by giving a discriminant for the type of performance function. The HMV numerical algorithm effectively integrates the AMV and CMV methods, making it more robust and efficient. In the following, the discriminant in the HMV method is described.

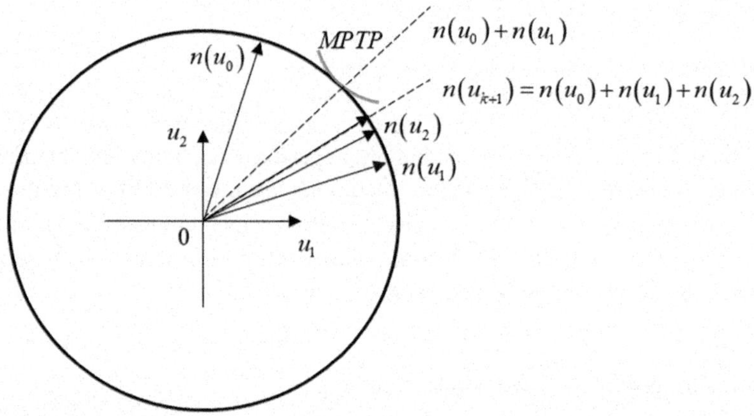

Fig. 7.6 Convergence process of CMV method for concave performance function adapted from Hao et al. [7]

The HMV method proposes a discriminant of function type based on the steepest descent direction at three consecutive iterations, expressed by

$$\xi^{(k+1)} = \left(n^{(k+1)} - n^{(k)}\right) \cdot \left(n^{(k)} - n^{(k-1)}\right) \tag{7.21}$$

When $\xi^{(k+1)}$ is larger than 0, the function is convex function at u_{HMV}^{k+1}. Otherwise, the function is concave function. Thus, once the type of performance function is identified, the next iteration of the update point can be performed by adaptively selecting the appropriate method. The process of HMV can be shown as follows.

Step 1: Set the number of iterations and the convergence parameter, and use the MV method to calculate the steepest direction of gradient descent $n\left(u_{HMV}^{(0)}\right)$ of the initial performance function in U-space.

Step 2: If the performance function is of convex type or $k < 3$, the next update point is calculated using the AMV method. Otherwise, if the performance function is concave and $k \geq 3$, CMV method is used to compute the next update point.

Step 3: In the new MPP $u_{HMV}^{(k+1)}$ calculate $G\left(u_{HMV}^{(k+1)}\right)$ and reliability index $\beta^{(k+1)}$. If the convergence criterion is satisfied then the computation is stopped and the optimal MPP is considered to be found, otherwise the next step is executed.

Step 4: Compute $\nabla_U G\left(u_{HMV}^{(k+1)}\right)$ and the performance function discriminant $\varsigma^{(k+1)}$, and set $k = k + 1$ to return Step 2.

For highly nonlinear performance functions, the efficiency of the HMV method is still unsatisfactory and may not even converge. Therefore, the HMV method is improved, and thus the enhanced Hybrid-Mean-Value method (HMV+) is proposed, which will be introduced in the next section.

4. **Enhanced HMV (HMV+) Method**

Based on HMV method, the arc interpolation method is introduced in the HMV+ method to obtain a more accurate output probability distribution $F_G(g)$. When the probability level $\beta_i > 0$, the value of the performance function decreases at the next search point. In the case when the value of the performance function increases at the next search point, the arc interpolation method is used, carried by using the performance function and its sensitivity values at the two search points $u_{HMV+}^{(k)}$ and $u_{HMV+}^{(k-1)}$. Hence, the performance function is interpolated along the arc region between these two search points to obtain the next update point $u_{HMV+}^{(k+1)}$.

Step 1: Set the number of iterations and the convergence parameter as for the probability level and make $u_{HMV+}^{(0)} = 0$;

Step 2: Compute the performance function $g\left(u_{HMV+}^{(k)}\right)$ and its sensitivity $\nabla g\left(u_{HMV+}^{(k)}\right)$;

Step 3: Check whether the Karush-Kuhn-Tucker condition satisfies the following equation, stop if it does, and consider that the final MPP has been found.

$$\left| \mathrm{sgn}(\beta_i) \cdot \frac{u_{HMV+}^{(k)}}{\left\| u_{HMV+}^{(k)} \right\|} \cdot n_{HMV+}^{(k)} - 1 \right| \leq \varepsilon \tag{7.22}$$

where $k \geq 2$, n is the normalized steepest ascent direction of $G_U(u)$, and $sgn(\beta_i)$ is the sign function whose value is -1 when $\beta_i < 0$ otherwise it is 1.

Step 4: If $k \geq 2$ and $sgn(\beta_i)\left| g\left(u_{HMV+}^{(k)}\right) - g\left(u_{HMV+}^{(k-1)}\right) \right| < 0$, the performance function is interpolated along the arc region between the two search points and obtain a new search point $u_{HMV+}^{(k+1)}$ by maximizing the approximate performance function. Otherwise, the HMV method is used to obtain a new search point $u_{HMV+}^{(k+1)}$ and set $k = k + 1$ and return to Step 2.

The HMV+ method can achieve significant results, and the numerical strategy actually plays a key role in the computational efficiency. The researchers further considered the numerical strategy to find numerical methods that make the PMA method more efficient and robust. Thus, the enriched PMA (PMA+) method was proposed, which will be introduced in the next section.

5. **Enriched PMA (PMA+) Method**

The numerical computation scheme is also decisive for the efficiency of the computational solution. The PMA+ (enriched PMA) method integrates four major improvements aimed at developing new efficient and robust methods for probabilistic constraint evaluation of RBDO without sacrificing numerical accuracy and stability. The four major improvements involved in PMA+ are summarized as follows:

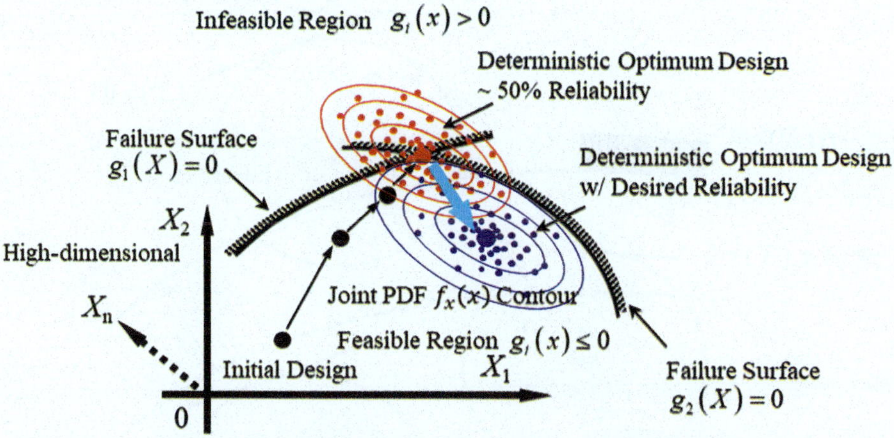

Fig. 7.7 The relative position of the DDO optimal solution to the RBDO optimal solution

(1) HMV+ method is applied in reliability analysis

HMV+ further solves the shortcomings of the HMV method and is a better numerical method from the comprehensive viewpoint of solution efficiency and stability. Therefore, the HMV+ method is used for reliability analysis.

(2) Carried RBDO based on DDO result

As shown in Fig. 7.7, the optimal solution of RBDO usually exists near the optimal solution of deterministic design optimization (DDO). Hence, the HMV+ method executes DDO at the initial design point first. Then the RBDO starting from the optimal solution of DDO, which has fewer iterations and higher numerical efficiency compared to executing RBDO solution from the initial design point.

(3) Probabilistic feasibility check

The reliability analysis of the current design point is required continuously during the iteration of the optimal solution of RBDO. And the PMA+ method proposes the concept of probabilistic feasibility checking, which aims to perform probabilistic feasibility judgment without performing the reliability analysis process completely while maintaining numerical accuracy. Thus, the numerical calculation efficiency is improved.

As shown in Fig. 7.8, a small positive number ε_f is introduced. If the probabilistic performance satisfies $-g_i < -\varepsilon_f$, the i-th probabilistic constraint is considered feasible at the current design point; if $-\varepsilon_f \leq -g_i \leq 0$, the i-th probabilistic constraint is considered active; if $-g_i > -\varepsilon_f$, the i-th probabilistic constraint is considered to violate the reliability requirement.

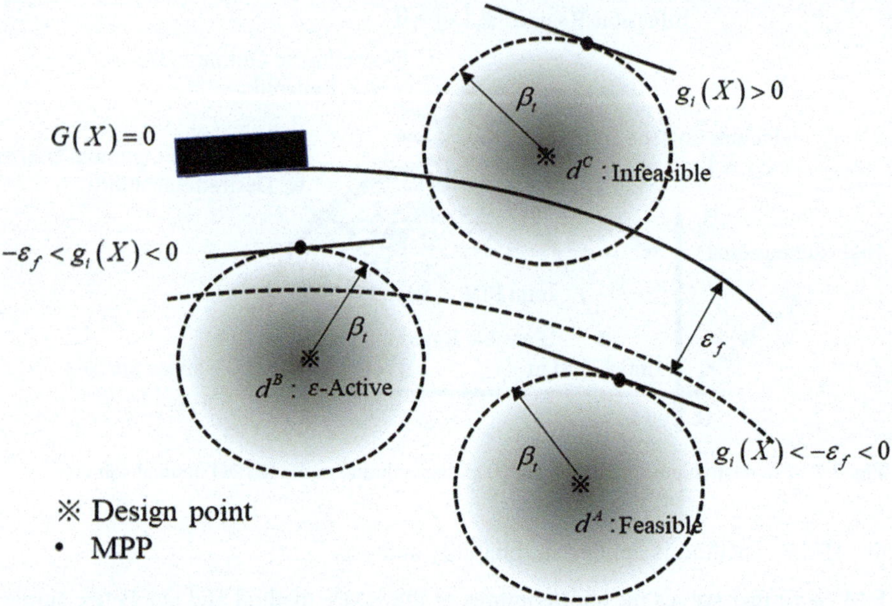

Fig. 7.8 Feasibility of probabilistic constraints in RBDO adapted from Youn et al. [19, 20]

(4) Rapid reliability analysis under design proximity conditions

The computational cost of RBDO mainly comes from the reliability analysis of the current design point. When the design points of the previous and subsequent iterations are close to each other, the reliability analysis can be performed by using the information obtained in the previous design iteration, which can effectively reduce the computational burden. Thus, the numerical calculation efficiency can be effectively improved when the iteration is near the end.

The proximity of the design points in X-space and the MPP sum in U-space between the previous and subsequent iterations is determined using the following equations:

$$\Delta d^{(k)} = \left\| \sqrt{\sum (X)^{-1}} \left[d^{(k)} - d^{(k-1)} \right] \right\| \le \varepsilon_d \tag{7.23}$$

$$\Delta x^{*(k-1)} = \left\| \left[x^{*(k-1)} - x^{*(k-2)} \right] \right\| \le \varepsilon_d \tag{7.24}$$

where $d^{(k)}$ is the design vector of the k-th iteration, $x^{*(k-1)}$ is the MPP point in the k-1th iteration, ε_d the constant of the design tightness criterion, and the diagonal components of the covariance matrix $\sum(X)$ of the random vector X corresponding to the design vector d are defined as

$$\Sigma^i = \sigma_{X_i}^2 = \int_{-\infty}^{\infty} (x_i - \mu_i)^2 f_{X_i}(x_i) dx_i$$
$$\text{where} \sqrt{\Sigma^i} = \sigma_{X_i}, \sqrt{\Sigma^i} = \text{diagonal}\left(\sqrt{\Sigma^i}\right) \quad (7.25)$$

Once it has been determined that the MPPs updated in the two previous and two iterations are close enough, the reliability analysis calculations can be performed using the HMV+ method, which is similar to the principle of performing RBDO at the optimal solution of DDO instead of starting from the initial design point. Performing the reliability analysis close to the optimal MPP reduces the number of reliability evaluations and thus increases the efficiency of numerical calculations.

7.3.4 Full Loop of MPP-Based RBDO

Figure 7.9 describes the complete numerical procedure of the proposed RBDO based on PMA+. On the left of Fig. 7.9 is the sub-optimization loop of the reliability analysis for evaluating a set of potential probabilistic constraints and their sensitivities. The sensitivity is obtained by taking derivative of the performance function. For explicit performance functions, the sensitivity can be directly obtained through derivation rules. For implicit performance functions, the derivative of the performance function obtained by difference method, including forward difference, backward difference, or central finite difference. Taking the commonly used central finite difference method as an example, the sensitivity of the performance function is calculated as follows:

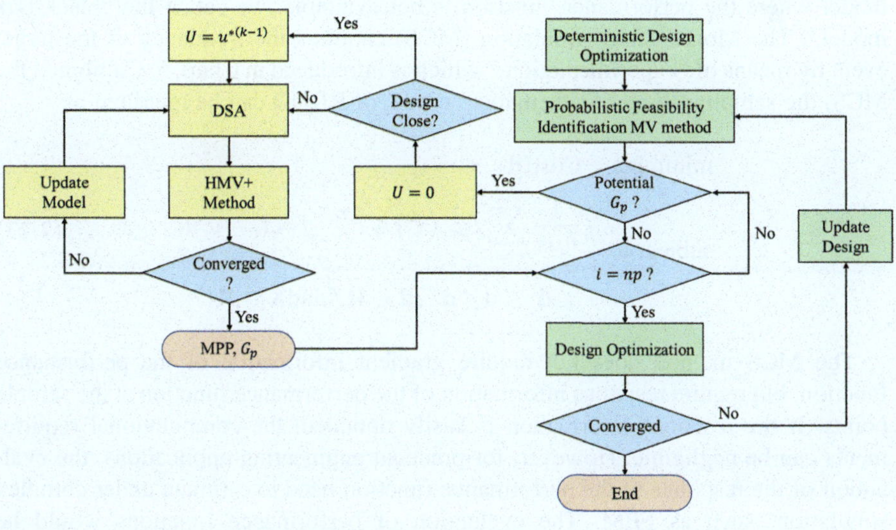

Fig. 7.9 Flow chart of PMA+-based RBDO adapted from Youn et al. [19, 20]

$$\frac{\partial G_i(\boldsymbol{x})}{\partial \boldsymbol{x}} \cong \frac{G_i(\boldsymbol{x}+\Delta) - G_i(\boldsymbol{x}-\Delta)}{2\Delta} \tag{7.26}$$

where $G_i(\cdot)$ represents the i-th performance function and Δ represents the step size of the difference method.

The right part of the loop of Fig. 7.9 is the main design optimization procedure where the RBDO undergoes design optimization. In the main optimization loop, a set of potential probabilistic constraints is identified by feasibility checks prior to the sub-optimization loop, thus reducing the computational burden throughout the RBDO process. In the sub-optimization loop, probability feasibility is identified to determine if fast reliability analysis should be used. The sequential quadratic programming method is used for design optimization and the HMV+ method is used for reliability analysis. The detailed description of other optimization procedure and methods will be introduced in Sect. 7.5.

7.4 Sampling-Based RBDO

7.4.1 Monte Carlo Simulation

The MPP-based RBDO method was proposed to solve the multiple integration problem involved in calculating the probability constraint in the RBDO model from the perspective of probabilistic resolution, and robust and efficient numerical algorithms were introduced. The common point of these numerical algorithms is that they all need to utilize the sensitivity of the performance function, which also contains many uncertainties in practical engineering applications. For the RBDO model where the performance function is not available, we call it the "black-box model." The Monte Carlo simulation (MCS) captures the frequency of the target event by means of "experimentation," which is introduced in Chap. 5. Combined the MCS, the sampling-based mathematical model of RBDO can be rewritten as

$$\begin{aligned} &\text{minimize} \quad \text{Cost}(\mathbf{d}) \\ &\text{subject to} \quad \frac{1}{n_{MC}} \sum_{k=1}^{n_{MC}} I_{\Omega_{F_j}}\left(x^k\right) \le P_{F_j}^{Tar}, j = 1, \ldots, nc \\ &\qquad\qquad \mathbf{d}^L \le \mathbf{d} \le \mathbf{d}^U, \mathbf{d} \in \mathbf{R}^{nd} \text{and} \mathbf{X} \in \mathbf{R}^{nr} \end{aligned} \tag{7.27}$$

The MCS method does not require gradient information of the performance function, but require response information of the performance function at the sample points. If the response information is easily obtained, the computational requirements can be negligible. However, for practical engineering applications, the evaluation of the response of the performance function need to estimate under complex simulation, such as FEM. The evaluation of performance functions would be extremely time-consuming or even unrealistic. Therefore, the MCS method cannot be directly used for solving the practical engineering RBDO problems yet.

The surrogate model can provide an approximate limit state bound (i.e., $G(x) = 0$). Besides, the computational effort of MCS based on surrogate models can be greatly reduced, and sufficiently accurate predictions of the performance function can be obtained with lower computational requirements. Next, the sampling-based RBDO approach is presented from the perspective of surrogate models in the next section.

7.4.2 Surrogate Model

There are many types of surrogate models, among which polynomial response surface modeling (PRSM) is the most used method due to its fast establishment, but it is limited in terms of global interpolation accuracy. The problem can be well solved by using chaotic polynomial (PCE), which corresponds to the response surface of a specific basis, but it is difficult to determine an appropriate DoE and the number of polynomials, and cross-validation is also required for accuracy assessment. Support vector machine (SVM) has shown many unique advantages in solving small sample, nonlinear and high-dimensional pattern recognition, and is also widely used in agent model building. In addition to the abovementioned common agent models, Kriging has attracted a lot of attention in the construction of surrogate models. Sachs-Ayer applied this method to approximate computer experiments in 1989. The details of the surrogate model can be referred in Chap. 3.

7.4.3 Stochastic Sensitivity Analysis Based on Surrogate Model

After the accurate surrogate model is established, the reliability analysis can be performed using MCS. An important step in design optimization is to obtain sensitivity derivatives, which can be used to study the effect of parametric modifications, to calculate search directions to find an optimum design. The sensitivity analysis discusses "how" and "how much" changes in the parameters of an optimization problem modify the optimal objective value and the point where the optimum is attained. On the one hand, the sensitivity of the probabilistic response can be obtained by the finite difference method (FDM). However, since the probability response is derived from the MCS, many sample points are required to obtain an accurate probability response sensitivity. In addition, using the FDM to solve the probability response sensitivity for optimization may also lead to infeasible points. On the other hand, the probability response sensitivity can be obtained from the constructed surrogate model. However, even if the surrogate model is very accurate, the sensitivity information it provides is inaccurate. Therefore, Lee et al. proposed a sampling-based stochastic sensitivity analysis method that uses a scoring function (score function) for the calculation of probabilistic response sensitivity. The method neither requires a transformation involving space (X-space to U-space) nor does the computational procedure have any approximation.

The probability of failure P_F in Eq. (7.27) can be defined as

$$\begin{aligned} P_F(\Psi) &\equiv P[\mathbf{X} \in \Omega_F] \\ &= \int_{\mathbf{R}^{nr}} I_{\Omega_F}(\mathbf{x}) f_{\mathbf{X}}(\mathbf{x};\Psi) d\mathbf{x} = E[I_{\Omega_F}(\mathbf{X})] \end{aligned} \tag{7.28}$$

where $\Psi$is a vector of distribution parameters, which is the mean of **X**, $\Omega_F$is the failure set; $f_{\mathbf{X}}(\mathbf{x};\Psi)$ is the joint probability density function (PDF) of X. Taking the partial derivative of the probability of failure in Eq. (7.28) with respect to the *i*-th design variable μ_i yields

$$\frac{\partial P_F(\psi)}{\partial \mu_i} = \frac{\partial}{\partial \mu_i} \int_{\boldsymbol{R}^{nr}} I_{\Omega_F}(\boldsymbol{x}) f_{\boldsymbol{X}}(\boldsymbol{x};\boldsymbol{\mu}) d\boldsymbol{x} \tag{7.29}$$

By utilizing the Leibniz's rule of differentiation, the differential and integral operators can be expressed as

$$\begin{aligned} \frac{\partial P_F(\psi)}{\partial \mu_i} &= \int_{\boldsymbol{R}^{nr}} I_{\Omega_F}(\boldsymbol{x}) \frac{\partial f_{\boldsymbol{X}}(\boldsymbol{x};\boldsymbol{\mu})}{\partial \mu_i} d\boldsymbol{x} \\ &= \int_{\boldsymbol{R}^{nr}} I_{\Omega_F}(\boldsymbol{x}) \frac{\partial \ln f_{\boldsymbol{X}}(\boldsymbol{x};\boldsymbol{\mu})}{\partial \mu_i} f_{\boldsymbol{X}}(\boldsymbol{x};\boldsymbol{\mu}) d\boldsymbol{x} \\ &= \left[I_{\Omega_F}(\boldsymbol{x}) \frac{\partial \ln f_{\boldsymbol{X}}(\boldsymbol{x};\boldsymbol{\mu})}{\partial \mu_i} \right] \end{aligned} \tag{7.30}$$

Since $I_{\Omega_F}(\boldsymbol{x})$ is not a function of μ_i, the sensitivity analysis only needs to consider the partial derivative of the log function of the joint PDF of Eq. (7.30) with respect to μ_i is known as the first-order score function for μ_i and is denoted as

$$s_{\mu_i}^{(1)}(\boldsymbol{x};\boldsymbol{\mu}) \equiv \frac{\partial \ln f_{\boldsymbol{X}}(\boldsymbol{x};\boldsymbol{\mu})}{\partial \mu_i} \tag{7.31}$$

Hence, to further derive sensitivities of the probability of failure in Eq. (7.31), the derive of the first-order score function of (31) can be carried out by independent input random variables and correlated input random variables, which will be introduced in the next section.

1. **Independent Input Random Variables**

Consider a random input $\boldsymbol{X} = \{X_1, \ldots, X_{nr}\}^T$ whose components are statistically independent random variables. Then the joint PDF of $\boldsymbol{X}$ is expressed as a multiplication of its marginal PDFs as

$$f_{X}(x;\mu) = \prod_{i=1}^{nr} f_{X_i}(x_i;\mu_i) \tag{7.32}$$

where $f_X(x_i, \mu_i)$ is the marginal PDF corresponding to the i-th random variable X_i. Therefore, for statistically independent random variables, the first-order score function for μ_i is expressed as

$$s_{\mu_i}^{(1)}(x;\mu) \equiv \frac{\partial \ln f_X(x;\mu)}{\partial \mu_i} = \frac{\partial \ln f_{X_i}(x_i;\mu_i)}{\partial \mu_i} \tag{7.33}$$

The marginal PDF and the cumulative distribution function (CDF) are available analytically as listed in Chap. 1.

2. **Correlated Input Random Variables**

Consider a bivariate correlated random input $X = \{X_i, X_j\}^T$. Then, the joint PDF of X is expressed as

$$\begin{aligned} f_X(x;\mu) &= \frac{\partial^2 C(u,v;\theta)}{\partial u \partial v} f_{X_i}(x_i;\mu_i) f_{X_j}(x_j;\mu_j) \\ &= C_{,uv}(u,v;\theta) f_{X_i}(x_i;\mu_i) f_{X_j}(x_j;\mu_j) \end{aligned} \tag{7.34}$$

where C is the copula function, $u = F_{X_i}(x_i, \mu_i)$ and $v = F_{X_j}(x_j;\mu_j)$ are marginal CDFs for X_i and X_j, respectively, and θ is the correlation coefficient between X_i and X_j. The partial derivative of the copula function with respect to the marginal CDFs u and v in Eq. (7.34) is called the copula density function and denoted as

$$c(u,v;\theta) \equiv \frac{\partial^2 C(u,v;\theta)}{\partial u \partial v} = C_{,uv}(u,v;\theta) \tag{7.35}$$

Accordingly, using Eq. (7.34), the first-order score functions in Eq. (7.33) for a correlated bivariate input are expressed as

$$s_{\mu_i}^{(1)}(x;\mu) \equiv \frac{\partial \ln f_X(x;\mu)}{\partial \mu_i} = \frac{\partial \ln c(u,v;\theta)}{\partial \mu_i} + \frac{\partial \ln f_{X_i}(x_i;\mu_i)}{\partial \mu_i} \tag{7.36}$$

The derivation of the first term in the right-hand side of Eq. (7.36) can be inferred in Table. 1, and the second term can be calculated similarly in Eq. (7.33). In Table 7.1, the partial derivative of the marginal CDF with respect to μ_i , $\partial \mu_i$ can be straightforwardly obtained from the analytic CDFs shown in Table 7.1. Even if several pairs of bivariate correlated random variables exist in $X = \{X_1, \ldots, X_{nr}\}^T$, the first-order score function for μ_i is the same as Eq. (7.36).

Table 7.1 Log-derivative of copula density function adapted from [9]

Copula type	$\frac{\partial \ln c(u,v;\theta)}{\partial \mu_i}$
Clayton	$\left(-\frac{1+\theta}{u}+\frac{(2\theta+1)u^{-(1+\theta)}}{u^{-\theta}+v^{-\theta}+1}\right)\frac{\partial u}{\partial \mu_i}$
AMH	$\left[\frac{\theta^2(1-v)+\theta(v+1)}{1-\theta^2(1-u)(1-v)-\theta(2-u-v-uv)}-\frac{3\theta(1-v)}{1-\theta(1-u)(1-v)}\right]\frac{\partial u}{\partial \mu_i}$
Frank	$\theta\left[\frac{2\left(e^{\theta(1+u)}-e^{\theta(u+v)}\right)}{e^{\theta}-e^{\theta(1+u)}-e^{\theta(1+v)}+e^{\theta(u+v)}}+1\right]\frac{\partial u}{\partial \mu_i}$
FGM	$\left[\frac{2\theta(2v-1)}{1+\theta(1-2u)(1-2v)}\right]\frac{\partial u}{\partial \mu_i}$
Gaussian	$\left[\frac{\Phi^{-1}(u)}{\phi(\Phi^{-1}(u))}+\frac{\theta\Phi^{-1}(u)-\Phi^{-1}(u)}{\phi(\Phi^{-1}(u))(1-\theta^2)}\right]\frac{\partial u}{\partial \mu_i}$
Independent	0

3. **Sensitivity Calculation of Probability of Failure**

Denote the surrogate model for the constraint function $G_j(\boldsymbol{X})$ as $\widehat{G}_j(\boldsymbol{X})$. Then the MCS is used to express the probabilistic constraints can be expressed by

$$P_{F_j} \equiv P\left[G_j(X) > 0\right] \cong \frac{1}{M}\sum_{m=1}^{M} I_{\widehat{\Omega}_{F_j}}\left(x^{(m)}\right) \leq P_{F_j}^{Tar} \tag{7.37}$$

where M is the sampling size of MCS, $x^{(m)}$ is the m-th realization of X, and the failure set $\widehat{\Omega}_{F_j}$ for the surrogate model is defined as $\widehat{\Omega}_{F_j} \equiv \left\{x : \widehat{G}_j(x) > 0\right\}$. The sensitivity of the probability of failure in Eq. (7.2) can be transmitted

$$\frac{\partial P_{F_j}}{\partial \mu_i} \cong \frac{1}{M}\sum_{m=1}^{M} I_{\widehat{\Omega}_{F_j}}\left(x^{(m)}\right) s_{\mu_i}^{(1)}\left(x^{(m)};\mu\right) \tag{7.38}$$

where $s_{\mu_i}^{(1)}\left(x^{(m)};\mu\right)$ can be obtained by Eq. (7.36). The aforementioned first-order score functions (Eqs. (7.33) and 7.36)) can be used to calculate the sensitivity of the probability of failure. To estimate of the accuracy of Eq. (7.38) can be rewritten as

$$\begin{aligned}&\frac{1}{M}\sum_{m=1}^{M} I_{\widehat{\Omega}_{F_j}}\left(x^{(m)}\right) s_{\mu_i}^{(1)}\left(x^{(m)};\mu\right)\\&=\frac{1}{M}\sum_{m_f=1}^{M_f} I_{\widehat{\Omega}_{F_j}}\left(x^{(m_f)}\right) s_{\mu_i}^{(1)}\left(x^{(m_f)};\mu\right)=\frac{M_f}{M}\mu_{s_f}=P_{F_j}\mu_{s_f}\end{aligned} \tag{7.39}$$

where M_f is the number of failed samples and μ_{s_f} is the mean value of the score function value for the failed samples. Hence, the accuracy for the MCS to compute Eq. (7.38) can be measured by $\varepsilon_{MCS}\mu_{s_f}$. Since μ_{s_f} changes based on the problems, the accuracy of the probability of failure varies according to problems or design points but not related to the surrogate model. Hence, the computation of the sensitivity using the score function does not include any approximation except the statistical noise due to MCS when the number of the sampling point is large enough. After the analysis of probability of failure, the design procedure can be carried, which will be introduced in the next section.

7.5 Double-Loop, Single-Loop, and Decoupled RBDO

7.5.1 Double-Loop RBDO

In RIA and PMA, the optimization model structure for reliability-based design optimization using either the RIA-based RBDO mathematical model or the PMA-based RBDO mathematical model is called the "double-loop structure," which can be referred in Sect. 7.3.4. The "double-loop" means that the "inner-loop" performs reliability analysis and the "outer-loop" performs design optimization, as shown in Fig. 7.10, the two steps (design optimization and reliability analysis).

Most of the computational effort in performing RBDO comes from the reliability analysis process. The characteristic of the two-cycle structure dictates that a large number of times of reliability analysis is required during the execution of RBDO. Therefore, the efficiency of RBDO with a two-cycle structure is low. To overcome these difficulties, subsequent researchers have proposed "single-loop structure" and "decoupled structure."

7.5.2 Single-Loop RBDO

The single-loop structure has only one loop in the design optimization process, which greatly improves the efficiency of the solution compared with the two-loop structure, but with the corresponding sacrifice of the accuracy of the results. The key idea of single-loop is to approximate the probabilistic constraints into deterministic

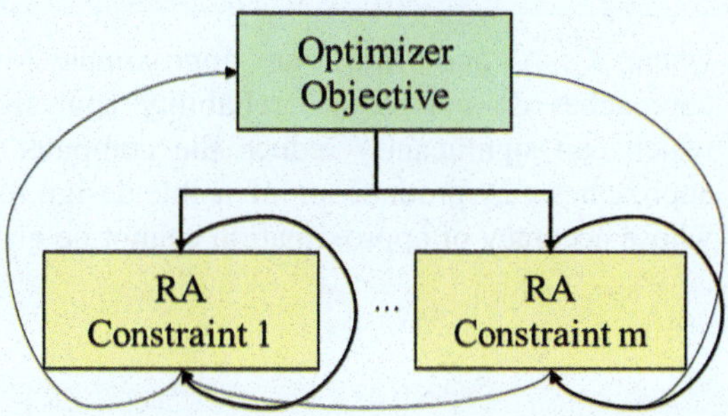

Fig. 7.10 Flowchart of double-loop RBDO adapted from [12]

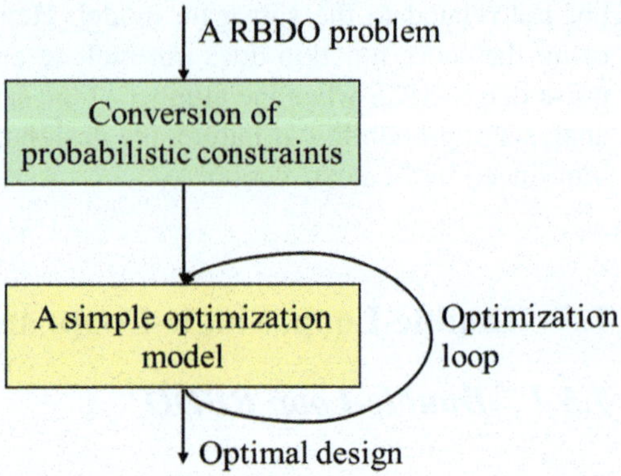

Fig. 7.11 Flowchart of single-loop RBDO adapted from Shan et al. [12]

ones without additional reliability analysis (as shown in Fig. 7.11). Chen et al. [2] proposed the single-loop single-variable (SLSV) method in 1997. However, the SLSL method is unstable due to the influence of conditions such as the selection of the initial point, the degree of nonlinearity of the constraint, etc. Liang et al. [10] developed the single-loop approach (SLA) based on SLSV, which approximates the most probable target point (MPTP) with active constraints by introducing the KKT condition.

The optimization model of the SLA method is as follows:

$$
\begin{aligned}
&\text{find } d, \mu_X \\
&\min \ \mathrm{f}(\mathrm{d}, \mu_X, \mu_P) \\
&\text{s.t.} \quad \mathrm{g}_c\left(d, \tilde{X}^{k+1}_{i,MPTP}, \tilde{P}^{k+1}_{i,MPTP}\right) \geq 0, c = 1, 2, \ldots, N \\
&\mathrm{d}_i^L \leq d_i \leq d_i^U, \mu_{X_j}^L \leq \mu_{X_j} \leq \mu_{X_j}^U \\
&\tilde{X}^{k+1}_{c,MPTP} = \mu_X^{k+1} - \sigma_X \beta_c^t \frac{\sigma_X \nabla g_c\left(\tilde{X}^k_{c,MPTP}\right)}{\left\|\sigma_X \nabla g_c\left(\tilde{X}^k_{c,MPTP}\right)\right\|} \\
&\tilde{P}^{k+1}_{c,MPTP} = \mu_P^{k+1} - \sigma_P \beta_c^t \frac{\sigma_P \nabla g_c\left(\tilde{P}^k_{c,MPTP}\right)}{\left\|\sigma_P \nabla g_c\left(\tilde{P}^k_{c,MPTP}\right)\right\|}
\end{aligned}
\tag{7.40}
$$

where $\tilde{X}^{k+1}_{MPTP}$ and P^{k+1}_{MPTP} are approximate MPTP points. Since the single-loop approaches do not require reliability analysis during the optimization process, which can significantly reduce the computer burden. However, the single-loop approaches may produce an infeasible design for highly nonlinear design problems whose accuracy of approximation cannot be guaranteed.

7.5.3 *Decoupled RBDO*

Decoupling in RBDO separates the reliability analysis and the optimization process. It applies the reliability analysis results to the optimization process by using a single-loop strategy. Compared to the double-loop RBDO, which conducts the reliability analysis for all design changes in the outer loop, the decoupled RBDO conducts the reliability analysis only once after the deterministic optimum design from the outer loop is achieved. That is, the outer loop (i.e., design optimization) may have several iterations but it does not call the inner loop (i.e., reliability analysis) each time, which reduces the number of the reliability analyses thus reduce the computational cost. The structure diagram is shown in Fig. 7.12.

The key idea is the usage of shifting vector s_i^{k+1}, which transforms deterministic constraints into shifted constraints, updated after every reliability analysis. As the design optimization proceeds, the difference between the shifted and probabilistic constraints diminishes. The optimization process can be expressed by

$$\begin{aligned} &\min\text{imize}\, f(\mathbf{d}) \\ &\text{subject to}\; G_j\left(\mathbf{d}, \mu_{\mathrm{x}} - \mathbf{s}_j^{k+1}\right) \le 0, j = 1, 2, \ldots, N_c \end{aligned} \tag{7.41}$$

The earliest decoupling structure was proposed by Li et al. [15] which was based on the RIA model. Li et al. solved the RBDO problem with the help of linear programming. They constructed an approximate probability constraint by linearly

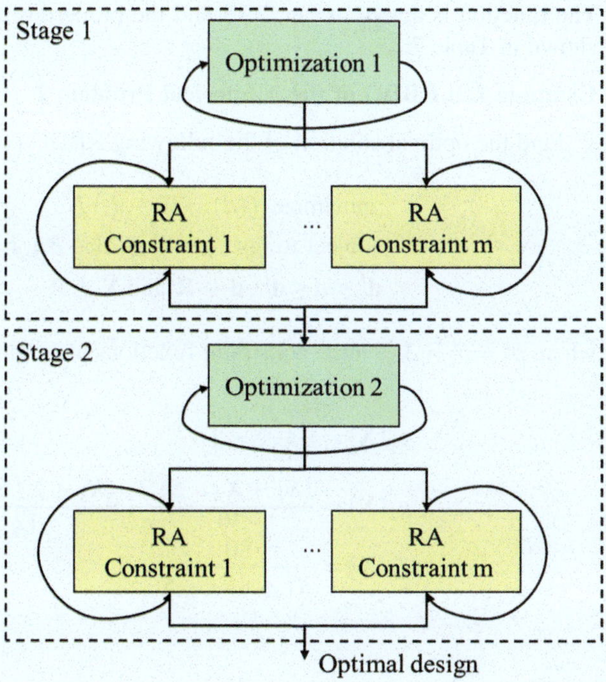

Fig. 7.12 Flowchart of decoupled RBDO adapted from [15]

approximating the reliability index based on the reliability results and sensitivity analysis results of the previous iteration. Later, Cheng et al. [3] proposed a sequential approximate programming (SAP) for the RBDO problem by adapting the traditional sequential approximate programming method. Zou and Mahadevan [23] proposed another direct decoupling method by expanding the approximation of the probability constraint with first-order Taylor technique based on the current failure probability and sensitivity information. One of the main concerns of the decoupled RBDO methods is that the calculated sensitivity of the probability of failure with respect to design variables may be inaccurate resulting in nonconvergence of the RBDO process or inaccurate RBDO optimum.

Example 7.1: RBDO of the Numerical Problem 1

Find the optimal solution of the following RBDO problem with 2 constraints.

$$\begin{aligned} &\text{min imize } f(d) = (d_1 - 3.7)^2 + (d_2 - 4)^2 \\ &\text{subject to } [g_i(X) < 0] \leq \Phi(-\beta_i^t), i = 1, 2 \\ &\mathbf{d}^L \leq \mathbf{d} \leq \mathbf{d}^U, \mathbf{d} \in \mathbf{R}^2 \text{and } \mathbf{X} \in \mathbf{R} \end{aligned} \tag{7.42}$$

where $\beta_1^t = \beta_2^t = 2.0$, two constraint functions are expressed as

$$\begin{aligned} g_1(X) &= -X_1 \sin(4X_1) - 1.1X_2 \sin(2X_2) \\ g_2(X) &= X_1 + X_2 - 3 \end{aligned} \tag{7.43}$$

Solution

The function is drawn in Fig. 7.13 and the properties of two random variables are shown in Table 7.2.

Example 7.2: RBDO of the Numerical Problem 2

Find the optimal solution of the following RBDO problem with 3 constraints.

$$\begin{aligned} &\text{min imize } f(d) = 10 - d_1\text{-}d_2 \\ &\text{subject to } [g_i(X) < 0] \leq \Phi(-\beta_i^t), i = 1, 2, 3 \\ &\mathbf{d}^L \leq \mathbf{d} \leq \mathbf{d}^U, \mathbf{d} \in \mathbf{R}^2 \text{and } \mathbf{X} \in \mathbf{R} \end{aligned}$$

where $\beta_1^t = \beta_2^t = 3.0$, three constraint functions are expressed as

$$\begin{aligned} g_1(\mathbf{X}) &= \frac{X_1^2 X_2}{20} - 1 \\ g_2(\mathbf{X}) &= \frac{(X_1 + X_2 - 5)^2}{30} + \frac{(X_1 + X_2 - 12)^2}{12} - 1 \\ g_3(\mathbf{X}) &= \frac{80}{X_1^2 + 8X_2 + 5} - 1 \end{aligned}$$

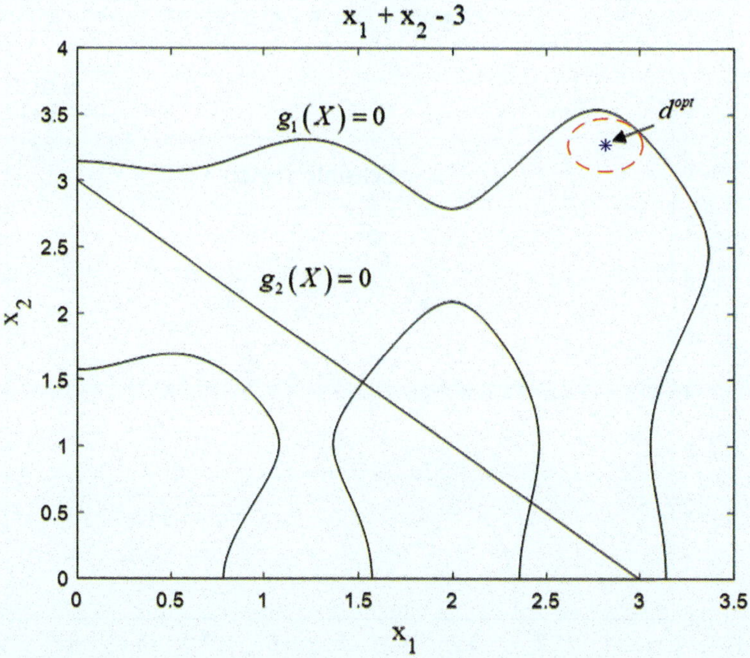

Fig. 7.13 RBDO of the numerical problem 1 with 2 constraints

Table 7.2 Parameter of RBDO of the numerical problem 1 with 2 constraints

Random variables	Distribution	d^L	d^0	d^U	Standard deviation
X_1	Normal	0.0	2.5	3.7	0.1
X_2	Normal	0.0	2.5	4.0	0.1

Solution

The function is drawn in Fig. 7.14; and the properties of two random variables are shown in Table 7.3.

Example 7.3: RBDO of the Numerical Problem 3

Find the optimal solution of the following RBDO problem with 3 constraints.

$$\begin{aligned} &\text{min imize } \mathrm{Cost}(\mathbf{d}) = -\frac{(d_1 + d_2 - 10)^2}{30} - \frac{(d_1 - d_2 + 10)^2}{120} \\ &\text{subject to } P\big(G_j(\mathbf{X}(\mathbf{d})) > 0\big) \le P_{F_j}^{Tar} = 2.275\%, j = 1, 2, 3 \\ &\mathbf{d}^L \le \mathbf{d} \le \mathbf{d}^U, \mathbf{d} \in \mathbf{R}^2 \text{ and } \mathbf{X} \in \mathbf{R} \end{aligned} \tag{7.44}$$

where three constraint functions are expressed as

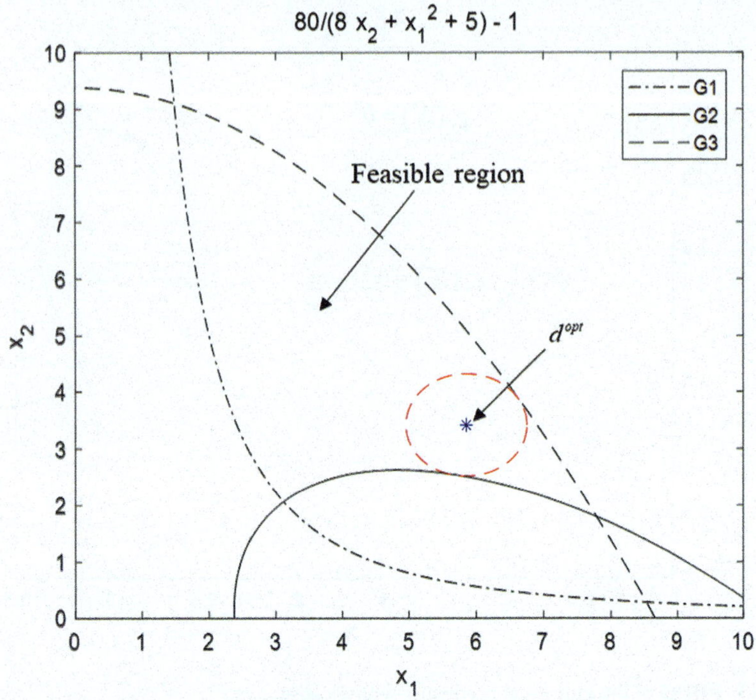

Fig. 7.14 RBDO of the numerical problem 2 with 3 nonlinear constraints

Table 7.3 Parameter of RBDO of the numerical problem 2 with 3 nonlinear constraints

Random variables	Distribution	d^L	d^0	d^U	Standard deviation
X_1	Normal	0.0	5	10	0.3
X_2	Normal	0.0	5	10	0.3

$$\begin{aligned} G_1(\mathbf{X}) &= 1 - \frac{X_1^2 X_2}{20} \\ G_2(\mathbf{X}) &= -1 + (Y-6)^2 + (Y-6)^3 - 0.6 \times (Y-6)^4 + Z \\ G_3(\mathbf{X}) &= 1 - \frac{80}{X_1^2 + 8X_2 + 5} \end{aligned} \tag{7.45}$$

where $\begin{Bmatrix} Y \\ Z \end{Bmatrix} = \begin{bmatrix} 0.9063 & 0.4226 \\ 0.4226 & -0.9063 \end{bmatrix} \begin{Bmatrix} X_1 \\ X_2 \end{Bmatrix}$

Solution

The functions are drawn in Fig. 7.15, and the properties of two random variables are shown in Table 7.4, and they are correlated with the Clayton copula ($\tau = 0.5$). The target probability of failure (P_F^{Tar}) is 2.275% for all constraints.

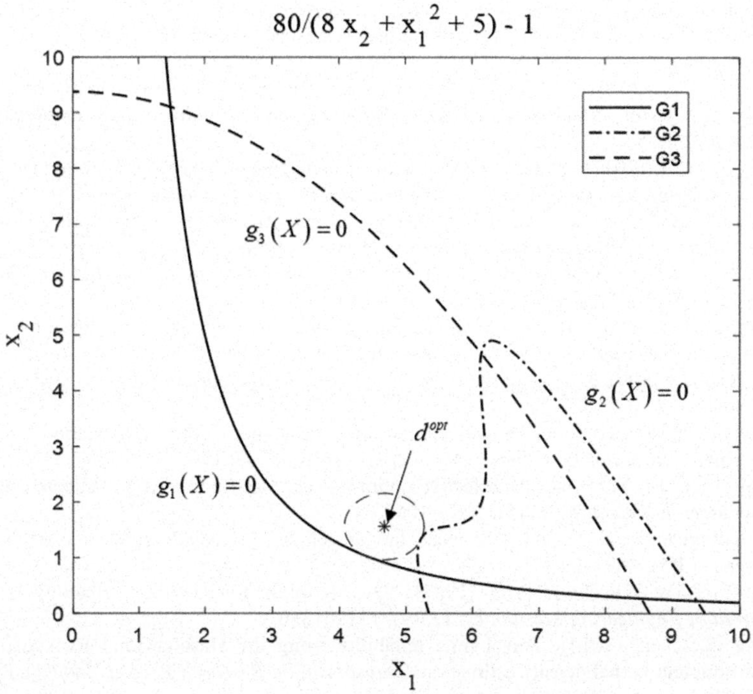

Fig. 7.15 RBDO of the numerical problem 3 with 3 nonlinear constraints

Table 7.4 Parameter of RBDO of the numerical problem 3 with 3 nonlinear constraints

Random variables	Distribution	d^L	d^0	d^U	Standard deviation
X_1	Normal	0.0	5.0	10.0	0.3
X_2	Normal	0.0	5.0	10.0	0.3

References

1. Breitung, K. (1984). Asymptotic approximations for multinormal integrals. *Journal of Engineering Mechanics, **110***(3), 357–366.
2. Chen, X., et al. (1997). Reliability based structural design optimization for practical applications. In *Proceeding of 38th Structures, structural dynamics, and materials conference: 1403.*
3. Cheng, G., et al. (2006). A sequential approximate programming strategy for reliability-based structural optimization. *Computers & Structures, **84***(21), 1353–1367.
4. Cheng, J., et al. (2019). Hybrid reliability-based design optimization of complex structures with random and interval uncertainties based on ASS-HRA. *IEEE Access, **7***, 87097–87109.
5. Ditlevsen, O., & Bjerager, P. (1986). Methods of structural systems reliability. *Structural Safety, **3***(3-4), 195–229.
6. Fang, J., et al. (2022). Wind turbine rotor speed design optimization considering rain erosion based on deep reinforcement learning. *Renewable & Sustainable Energy Reviews, **168***, 112788.

7. Hao, P., et al. (2019). An augmented step size adjustment method for the performance measure approach: Toward general structural reliability-based design optimization. *Structural Safety, **80***, 32–45.
8. Hu, W., et al. (2016). Reliability-based design optimization of wind turbine blades for fatigue life under dynamic wind load uncertainty. *Structural and Multidisciplinary Optimization, **54***, 953–970.
9. Lee, I., et al. (2011). Sampling-based RBDO using the stochastic sensitivity analysis and dynamic Kriging method. *Structural and Multidisciplinary Optimization, **44***, 299–317.
10. Liang, J., et al. (2004). A single-loop method for reliability-based design optimization. *International Design Engineering Technical Conferences, **46946***, 419–430.
11. Liu, P.-L., & Der Kiureghian, A. (1991). Optimization algorithms for structural reliability. *Structural Safety, **9***(3), 161–177.
12. Shan, S., et al. (2008). Reliable design space and complete single-loop reliability-based design optimization. *Reliability Engineering & System Safety, **93***(8), 1218–1230.
13. Tu, J., et al. (1999). A new study on reliability-based design optimization. *Journal of Mechanical Design, **121***(4), 557–564.
14. Tvedt, L. (1990). Distribution of quadratic forms in normal space—Application to structural reliability. *Journal of Engineering Mechanics, **116***(6), 1183–1197.
15. Weiji, L., & Li, Y. (1994). An effective optimization procedure based on structural reliability. *Computers & Structures, **52***(5), 1061–1067.
16. Wirsching, P., et al. (1991). Advanced fatigue reliability analysis. *International Journal of Fatigue, **13***(5), 389–394.
17. Wu, Y.-T., & Wirsching, P. H. (1987). New algorithm for structural reliability estimation. *Journal of Engineering Mechanics, **113***(9), 1319–1336.
18. Yang, S., et al. (2014). Travel time reliability using the Hasofer–Lind–Rackwitz–Fiessler algorithm and kernel density estimation. *Transportation Research Record, **2442***(1), 85–95.
19. Youn, B. D., et al. (2005a). Adaptive probability analysis using an enhanced hybrid mean value method. *Structural and Multidisciplinary Optimization, **29***, 134–148.
20. Youn, B. D., et al. (2005b). Enriched performance measure approach for reliability-based design optimization. *AIAA Journal, **43***(4), 874–884.
21. Youn, B. D., et al. (2003). Hybrid analysis method for reliability-based design optimization. *Journal of Mechanical Design, **125***(2), 221–232.
22. Zhu, S.-P., et al. (2021). Reliability-based structural design optimization: Hybridized conjugate mean value approach. *Engineering with Computers, **37***, 381–394.
23. Zou, T., & Mahadevan, S. (2006). A direct decoupling approach for efficient reliability-based design optimization. *Structural & Multidisciplinary Optimization, **31***, 190–200.

Chapter 8
Robust Design Optimization

Nomenclature

D	A real number
F	The original optimization objective
f	The new optimization objective function
$\mathbf{g}$	Constraint vector
g_i	The value of the design metric
$\overline{g_i}$	The class-function that need to be minimized for the metric
$\boldsymbol{N}$	The Nadir point where each objective has been maximized
$\boldsymbol{n}_{sc}$	The number of design metrics
$\widehat{\boldsymbol{n}}$	The normal unit direction to the convex hull of individual minima
o	The parent population
$\boldsymbol{p}$	Engineering system constant parameter vector
p	The specific metric
$\boldsymbol{R}$	The reliability vector specified for the constraint vector
$\mathbf{r}$	The n-dimension objective function vector
s_o	The size of the parent population
$\mathbf{U}$	The Utopia point where each objective has been minimized
$\boldsymbol{w}$	Weight vector using in normal boundary intersection method
$\boldsymbol{x}$	Design variable vector
$\boldsymbol{x}^L$	Lower bounds of design variable
$\boldsymbol{x}^U$	Upper bounds of design variable
$\boldsymbol{x}^*$	Optimal solution for design variables
β	A positive value
μ_f	The mean of the original optimization objective
μ_f^*	The ideal solution of function μ_f
σ_f	The standard deviation of the original optimization objective

W. Hu, *Design Optimization Under Uncertainty*,
https://doi.org/10.1007/978-3-031-49208-2_8

σ_f^* The ideal solution of function σ_f^*
$\mathbf{\Phi}$ The payoff matrix
$\overline{\mathbf{\Phi}}$ The normalized payoff matrix

8.1 Introduction

With the advancement of science and technology, the demands for industrial products with better performance, higher reliability, and lower cost have never been greater than it is now. However, due to the inevitable presence of uncertainties in the lifecycle of products, e.g., tolerances on manufactured dimensions or material properties [1, 2], product performance can fluctuate significantly, or even deviate substantially from the original value, leading to functional failures and accidents. Thus, it is essential to take uncertainty into account in the process of product analysis and design for robust and reliable performance while asking for a cost-effective solution. Robust design optimization (RDO) is such a method developed to achieve the goal.

Different from the reliability-based design optimization (RBDO) mentioned in the earlier chapter that only considers uncertainty in the probabilistic constraints, the RDO strategy accounts for the effects of variation and uncertainties by simultaneously optimizing the objective function and minimizing performance parameter variations [3]. The RDO does not expunge the uncertainties rather it makes the product performance insensitive to these variations. Our discussion of RDO will cover the problem statement and formulation of RDO as well as its main procedure and several optimization approaches.

8.2 Problem Statement and Formulation

It has been widely recognized that RDO is one of the most common methods for incorporating the impact of uncertainty into the design optimization formulation [4–8]. It aims at developing low-cost and high-quality products whose performances are insensitive to the effects of various natural variations in their manufacturing and operational environments. RDO is developed based on the philosophy of robust design proposed by Taguchi [9, 10]. As a pioneer of robust design, Taguchi presents the concept of RDO as a method of dealing with geometric deviations arising in the manufacturing process in the 1950s, which is widely known as Taguchi method. The core component of Taguchi method is design of experiments (DoE) [4, 11], which involves evaluating various designs to identify factors that impact product quality and optimize their normal levels. Meanwhile, the concept of robustness is also first defined by Taguchi as being insensitive to variations in both the system itself and the environment. Although Taguchi method is beneficial for the improvement of product quality, it usually encounters the problem of curse of dimensionality and is incapable of handling design problems in a continuous space with several design constraints due to the DoE. Hence RDO based on optimization techniques is proposed to solve the above phenomena.

In general, an RDO problem for an engineered system can be formulated as follows:

$$\begin{gathered}\text{Find } \boldsymbol{x} \in R^d \\ \text{Min } F(\mu_f(\boldsymbol{x},\boldsymbol{p}), \sigma_f(\boldsymbol{x},\boldsymbol{p})) \\ \text{S.t. } \boldsymbol{g}(\boldsymbol{x},\boldsymbol{p}) \leq 0 \\ \boldsymbol{x}^L \leq \boldsymbol{x} \leq \boldsymbol{x}^U\end{gathered} \tag{8.1}$$

where $\boldsymbol{x}$ is design variable vector, $\boldsymbol{p}$ is engineering system constant parameter vector, both $\boldsymbol{x}$ and $\boldsymbol{p}$ could be uncertain, μ_f and σ_f are the mean and standard deviation of the original optimization objective f(.), which represent general performance and robustness assessments respectively, F(.) is the new optimization object function with respect to μ_f and σ_f, for example, weighted sum of the mean and standard deviation, $\boldsymbol{g}$(.) denotes unequal constraint vector which is also replaced by the equal constraints, and $\boldsymbol{x}^L$ and $\boldsymbol{x}^U$ are lower bounds and upper bounds of design variable which define the boundaries of design space. It is apparent that system sensitivity to uncertainties can be minimized by incorporating σ_f into the objective function.

To realistically and vividly describe the concept of RDO, the graphical illustration of RDO is shown in Fig. 8.1. Let the horizontal axis represent the uncertain parameters, including random design variables and other system constant parameters, and the vertical axis represent an optimization objective function. There are two

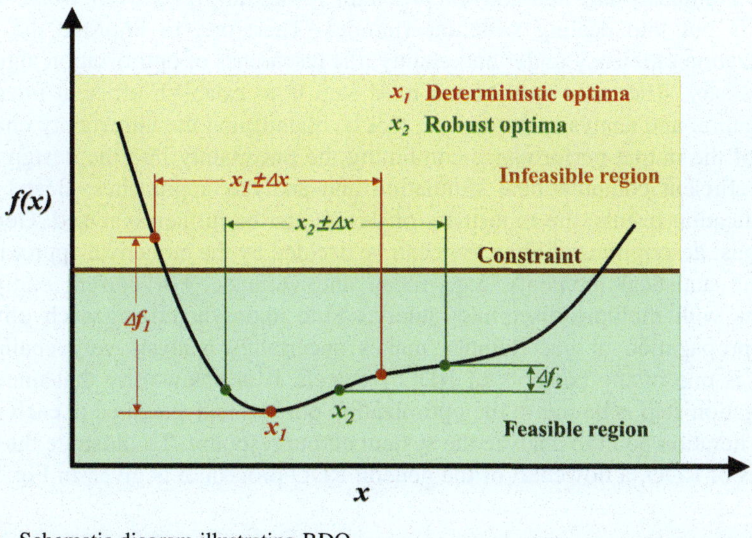

Fig. 8.1 Schematic diagram illustrating RDO

optimal solutions, among which x_2 is considered as more robust compared with x_1 because the former solution does not alter the objective function too much. Hence x_1 is not recommended as a design in practice even though it achieves better performance (lower objective function value).

8.3 Main Procedure of RDO

To effectively solve the RDO problems formulated by Eq. (8.1), a general RDO solving process composed of three important steps is presented to provide an overall understanding and a reference in this section. The first step, i.e., uncertain system modeling, consists of system modeling and uncertainty modeling. System modeling mainly refers to the mathematical expression of optimization problems, for example, design variables, optimization objectives, constraints, design space, and so on. Uncertainty modeling is identification, classification, and quantification of uncertainties involved in the system design. There are many mathematical theories and methods developed for engineering problems to model uncertainties so far, such as probabilistic approaches [12], possibilistic approaches [13], hybrid possibilistic-probabilistic approaches [14, 15], information gap decision theory (IGDT) [16], clouds theory [17], etc. Considering that there exists a great deal of uncertainties throughout the lifecycle of products in engineering practices, it will inevitably lead to an unbearable computational burden. Thus, sensitivity analysis is required to screen out some uncertainties that a negligible effect on engineering system design for simplifying RDO problems. The second step focuses on adopting proper optimization algorithms to optimize the robust objective under uncertainties. Due to the incapability of the deterministic global optimization in solving the large-scale, highly nonlinear, and non-convex problem, it naturally gets even worse if extra effort is put into dealing with uncertainties. Therefore, to improve the overall optimization efficiency under uncertainty, the researches of optimization algorithms is necessary. The main goal of the third step is to establish efficient uncertainty propagation and analysis approaches, that is, quantifying the uncertainty characteristics of the output performance combining the uncertainty into the design system using efficient computational simulation and analysis approaches. Based on the quantification results, the robustness of design can be further analyzed. Generally, the robust descriptions of objectives can be decided by the numerical approximation of their statistical moments, e.g., mean and variance. For current engineering systems with multiple disciplines intercrossing and syncretizing each other, the cross propagation of uncertainties makes uncertainty analysis very challenging, which is one of the hotspots of RDO research. It is noteworthy that uncertainty propagation step is bound to the optimization process and executed at each optimization iteration point to analyze the system output response. To illustrate the overall process of RDO, a flowchart of the general RDO procedure is given in Fig. 8.2.

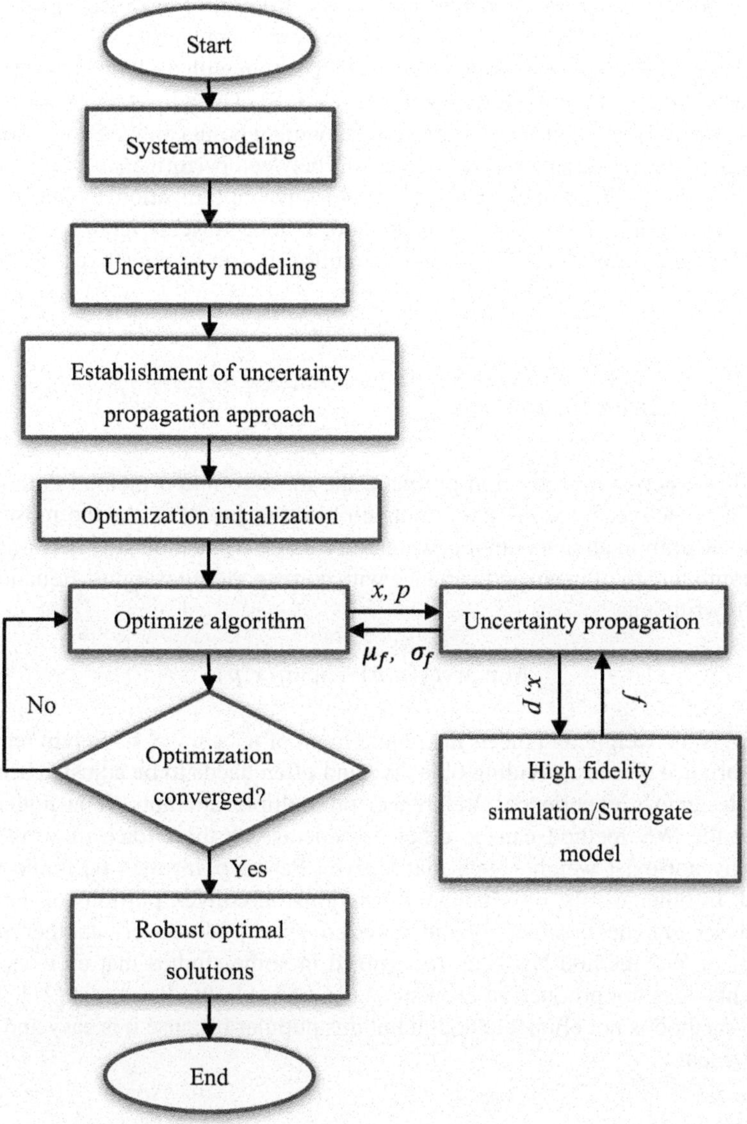

Fig. 8.2 General flowchart of the RDO process

8.4 RDO Methods

Considering the particular discussion and illustration of uncertainty modeling and propagation methods in the former chapters, the focus of the section is to introduce the optimization methods used for RDO. RDO is essentially a typical multi-objective optimization problem and often involves two conflicting objectives, i.e., reducing the mean and variance at the same time is very difficult or even impossible. Some studies have indicated that these expected properties, including high performance and performance robustness, are always in competition with each other, for example, the better robustness but the lower performance [18]. To address the issue, there are efficient methods developed to obtain all the potential points on the Pareto frontier of RDO and enable designers to reach a trade-off between performance and robustness. In this section, some general used multi-objective optimization methods will be introduced, including weighted sum method, compromise programming method, physical programming methods, and evolutionary multi-objective optimization methods.

8.4.1 Weighted Sum Method

For multi-objective optimization problem, the most common method is to convert the multi-objective functions into a single objective function. As the most widely used robust optimization technique, weighted sum (WS) method [19, 20] optimizes a linear combination of mean and variance with corresponding weights. Equation (8.1) can be rewritten as:

$$\text{Min } w_1\mu_f(\boldsymbol{x},\boldsymbol{p}) + w_2\sigma_f(\boldsymbol{x},\boldsymbol{p}) \tag{8.2}$$

where w_i is the weight and meets the requirement of $w_1 + w_2 = 1$. Weight represents the importance of corresponding objective and often needs to be adjusted according to the designer's experience, preference, or multiple attempts. It is noticed that whether the WS method can be efficient depends on the choice of weights and there may not exist weights such that a given Pareto point may be found by WS method. In other words, only potential solutions on convex portions of the Pareto optimal set can be obtained in the criterion space. Moreover, another existing problem of WS method has been recognized in some studies that an even spread of weights does not produce an even spread of points in the Pareto set [21]. To sum up, WS method is not efficient enough but it is popular because it is easy and simple to implement.

8.4.2 Compromise Programming Method

To compensate the shortcomings of the WS method, compromise programming (CP) method [22] is developed for solving multi-objective optimization. As a modified version of goal programming [23], the fundamental principle of CP is to identify an ideal solution as a point at which each attribute reaches its optimum value and then to find a solution that is as near to that point as possible. The point is also called a Utopia point. Since to minimize the distance between the efficient point and the Utopia point, different metrics can be used. Therefore, the CP method for multi-objective optimization is formulated as the following:

$$\text{Min } \|\mathbf{r}-\mathbf{U}\|_p^{\mathbf{w}} = \left(\sum_{i=1}^{n} (w_i|r_i - u_i|)^p \right)^{1/p} \tag{8.3}$$

where $\mathbf{r} = (r_1, r_2, \ldots, r_n) = (f_1, f_2, \ldots, f_n)$ is the n-dimension objective function vector. $\boldsymbol{U}$ represents the Utopia point where each objective has been minimized, which equals $(f_1^*, f_2^*, \ldots, f_n^*)$ for this minimization optimization problem. $p = \{1, 2, \ldots\} \cup \{\infty\}$ defines the specific metric and $w_i \geq 0$, $i = 1, 2, \ldots, n$. The designers first determine the value of p and then adjust the weight vector $\boldsymbol{w}$ to find the Pareto optimal solutions. Notice that for $\boldsymbol{x} \in R^d$, $u_i \leq f_i(\boldsymbol{x})$, $i = 1, 2, \ldots, n$, the symbol of absolute value in Eq. (8.3) can be removed. In particular, $p = 1$ stands for the Manhattan distance metric, that is the WS method. The $p = 2$ is Euclidean distance metric and $p = \infty$ is the Tchebycheff distance metric. In order to capture certain solutions located on the non-convex portions of the Pareto optimal set, a relatively large value of p may be required. When Tchebycheff distance metric is applied, Eq. (8.3) can be transformed as a min-max problem:

$$\min_{\boldsymbol{x} \in X} \max_{i=1,\ldots,n} \{w_i(f_i(\boldsymbol{x}) - u_i)\} \tag{8.4}$$

which is equivalent to the following β-problem:

$$\begin{gathered} \text{Min } \beta \\ w_i(f_i(\boldsymbol{x}) - u_i) \leq \beta, i = 1, \ldots, n \end{gathered} \tag{8.5}$$

where β is a positive value. After substituting two objectives of RDO (mean and variance) into the above equation, the formula of RDO is given by

$$\begin{gathered} \text{Find } \boldsymbol{x} \in R^d \\ \text{Min } \beta \\ \text{S.t. } w_1\left(\frac{\mu_f}{\mu_f^*} - 1.0 + \epsilon_1\right) \leq \beta,\ w_2\left(\frac{\sigma_f}{\sigma_f^*} - 1.0 + \epsilon_2\right) \leq \beta \\ g(\boldsymbol{x}, \boldsymbol{p}) \leq 0,\ \boldsymbol{x}^L \leq \boldsymbol{x} \leq \boldsymbol{x}^U \end{gathered} \tag{8.6}$$

where μ_f^* and σ_f^* are the ideal solution of function μ_f and σ_f respectively. Based on the above CP method using Tchebycheff distance metric, all Pareto optimal solutions of the RDO problem can be guaranteed to be found by adjusting weights. To make a better comparison between the WS method and the CP method, the solutions

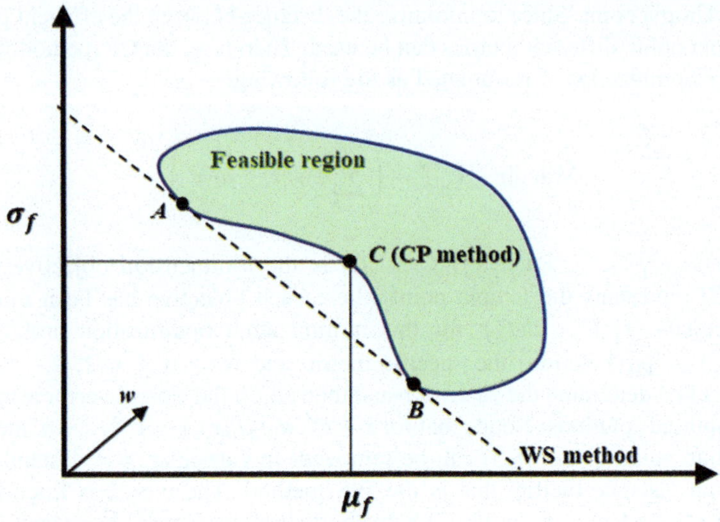

Fig. 8.3 Generating Pareto solutions by WS and CP methods

obtained from two methods are plotted in the objective space, as shown in Fig. 8.3. It can be observed that the efficient Pareto solutions located on the arc between A and B could not be found by the WS method, e.g., point C, but they are accessible for the CP method based on Tchebycheff distance metric.

8.4.3 *Physical Programming Method*

As another method employed for multi-objective optimization problem, physical programming (PP) method was developed by Messac et al. [24] and has already been used in RDO [25, 26]. By constructing the so-called class-function reflecting the degree of designer's preference, the PP method allows designers to make a qualitative and quantitative physical description of their preferences without conjuring corresponding weights. Therefore, compared with CP method, PP method is more flexible in expressing designer's preference to each design metric and more capable of capturing complete Pareto optimal set.

When adopting PP method, the designers firstly choose a set of design parameters and design metrics based on the current level of knowledge regarding the desired design. After that, they develop a mapping between design parameters and design metrics and construct a set of class-functions according to design metrics. Each class contains two cases, hard and soft, referring to the sharpness of the preference. All soft class-functions will become constituent components of the objective function and all the hard classes would simply become constraints. In detail, they are subdivided into four situations: smaller is better, larger is better, value is better,

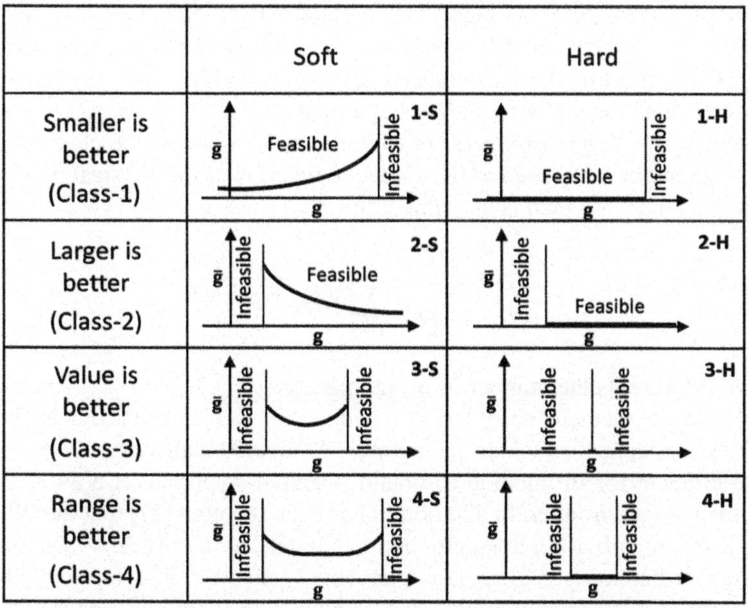

Fig. 8.4 Classification of class-function [24]

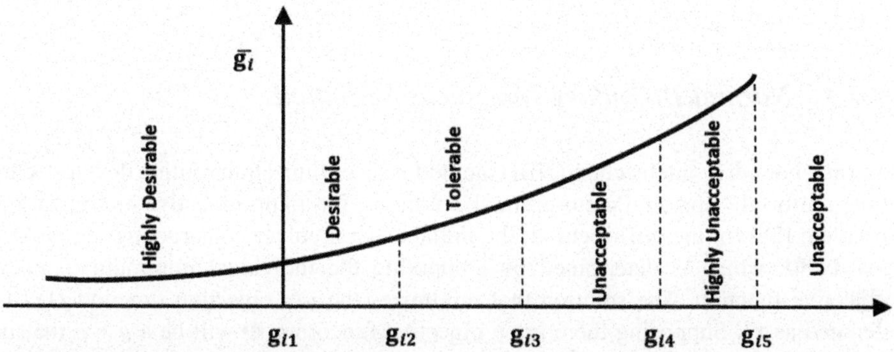

Fig. 8.5 Range division of Class-1S function [24]

and range is better. Figure 8.4 shows the qualitative meaning of each class. The horizontal axis represents the value of the design metric under consideration, g_i, while the vertical axis represents the class-function, $\overline{g_i}$, which need to be minimized for the responding metric. Furthermore, the PP method defines the preference range for each class design metrics in order to more precisely and flexibly express preferences for each metric rather than just using the terms minimize, maximize, greater than, less than, or equal to. Take the case of Class-1 as example, the arranged range are defined as Fig. 8.5, which are highly desirable range, desirable range, tolerable range, undesirable range, highly undesirable range, and unacceptable range in order of decreasing preference. The parameters g_{i1} through g_{i5} are physically meaningful values specified by designers to quantify the preference for the i-th design metric. In particular, the objective space optimization direction is determined by the class-function value for each design metric.

Generally, the common logarithm of the average value of the class-function of each design metric is selected as the objective function of the physical programming optimization model, which is given as follows:

$$\widehat{g} = \log\left\{\frac{1}{n_{sc}}\sum_{i=1}^{n_{sc}} \overline{g}_i[g_i(\boldsymbol{x},\boldsymbol{p})]\right\} \tag{8.7}$$

where n_{sc} represents the number of design metrics.

With the PP method, designers can judge and weigh mean and deviation by specifying the range for each design metrics. Compared with WS and CP method, it is more efficient for PP method to obtain potential solutions on both convex and non-convex Pareto frontier. In addition, it has been reported [27] that the PP method can be used to provide all Pareto optimal points as a sufficient and necessary condition for Pareto optimality. However, due to the need to determine the design metrics sets for different preferences, more knowledge of optimization is required for PP method, leading to more complex in initializing set of optimization formulation than WS and CP methods.

8.4.4 Normal Boundary Intersection Method

Normal boundary intersection (NBI) method is an optimization routine developed to find uniformly spread Pareto-optimal solutions for multi-objective optimization problem [28]. In the NBI method, the initial stage involves constructing the payoff matrix $\boldsymbol{\Phi}$, which is determined by computing the individual minimum of each objective function. The solution that minimizes the i-th objective function $f_i(\boldsymbol{x})$ is denoted as $\boldsymbol{x}_i^*$. Supposing there are n objective functions, $\boldsymbol{\Phi}$ will be a $n \times n$ matrix where each row consists of i-th objective function value by substituting n individual optima points $\{\boldsymbol{x}_1^*, \boldsymbol{x}_2^*, \ldots, \boldsymbol{x}_n^*\}$ respectively.

$$\boldsymbol{\Phi} = \begin{pmatrix} f_1^*(\boldsymbol{x}_1^*) & \ldots & f_1(\boldsymbol{x}_i^*) & \ldots & f_1(\boldsymbol{x}_n^*) \\ \vdots & \ddots & & & \vdots \\ f_i(\boldsymbol{x}_1^*) & \ldots & f_i^*(\boldsymbol{x}_i^*) & \ldots & f_i(\boldsymbol{x}_n^*) \\ \ldots & & & \ddots & \vdots \\ f_n(\boldsymbol{x}_1^*) & \ldots & f_n(\boldsymbol{x}_i^*) & \ldots & f_n^*(\boldsymbol{x}_n^*) \end{pmatrix} \tag{8.8}$$

All objective functions need to be normalized based on their minimum and maximum values to avoid differences in magnitude and order of magnitude between them. By joining the minimum and maximum values of each objective function, two

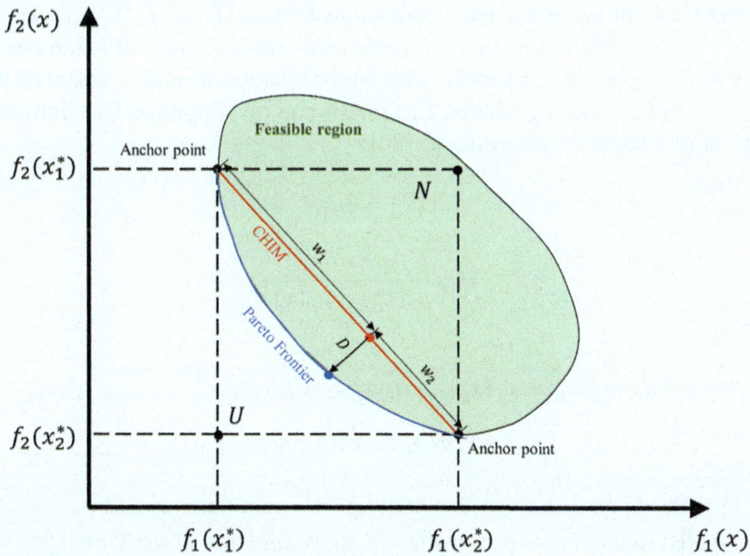

Fig. 8.6 Graphical description of NBI method for RDO

vectors can be obtained, called the Utopia point $\boldsymbol{U} = \left(f_1^U, f_2^U, \ldots, f_n^U\right) = \left(f_1^*(\boldsymbol{x}_1^*), f_2^*(\boldsymbol{x}_2^*), \ldots, f_n^*(\boldsymbol{x}_n^*)\right)$ and the Nadir point $\boldsymbol{N} = \left(f_1^N, f_2^N, \ldots, f_n^N\right)$. Then, the normalization of the objective functions can be achieved using the above two vectors. Take the i-th objective function as an example:

$$\overline{f}_i(\boldsymbol{x}) = \frac{f_i(\boldsymbol{x}) - f_i^U}{f_i^N - f_i^U}, \quad i = 1, 2, \ldots, n \tag{8.9}$$

As a result of this normalization process, the normalized payoff matrix $\overline{\boldsymbol{\Phi}}$ is obtained in which each element is calculated by Eq. (8.9).

Figure 8.6 illustrates how the convex combination of each row in the payoff matrix forms the convex hull of individual minima (CHIM) [29] in an arbitrary bi-objective problem. The endpoint of the Utopia line is also called the anchor point, corresponding to the solution of the single-optimization problem. Considering a convex weighting $\boldsymbol{w}$, any point in the normalized space, on the Utopia line, is expressed as $\overline{\boldsymbol{\Phi}}\boldsymbol{w}$. Let $\widehat{\boldsymbol{n}}$ denote the normal unit direction to the CHIM at the point $\overline{\boldsymbol{\Phi}}\boldsymbol{w}$ towards the origin, then $\overline{\boldsymbol{\Phi}}\boldsymbol{w} + D\widehat{\boldsymbol{n}}, D \in R$ represents the set of points on that normal. The intersection point between the normal vector and the boundary of the feasible region closest to the origin corresponds to maximizing the distance between the Utopia line and the Pareto frontier. The multi-objective optimization problem then can be formulated by

$$\begin{aligned} &\text{Max } D \\ \text{s.t. } &\overline{\boldsymbol{\Phi}}\boldsymbol{w} + D\widehat{\boldsymbol{n}} = \overline{\boldsymbol{F}}(\boldsymbol{x}) \end{aligned} \tag{8.10}$$

where $\overline{F}(x)$ is a normalized objective vector as $\overline{F}(x) = (\overline{f_1}(x),\overline{f_2}(x), \ldots,\overline{f_n}(x))$. By iteratively solving this optimization problem for various values of parameter w, it is possible to generate a uniformly distributed Pareto frontier. Since RDO is a bi-objective optimization problem, Eq. (8.10) can be simplified by eliminating the conceptual parameter D according to [30]:

$$\begin{gathered}\text{Find } x \in R^d \\ \text{Min}\frac{\mu_f - \mu_f(x_\mu^*)}{\mu_f(x_\sigma^*) - \mu_f(x_\mu^*)} \\ \text{s.t.}\frac{\mu_f - \mu_f(x_\mu^*)}{\mu_f(x_\sigma^*) - \mu_f(x_\mu^*)} - \frac{\sigma_f - \sigma_f(x_\sigma^*)}{\sigma_f(x_\mu^*) - \sigma_f(x_\sigma^*)} + 2w - 1 = 0 \\ g(x,p) \le 0,\ x^L \le x \le x^U\end{gathered} \tag{8.11}$$

where x_μ^* and x_σ^* the solutions that individually minimize μ_f and σ_f, respectively. While the NBI method produces a smooth approximation of the Pareto boundary, it is important to note that some points obtained through the NBI method may not necessarily be Pareto optimal solutions. It is worth noting that despite the high efficiency and well-defined mathematical significance of the four RDO methods (WS, CP, PP, and NBI) discussed so far, it is important to acknowledge that these methods rely on prior knowledge and experience. For instance, the selection of weights and preferences requires a level of expertise and understanding.

8.4.5 Evolutionary Multi-objective Optimization Method

Due to the complexity and multimodal characteristics of realistic optimization problems, it is difficult for traditional gradient-based algorithms to deal with multi-objective problems unless a multi-objective problem is converted into multiple single-objective problems, which raises the bar for beginners. In addition, gradient-based algorithms use local gradient information as the optimization direction and start from a single initial point to search for the optimal solution iteratively, leading to its vulnerability to the impact of initial point and precision of gradient information and easy to fall into the local optimum. Conversely, stochastic search algorithms with stochastic and gradient-free nature are able to deal with linear or nonlinear multi-objective optimization problems directly and more robust and faster to converge to the global optimal result. As the most representative of stochastic search algorithms, evolutionary algorithms have been widely used to address robust design optimization problems.

Evolutionary multi-objective optimization (EMO) method finds a set of non-dominated solutions for multi-objective optimization problems by evolving a population, which simulates natural evolution mechanism. The solution can be taken to be the optimal solution if there does not exist another solution that dominates it. The definition of domination is that, for $\boldsymbol{x}_1$, $\boldsymbol{x}_2 \in R^d$, $\forall i \in \{1, 2, \ldots, n\}$, $f_i(\boldsymbol{x}_2) \geq f_i(\boldsymbol{x}_1)$, and $\exists \mathrm{j} \in \{1, 2, \ldots, n\}$, $f_j(\boldsymbol{x}_2) > f_j(\boldsymbol{x}_1)$, in such scenario $\boldsymbol{x}_1$ is said to dominate $\boldsymbol{x}_2$. Generally, the EMO method starts with the initialization of a parent population o with the size of s_o, including the evaluations of the objectives and constraints of each candidate in the population. The fitness of each candidate is assigned based on the objective function and constraint violations. If a candidate violates the constraint, it will be punished, resulting in its worse fitness than other unviolated solutions. After the parent population is placed in the mating pool, three popular operators, namely selection, crossover, and mutation, are usually used to generate the offspring population.

Despite their popularity, one of the main obstacles encountered by EMO methods in dealing with robust design problems is the high computational cost, which is the result of the large number of objective functions and constraint evaluation required by the majority of existing EMO methods. In order to address the above expensive problems, three main schemes have been proposed by [31], which are problem approximation, functional approximation, and evolutionary approximation, respectively. In detail, the idea of problem approximation is to substitute a less expensive problem statement for the original one. For example, in finite element simulation process, coarse grid is used rather than the fine grid. For functional approximation, surrogate model is adopted to replace the original system model and to predict the objectives and constraints of iteration point in the optimization process. In spite of extensive development of surrogate model technology, high-fidelity surrogate models are still required when faced with high-dimensional and highly nonlinear design problems. As a unique method for evolutionary algorithms, evolutionary approximation aims to estimate the fitness of individuals by using information from other (similar) individuals (such as clustering and fitness inheritance), thus saving computing cost. It is worth noting that neither the clustering nor the fitness inheritance strategies demonstrate the required convergence in practice.

Compared with WS, CP, and PP methods, EMO method requires designers to have less priori knowledge and experience, which allows the designers to easily and conveniently identify the desired or best trade-offs between high performance and performance robustness from the Pareto frontier. Besides, EMO method is capable of finding optimal trade-offs for RDO problem in a single run [32], while WS, CP, and PP methods must carry out a sequence of separate executions to search for all Pareto frontier over the entire design space. However, under the limited computational cost, WS, CP, and PP methods may be more suitable for searching for a satisfactory solution than EMO methods.

8.5 Reliability-Based Robust Design Optimization

To improve product design in both robustness and reliability, a hybrid algorithm called reliability-based robust design optimization (RBRDO) has been developed to combine the principles of both RDO and RBDO [33]. By integrating RDO and RBDO techniques, RBRDO enhances the search for robust optima while obeying reliability type of constraints. In RBRDO, the mean and variance of a performance function are simultaneously minimized subject to probabilistic constraints. A typical RBRDO problem is defined as follows [33]:

$$\begin{aligned} &\text{Min } F\left(\mu_f(\boldsymbol{x},\boldsymbol{p}), \sigma_f(\boldsymbol{x},\boldsymbol{p})\right) \\ &\text{S.t.} P\{g(\boldsymbol{x},\boldsymbol{p}) \le 0)\} \ge \boldsymbol{R} \\ &\boldsymbol{x}^L \le \boldsymbol{x} \le \boldsymbol{x}^U \end{aligned} \tag{8.12}$$

where *P{.}* is the probability of the statement within the braces to be true, and $\boldsymbol{R}$ is the reliability vector specified for the constraint vector.

The concept of RBRDO is not novel, and numerous RBRDO methods have already been developed. Du et al. [34] utilized an inverse strategy, employing percentile performance to evaluate the robustness of the objective function and probabilistic constraints. Youn et al. [35] applied eigenvector dimension reduction (EDR) method to conduct probability analysis in the RBRDO problem. Rathod et al. [36] conducted a comparative study of different formulations of the RBRDO model and proposed an evolutionary genetic algorithm to optimize the RBRDO model based on a hybrid quality loss function. However, these works have not covered the problem of multi-objective RBRDO. Gonçalo et al. [36] proposed a new approach for the multi-objective optimization of composite structures under the effects of uncertainty in mechanical properties, structural parameters, and external loads. More advanced methods will be proposed in the future.

Exercises

8.1 Explain the definition of RDO. Provide examples of its engineeering applications.

8.2 Explain the main differences between RDO and RBDO.

8.3 Describe the three main steps involved in the RDO method in brief.

8.4 List the advantages and disadvantages of the WS method, the PP method, and the CP method used for RDO.

8.5 Suppose you are designing a cantilever beam and attempting to minimize its deflection by adjusting three parameters: width (*b*), height (*h*), and length (*L*). Each parameter has two levels as defined in the Taguchi method [9, 10]. The following experimental data was obtained after conducting a series of experiments. Use the Taguchi method to determine the optimal combination of parameters.

Experiment	b	h	L	Deflection
1	–	–	–	0.02
2	–	–	+	0.03
3	–	+	–	0.04
4	–	+	+	0.03
5	+	–	–	0.02
6	+	–	+	0.04
7	+	+	–	0.05
8	+	+	+	0.06

References

1. Cheng, J., Lu, W., Hu, W., Liu, Z., Zhang, Y., & Tan, J. (2019). Hybrid reliability-based design optimization of complex structures with random and interval uncertainties based on ASS-HRA. *IEEE Access, 7*, 87097–87109.
2. Hu, W., Choi, K., & Cho, H. (2016). Reliability-based design optimization of wind turbine blades for fatigue life under dynamic wind load uncertainty. *Structural Multidisciplinary Optimization, 54*, 953–970.
3. Huan, Z., Zhenghong, G., Fang, X., & Yidian, Z. (2019). Review of robust aerodynamic design optimization for air vehicles. *Archives of Computational Methods in Engineering, 26*, 685–732.
4. Beyer, H.-G., & Sendhoff, B. (2007). Robust optimization – A comprehensive survey. *Computer Methods in Applied Mechanics Engineering, 196*(33–34), 3190–3218.
5. Chen, W., Allen, J. K., Tsui, K.-L., & Mistree, F. (1996). A procedure for robust design: Minimizing variations caused by noise factors and control factors. *Journal of Mechanical Design, Transactions of the ASME, 118*, 478.
6. Gabrel, V., Murat, C., & Thiele, A. (2014). Recent advances in robust optimization: An overview. *European Journal of Operational Research, 235*(3), 471–483.
7. Zang, C., Friswell, M., & Mottershead, J. (2005). A review of robust optimal design and its application in dynamics. *Computers Structures, 83*(4–5), 315–326.
8. Chatterjee, T., Chakraborty, S., & Chowdhury, R. (2019). A critical review of surrogate assisted robust design optimization. *Archives of Computational Methods in Engineering, 26*(1), 245–274.
9. Taguchi, G. (1987). *System of experimental design: Engineering methods to optimize quality and minimize costs*. UNIPUB/Kraus International Publications.
10. Taguchi, G., & Asian Productivity Organization. (1986). *Introduction to quality engineering: Designing quality into products and processes*. Asian Productivity Organization.
11. Yao, W., Chen, X., Luo, W., Van Tooren, M., & Guo, J. (2011). Review of uncertainty-based multidisciplinary design optimization methods for aerospace vehicles. *Progress in Aerospace Sciences, 47*(6), 450–479.
12. Dantzig, G. B. (1955). Linear programming under uncertainty. *Management Science, 1*(3-4), 197–206.
13. Klir, G., & Yuan, B. (1995). *Fuzzy sets and fuzzy logic*. Prentice Hall.
14. Aien, M., Rashidinejad, M., & Fotuhi-Firuzabad, M. (2014). On possibilistic and probabilistic uncertainty assessment of power flow problem: A review and a new approach. *Renewable and Sustainable Energy Reviews, 37*, 883–895.
15. Soroudi, A., & Ehsan, M. (2011). A possibilistic–probabilistic tool for evaluating the impact of stochastic renewable and controllable power generation on energy losses in distribution networks—A case study. *Renewable and Sustainable Energy Reviews, 15*(1), 794–800.
16. Ben-Haim, Y. (2006). *Info-gap decision theory: Decisions under severe uncertainty*. Elsevier.

17. Neumaier, A., Fuchs, M., Dolejsi, E., Csendes, T., Dombi, J., Bánhelyi, B., Gera, Z., & Girimonte, D. (2007). *Application of clouds for modeling uncertainties in robust space system design*. ACT Ariadna Research ACT-RPT-05- European Space Agency.
18. Asafuddoula, M., Singh, H. K., & Ray, T. (2014). Six-sigma robust design optimization using a many-objective decomposition-based evolutionary algorithm. *IEEE Transactions on Evolutionary Computation, 19*(4), 490–507.
19. Lee, S. W., & Kwon, O. J. (2006). Robust airfoil shape optimization using design for six sigma. *Journal of Aircraft, 43*(3), 843–846.
20. Tang, Z., & Périaux, J. (2012). Uncertainty based robust optimization method for drag minimization problems in aerodynamics. *Computer Methods in Applied Mechanics Engineering, 217*, 12–24.
21. Shukla, P. K., & Deb, K. (2007). On finding multiple Pareto-optimal solutions using classical and evolutionary generating methods. *European Journal of Operational Research, 181*(3), 1630–1652.
22. Chen, W., Wiecek, M. M., & Zhang, J. (1998). Quality utility: A compromise programming approach to robust design. In *Proceeding of the international design engineering technical conferences and computers and information in engineering conference* (p. V002T002A032). American Society of Mechanical Engineers.
23. Tamiz, M., Jones, D., & Romero, C. (1998). Goal programming for decision making: An overview of the current state-of-the-art. *European Journal of Operational Research, 111*(3), 569–581.
24. Messac, A. (1996). Physical programming-effective optimization for computational design. *AIAA Journal, 34*(1), 149–158.
25. Chen, W., Sahai, A., Messac, A., & Sundararaj, G. J. (2000). Exploration of the effectiveness of physical programming in robust design. *Journal of Mechanical Design, 122*(2), 155–163.
26. Messac, A., & Ismail-Yahaya, A. (2002). Multiobjective robust design using physical programming. *Structural Multidisciplinary Optimization, 23*, 357–371.
27. Messac, A., & Mattson, C. A. (2002). Generating well-distributed sets of Pareto points for engineering design using physical programming. *Optimization Engineering, 3*, 431–450.
28. Das, I., & Dennis, J. E. (1998). Normal-boundary intersection: A new method for generating the Pareto surface in nonlinear multicriteria optimization problems. *SIAM Journal on Optimization, 8*(3), 631–657.
29. Vahidinasab, V., & Jadid, S. (2010). Normal boundary intersection method for suppliers' strategic bidding in electricity markets: An environmental/economic approach. *Energy Conversion Management Science, 51*(6), 1111–1119.
30. Lopes, L. G. D., Brito, T. G., Paiva, A. P., Peruchi, R. S., & Balestrassi, P. P. (2016). Robust parameter optimization based on multivariate normal boundary intersection. *Computers Industrial Engineering, 93*, 55–66.
31. Jin, Y. (2005). A comprehensive survey of fitness approximation in evolutionary computation. *Soft Computing, 9*(1), 3–12.
32. Jin, Y., & Sendhoff, B. (2003). Trade-off between performance and robustness: An evolutionary multiobjective approach. In Proceeding of the evolutionary multi-criterion optimization: Second international conference, EMO 2003, Faro, Portugal, April 8–11, 2003. Proceedings 2, Springer, pp. 237–251.
33. Shahraki, A. F., & Noorossana, R. (2014). Reliability-based robust design optimization: a general methodology using genetic algorithm. *Computers Industrial Engineering, 74*, 199–207.
34. Du, X., Sudjianto, A., & Chen, W. (2004). An integrated framework for optimization under uncertainty using inverse reliability strategy. *Journal of Mechanical Design, 126*(4), 562–570.
35. Youn, B. D., & Xi, Z. (2009). Reliability-based robust design optimization using the eigenvector dimension reduction (EDR) method. *Structural Multidisciplinary Optimization, 37*, 475–492.
36. Rathod, V., Yadav, O. P., Rathore, A., & Jain, R. (2013). Optimizing reliability-based robust design model using multi-objective genetic algorithm. *Computers Industrial Engineering, 66*(2), 301–310.

Chapter 9
Physics-Informed Neural Networks for Design Optimization Under Uncertainty

Nomenclature

b	Boundary conditions
c	Wave speed
$\boldsymbol{d}$	Design variable
$\boldsymbol{G}$	Performance function
$\boldsymbol{i}$	Initial condition
$\boldsymbol{j}$	Network model loss
$\boldsymbol{L}$	Network layer
N	Neutral network
$\boldsymbol{\mathcal{N}}$	Nonlinear operator
$\boldsymbol{N_f}$	Number of sampling points
$\boldsymbol{N_{MCS}}$	Sampling size of Monte Carlo simulation
$\boldsymbol{P_F}$	Probability of failure
$\boldsymbol{P_{Fj}^{Tar}}$	Target probability of failure
$\boldsymbol{t}$	Time
$\boldsymbol{u}$	Solution of nonlinear equations
$\boldsymbol{u_d}$	Threshold value
v	System's viscosity
$\boldsymbol{X}$	Input variable
$\boldsymbol{x}$	Spatial coordinate
$\boldsymbol{Y}$	Network output
Z	Decay rate
z	Network layer output
$\boldsymbol{\delta}$	Perturbation
$\boldsymbol{\theta}$	Network model parameters
$\boldsymbol{\lambda}$	Lagrange multipliers
$\boldsymbol{\sigma}$	Nonlinear activation function
$\boldsymbol{\Omega}$	Domain(math)

W. Hu, *Design Optimization Under Uncertainty*,
https://doi.org/10.1007/978-3-031-49208-2_9

9.1 Introduction

Reliability analysis (RA) is a method that uses the observed values of the system variables to calculate the probability of system failure [1]. The traditional method of reliability analysis has been introduced detailly in Chaps. 5 and 6. The surrogate models are always utilized to regress the limit-state function and predict the probability of failure [2], which have been introduced in Chap. 3. However, it is hard to obtain enough data for establishing a well-trained surrogate model through simulations in high-dimensional nonlinear problems.

In general, system response is governed by complex ordinary/partial differential equation (ODE/PDE), which take expensive computational costs by their repeatedly executing relevant experiments or simulations. Recently physics-informed machine learning (PIML) methods are proposed to solve PDE by incorporating prior physical knowledge into common neutral networks during the training stage. PIML can be seen as a way of integrating data-driven machine learning models with physics-based mathematical models, such as partial differential equations (PDEs), conservation laws, or symmetries. There are different approaches for implementing PIML, depending on how the physical prior knowledge is represented and incorporated into the machine learning model, including loss function regularization, network architecture design, data augmentation, and model-based optimization [3]. By doing so, PIML can guide the machine learning model towards solutions that are physically plausible, improving accuracy and efficiency even in uncertain and high-dimensional contexts.

To address the problem of the large demand for real response data during the construction of surrogate model, physics-informed neural network (PINN) as well as PINN-based RA is introduced. The govern function of system response in form of ODE/PDE will be converted in loss function of the neutral network to embedding priori knowledge. The neural network is trained to fulfill the PDE (i.e., physical laws) with the minimum loss of the equation and then can be used to predict the system response. The loss values in network training are calculated directly from the current model output and the automatic differentiation (AD) technique, so that the training data of a PINN model can be obtained without relying on costly and time-consuming high-fidelity simulations. In RA process, state parameters and stochastic variables are set as the input of the PINN model and use the network model to estimate the system response.

PINNs are a type of artificial intelligence that combines data and physical laws to solve forward and inverse PDEs [4]. Cuomo et al. [5] provided a comprehensive review of PINNs, with the main objective of describing these networks and their associated strengths and weaknesses and then attempted to incorporate publications on a broader range of collocation-based physics-informed neural networks. Considering the slow convergence and underfitting of PINN model, Nabian et al. [6] presented an active training scheme based on importance sampling. Wu et al. [7] summarized commonly used sampling methods and introduce two residual-based adaptive sampling methods for PINN to generate training points. For reliability applications, Xu et al. [8] provided a comprehensive survey of PIML methods for reliability and system safety applications; Souvik et al. [9] used a PINN model to predict the complex system response governed by PDE and proposed a simulation-free framework for reliability analysis. Based on this framework, Zhang et al. [10] then presented an active learning approach that adaptively trains the PINN model on regions of high importance for failure probability characterization to enhance the training efficiency and accuracy of RA. The PINN methods also have been applied to different reliability assessment objects, such as multi-state system [11], small failure probabilities system [12], uncertainty quantification & propagation [13], and realistic industrial problems [14].

In conclusion, PINNs can be used to efficiently and accurately solve some reliability analysis problems due the inherent advantages of PINNs that include the physical theories and do not require computationally expensive simulations. In addition, PINNs could possibly carry out uncertainty quantification and propagation, measurement data fusion, and system reliability assessment.

In this chapter, a physics-informed method is introduced to replace traditional surrogate models in reliability analysis problems, including Sect. 9.2, the basis introduction of PINNs; Sect. 9.3, the construction of the PINN model used in RA and several numerical examples; and Sect. 9.4, a brief introduction to PINN-based design optimization under uncertainty.

9.2 Basis of Physics-Informed Neural Network

Physics-informed methods are a way of integrating data and mathematical physics models to improve machine learning performance on tasks that involve a physical mechanism [15]. They can leverage physical prior knowledge, such as partial differential equations (PDEs), symmetries, conservation laws, or intuitive physics, to guide the machine learning model towards solutions that are physically plausible and accurate.

9.2.1 Basic Structure of Multi-layer Perceptron

The key idea of PINN is using a neural network model as a high precision function approximator. The network model structure refers to the input and output of practical applications (e.g., finite analysis inputs and results). Here, a fully connected neural network (FCN) is taken as an example. This FCN includes an input layer and model output layer. The output of each layer is used as the input of the next layer after affine transformation and activation function. The basic structure of FCN is shown in Fig. 9.1.

Assuming that the number of neurons contained in l-th layer is n_l, the weighted input z_i^l for the i-th neuron of the layer can be expressed as Eq. (9.1):

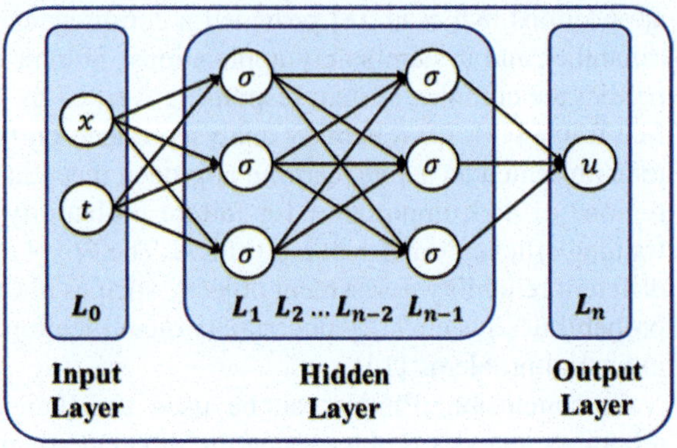

Fig. 9.1 The structure diagram of the feed-forward neural network

$$z_i^l = \sigma_{l-1}\left[\sum_{k=1}^{n_{l-1}} (W_{i,k}^l z_k^{l-1} + b_i^l)\right] \tag{9.1}$$

where W_i^l and b_i^l represent the corresponding weights and biases, respectively, and σ denotes the nonlinear activation function between the layers of the network. The output of each layer of the network can be represented by Eq. (9.2):

$$\begin{aligned} u^L &= \sigma_L(\mathbf{W}^{L+1} z^L + \boldsymbol{b}^L) \\ z^L &= \sigma_{L-1}(\mathbf{W}^L z^{L-1} + \boldsymbol{b}^L) \\ z^{L-1} &= \sigma_{L-2}(\mathbf{W}^{L-1} z^{L-2} + \boldsymbol{b}^{L-1}) \\ &\vdots \\ z^1 &= \sigma_0(\mathbf{W}^1 \boldsymbol{X} + \boldsymbol{b}^1). \end{aligned} \tag{9.2}$$

To simplify the formulation of the model, the output of the entire network is represented by u in Eq. (9.3)

$$u = N(\boldsymbol{X}, \boldsymbol{\theta}) \tag{9.3}$$

where $\boldsymbol{X}$ is the network model input, and $\boldsymbol{\theta}$ represents the current network parameters $[\boldsymbol{W}, \boldsymbol{b}]$. In the actual application process of the neural network model, the network model parameters $\boldsymbol{\theta}$ need to be continuously adjusted by minimizing the loss function.

The traditional neural network model training relies on the loss function to match the actual observation or experimental data, so it is also called data-driven NN.

9.2.2 Loss Construction Based on Priori Knowledge

To overcome the data dependency of traditional neural networks, the concept of PINN has been proposed in [4]. The PINN model can automatically satisfy the predefined physical constraints by embedding prior knowledge (i.e., partial differential equations) during training process. Figure 9.2 describes the basic structure of a PINN model, which reconstructs the loss function of network training.

Priori physical knowledge is usually expressed in partial differential equations, which is defined as Eq. (9.4):

$$u_t + \mathcal{N}[u, u_x, u_{xx}, \ldots] = 0 \tag{9.4}$$

where $u(x, t)$ is the implicit solution of nonlinear equations; $\mathcal{N}$ is a nonlinear operator containing partial differential components. The subscript in Eq. (9.5) is the partial differential of $u(x, t)$ to the x and t.

$$u_t = \frac{\partial u}{\partial t},\ u_x = \frac{\partial u}{\partial x},\ u_{xx} = \frac{\partial^2 u}{\partial x^2}, \ldots \tag{9.5}$$

The AD is used in the neural network to generate the partial differential components by backpropagation. Based on the equations mentioned above, the neural network approximation of the implicit solutions of the partial differential equations is obtained, as shown in Eq. (9.6). In addition, Eqs. (9.7) and (9.8) are the neural network expressions of Eqs. (9.4) and (9.5), respectively:

$$u(x, t) \approx \widehat{u}(x, t, \boldsymbol{\theta}) \tag{9.6}$$

$$\widehat{u}_t + \mathcal{N}[\widehat{u}, \widehat{u}_x, \widehat{u}_{xx}, \ldots] = 0 \tag{9.7}$$

$$\widehat{u}_t = \frac{\partial \widehat{u}}{\partial t}, \widehat{u}_x = \frac{\partial \widehat{u}}{\partial x}, \widehat{u}_{xx} = \frac{\partial^2 \widehat{u}}{\partial x^2}, \ldots \tag{9.8}$$

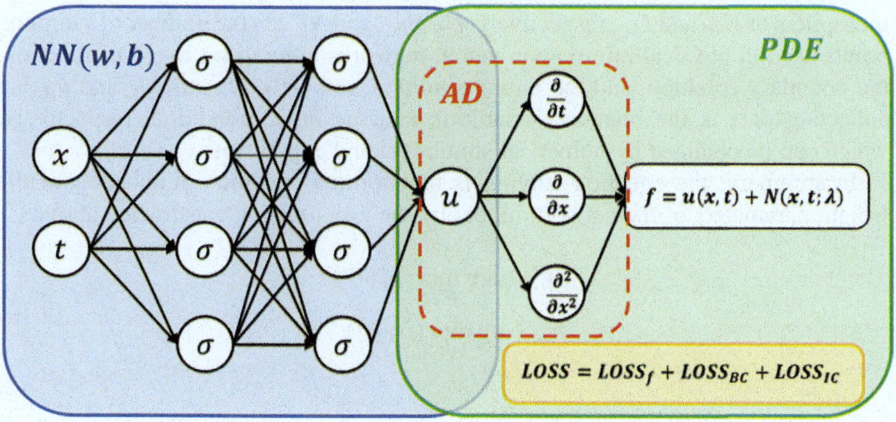

Fig. 9.2 The structure diagram of the physics-informed neural network

where $\boldsymbol{\theta}$ is the training parameters of each layer in neural network, which are the weight and bias coefficients in Eq. (9.2). The nonlinear activation function is introduced for the network model to approximate nonlinear input and output. For the nonlinear partial differential expression in Eq. (9.7), the equivalent constraint is defined as shown in Eq. (9.9). Thus, the problem of solving the equations is transformed into a coefficient optimization problem. Further, the neural network can approximate the partial differential equation (PDF) solution by training the weight coefficient θ of each layer in the neural network to make $f(x, t)$ tend to zero in their domain.

$$f(x,t) = \widehat{u}_t + \mathcal{N}[\widehat{u}, \widehat{u}_x, \widehat{u}_{xx}, \ldots; \lambda] \tag{9.9}$$

Based on the above method, the equivalent constraint equations, including partial differential equations, initial conditions, and boundary conditions, could be constructed. Then the loss function for neural network training can be built and expressed as:

$$L_p = L_f + \lambda_{BC} \cdot L_{BC} + \lambda_{IC} \cdot L_{IC} \tag{9.10}$$

$$L_f = \frac{1}{N_f} \sum_{i=1}^{N_f} |f(x_f^i, t_f^i)|^2 \tag{9.11}$$

$$L_{BC} = \frac{1}{N_{BC}} \sum_{i=1}^{N_{BC}} |\widehat{u}(x_{BC}^i, t_{BC}^i) - u_{BC}{}^i|^2 \tag{9.12}$$

$$L_{IC} = \frac{1}{N_{IC}} \sum_{i=1}^{N_{IC}} |\widehat{u}(x_{IC}^i, 0) - u_{IC}{}^i|^2 \tag{9.13}$$

where L_p, L_f, L_{BC}, L_{IC} are loss function of PINN, partial differential equation, boundary condition, and initial condition, respectively; λ_{BC} and λ_{IC} are Lagrange multipliers of L_{BC} and L_{IC}, respectively. N_f, N_{BC}, and N_{IC} are the number of sampling points used in physical information neural networks training for the PDE function, the boundary condtion, and the initial condition, respectively; and u_{BC} and u_{IC} are collecting data at the boundary condition and the initial condition, respectively, which can be obtained by numerical simulation and experiment.

In summary, the equation solving is transformed into the optimization of the weight parameters of the network model by the loss function constructed above.

$$\begin{aligned} \boldsymbol{W}^* &= \arg\min_{\boldsymbol{W} \in \theta} (L_p(\boldsymbol{W})) \\ \boldsymbol{b}^* &= \arg\min_{\boldsymbol{b} \in \theta} (L_p(\boldsymbol{b})) \end{aligned} \tag{9.14}$$

9.2.3 A Basic Example of PINN

To give a basic understanding of the performance of PINN, a simple propagation model of 1D sinusoidal wave is built, which can be defined as Eq. (9.15)

$$u = \begin{cases} \sin[\pi(t-x)] & t \leq x \\ 0 & t > x \end{cases} \tag{9.15}$$

And the propagation diagram of this sinusoidal wave can be seen in Fig. 9.3. The propagation of sinusoidal wave is governed by wave equation (Eq. 9.16)

$$\frac{\partial^2 u}{\partial t^2} - c^2 \frac{\partial^2 u}{\partial x^2} = 0 \tag{9.16}$$

where c is the wave speed, and $c = 1$ in this case.

The wave equation can be used as priori knowledge in construct the loss function of the PINN model, which can be expressed as Eqs. (9.17 and 9.18)

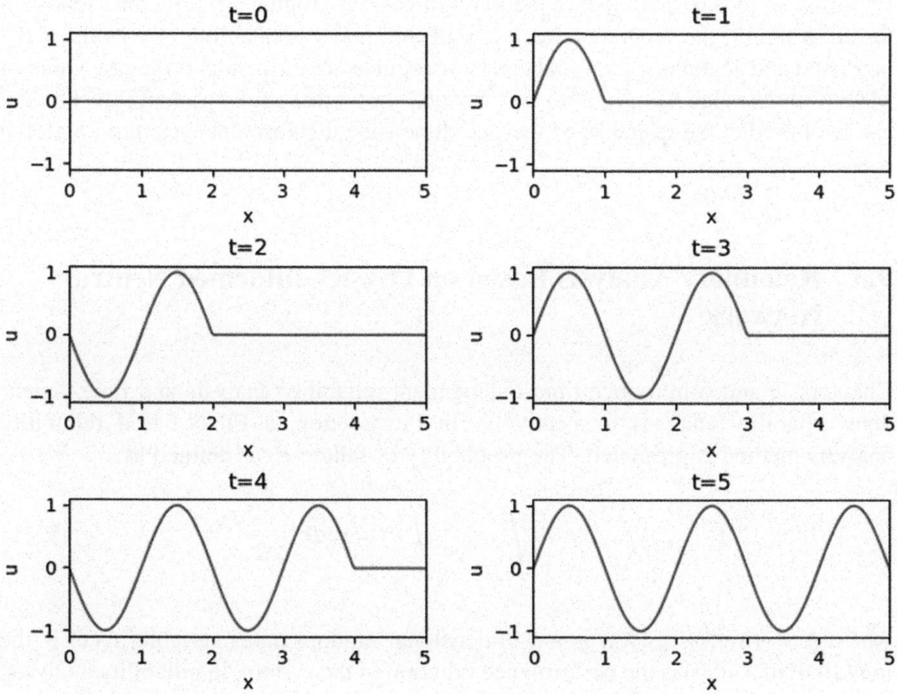

Fig. 9.3 Diagram of sinusoidal wave propagation

$$Loss = MSE_f + MSE_{ic} + MSE_b \tag{9.17}$$

$$\begin{cases} f = \dfrac{\partial^2 u}{\partial t^2} - c^2 \dfrac{\partial^2 u}{\partial x^2} & x \in [0, L], t \in [0, T] \\ ic_1 = u & x \in [0, L], t = 0 \\ ic_2 = \dfrac{\partial u}{\partial t} & x \in [0, L], t = 0 \\ b_i = u - \sin(\pi t) & x = 0, t \in [0, T] \\ b_o = u & x = 0, t \in [0, T] \end{cases} \tag{9.18}$$

where f corresponds to the residuals of the partial differential equation; ic is the initialization condition, which specifies the displacement and velocity of each point at the initial time, b imposes the boundary conditions on both ends of the domain. On the left end, a sinusoidal oscillation is applied to drive the system, while on the right end, the boundary is fixed to zero. The loss function consists of the above three components and is calculated as the sum of the mean square error (MSE) of the sampling points in the respective regions.

For training the network, 1000 training points have been generated using the Latin hypercube sampling. The Adams optimizer is run for 1500 iterations.

The result of the prediction at each time is shown in Fig. 9.4. In each time, the prediction of neural network (represented by a solid line) along with the exact values (represented by a dashed line) in the left figure show a high degree of consistency. In the right figure, the prediction accuracy of the model is quantified by MSE of the predicted and actual results. What can be recognized in the results is that the value of MSE is in the range from 10^{-6} to 10^{-5}, which shows that a fully trained PINN model can well predict the response of the one-dimensional sinusoidal system on a global scale.

9.3 Reliability Analysis Based on Physics-Informed Neutral Network

Chapters. 5 and 6 introduce time-independent reliability analysis and time-dependent reliability analysis, respectively. In this section, a PINN-based reliability analysis method is presented. The probability of failure P_F is defined as:

$$P_F = \int \cdots \int_{G(\cdot) < 0} f_{\mathbf{X}}(\mathbf{x}) d\mathbf{x} \tag{9.19}$$

where $\mathbf{X} = (X_1, X_2, \ldots, X_d)$ is a d-dimensional random input variable, $f_X(\mathbf{x})$ is the joint PDF of $\mathbf{X}$, $G(\cdot)$ is the performance function of the system. In reliability analysis, equating the system performance function $G(\cdot)$ to zero, a failure event can be recognized when $G(\cdot) < 0$.

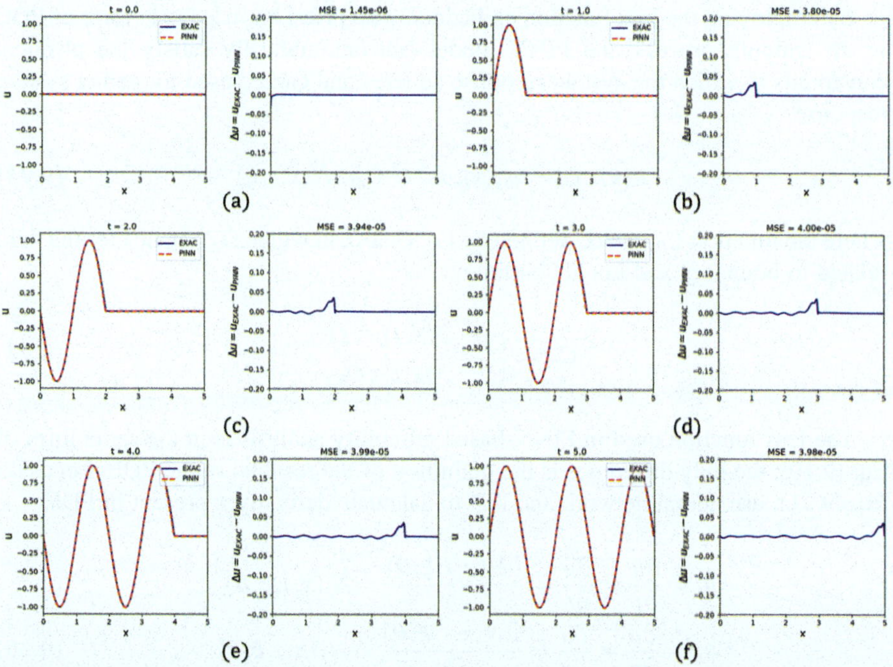

Fig. 9.4 Result of wave propagation process prediction use PINN. (**a**) $t = 1$, (**b**) $t = 2$, (**c**) $t = 3$, (**d**) $t = 4$, (**e**) $t = 5$, and (**f**) $t = 6$

The PINN method introduced in Sect. 9.2 is extended for solving reliability analysis problems here. The stochastic variables are set as inputs to the PINN model, and the output of the PINN model is the system response, from which a performance function of the system *G(X)* can be constructed based on system response. Consider the performance function *G(X)* is represented as

$$G(\boldsymbol{X}) = u(\boldsymbol{\xi}) - u_0 \tag{9.20}$$

where $u(\boldsymbol{\xi})$ denotes the system response, and u_0 indicates the threshold value to identify the failure. It is also assumed that $u(\boldsymbol{\xi})$ can be obtained by solving a stochastic PDE in the form of

$$u_t + \mathcal{N}[u, u_x, u_{xx}, \ldots; \boldsymbol{\xi}] = 0 \tag{9.21}$$

where $\boldsymbol{\xi}$ is the uncertainty of the system. The stochastic response can be described using PINN model as

$$u(\boldsymbol{x}, t, \boldsymbol{\xi}) \approx \widehat{u}(\boldsymbol{x}, t, \boldsymbol{\xi}) = N(\boldsymbol{x}, t, \boldsymbol{\xi}; \boldsymbol{\theta}) \tag{9.22}$$

where stochastic variable $\boldsymbol{\xi}$ is set as an input of neutral network.

By employing the loss function embedded with priori knowledge in form of PDE in the training process, the PINN model can automatically satisfy the physical constraints to which the system response adheres and can be used to predict system response.

$$\widehat{u}(\boldsymbol{x}, t, \boldsymbol{\xi}) = u_{b,i}(\boldsymbol{x}_b, t_0) + C_{bi} \cdot N(\boldsymbol{x}, t, \boldsymbol{\xi}) \tag{9.23}$$

where the function C_{bi} can be defined by Eq. (9.24), in which $\Omega_{b,\,i}$ stands for the field related to boundary and initial condition.

$$C_{bi} = \begin{cases} 0 & (x,t) \in \Omega_{b,i} \\ 1 & (x,t) \notin \Omega_{b,i} \end{cases} \tag{9.24}$$

The loss function used in PINN-based reliability analysis is in the same form of Eq. (9.10), the only difference is the definition of the residual of the PDE. Focus on Eq. (9.21), use neural network and AD to calculate derivatives present in PDE.

$$\begin{aligned} u_t &= \frac{\partial u}{\partial t} \approx \frac{\partial \widehat{u}}{\partial t} = \frac{\partial N(\boldsymbol{x}, t, \boldsymbol{\xi}; \boldsymbol{\theta})}{\partial t} = N_t(\boldsymbol{x}, t, \boldsymbol{\xi}; \boldsymbol{\theta}) \\ u_x &= \frac{\partial u}{\partial \boldsymbol{x}} \approx \frac{\partial \widehat{u}}{\partial \boldsymbol{x}} = \frac{\partial N(\boldsymbol{x}, t, \boldsymbol{\xi}; \boldsymbol{\theta})}{\partial \boldsymbol{x}} = N_x(\boldsymbol{x}, t, \boldsymbol{\xi}; \boldsymbol{\theta}) \\ u_{xx} &= \frac{\partial^2 u}{\partial \boldsymbol{x}^2} \approx \frac{\partial^2 \widehat{u}}{\partial \boldsymbol{x}^2} = \frac{\partial^2 N(\boldsymbol{x}, t, \boldsymbol{\xi}; \boldsymbol{\theta})}{\partial \boldsymbol{x}^2} = N_{xx}(\boldsymbol{x}, t, \boldsymbol{\xi}; \boldsymbol{\theta}) \end{aligned} \tag{9.25}$$

And the PDE residual can be defined as

$$R_{PDE} = N_t(\boldsymbol{x}, t, \boldsymbol{\xi}; \boldsymbol{\theta}) + \mathcal{N}[N(\boldsymbol{x}, t, \boldsymbol{\xi}; \boldsymbol{\theta}), N_x(\boldsymbol{x}, t, \boldsymbol{\xi}; \boldsymbol{\theta}), N_{xx}(\boldsymbol{x}, t, \boldsymbol{\xi}; \boldsymbol{\theta}), \ldots] \tag{9.26}$$

The network training process can be fully referred to the general form of PINN.

1. Generate collocation points, $D = \{\boldsymbol{x}_i, t_i, \boldsymbol{\xi}_i\}_{i=1}^{N_c}$ are generated by using some suitable DOE scheme.
2. Formulate the loss-functions as

$$L(\boldsymbol{\theta}) = \frac{1}{N_c} \sum_{k=1}^{N_c} [R_{PDE}] \tag{9.27}$$

3. Compute $\boldsymbol{\theta}$ by minimizing the loss-function

$$\boldsymbol{\theta}^* = \arg\min_{\boldsymbol{\theta}} L(\boldsymbol{\theta}) \tag{9.28}$$

When the training converges and the model parameters $\boldsymbol{\theta}$ are determined, $N(\boldsymbol{x}, t, \boldsymbol{\xi}; \boldsymbol{\theta})$ can predict the system response at any location point $(\boldsymbol{x}^*, t^*, \boldsymbol{\xi}^*)$.

Finally, the performance function $G(\boldsymbol{x})$ can be constructed from the system response $\widehat{u}(\boldsymbol{x}, t, \xi)$ predicted by PINN and reliability analysis can be conducted by calculating the probability of failure through Monte Carlo simulation (MCS).

$$G(\boldsymbol{X}) \approx \widehat{G}(\boldsymbol{x}) = \widehat{u}(\boldsymbol{\xi}) - u_0 \tag{9.29}$$

$$P_F \approx \widehat{P}_F = \frac{\sum_{j=1}^{N_{MCS}} I(\widehat{G}(\boldsymbol{X}))}{N_{MCS}} \tag{9.30}$$

where N_{MCS} is the sampling size of MCS.

Exercises for PINN-Based RA

Problem 1: Reliability analysis with a limit-state function in form of ODE

Given a simple stochastic ODE (Eq. 9.31),

$$\frac{du}{dt} = -Zu \tag{9.31}$$

where Z is the decay rate and is considered to be a stochastic variable $Z \sim \mathcal{N}(\mu, \sigma^2)$. Z is considered to be following normal distribution with $\mu = -2$ and $\sigma = 1$.

ODE (Eq. 9.32) follows initial condition,

$$u(t=0) = u_0 \tag{9.32}$$

where the initial value $u_0 = 1$. The limit-state function for reliability analysis can be defined as

$$G(u) = u - u_d \tag{9.33}$$

where $u_d = 0.5$ is the threshold value.

Solution (Table 9.1)

Table 9.1 Results of serveral RA methods in problem 1 adapted from [9]

Methods	P_f	β	N_S	$\epsilon = \frac{\lvert\beta_e - \beta\rvert}{\beta_e}$
Exact	**0.003539**	**2.6932**	—	—
MCS	0.0035	2.6949	10^6	0.06%
FORM	0.0036	2.6874	42	0.21%
SORM	0.0036	2.6874	44	0.21%
IS	0.0034	2.7074	1000	0.52%
DS	0.0034	2.7074	7833	0.52%
SS	0.0030	2.7456	1199	1.95%
PI-DNN	**0.0035**	**2.6949**	**0**	**0.06%**

Problem 2: Reliability analysis with a limit-state function in form of Burger's equation

Considering a viscous Burger's equation (Eq. 9.34)

$$u_t + uu_x = \nu u_{xx} \tag{9.34}$$

where $u(x, t)$ is the solution of Burger's equation, which can be interpreted as the system response; ν is the system's viscosity and $\nu > 0$; x is a variable of $[-1, 1]$. We can subject the given Burger's equation by following boundary conditions (Eq. 9.35)

$$\begin{aligned} u(x = -1) &= 1 + \delta \\ u(x = 1) &= -1 \end{aligned} \tag{9.35}$$

and the initial condition of the system

$$u(t = 0, x) = -1 + (1 - x)\left(1 + \frac{\delta}{2}\right) \tag{9.36}$$

where the δ stands for a little perturbation which is a uniformly distributed variable, and $\delta \sim \mathcal{U}(0, e), e \ll 1$.

The limit-state function can be defined as (Eq. 9.37)

$$J(z(\delta)) = -z(\delta) + z_0 \tag{9.37}$$

where z_0 is the threshold. For this example, we consider $e = 0.1$, $z_0 = 0.45$.

Solution (Table 9.2)

Table 9.2 Results of serveral RA methods in problem 2 adapted from [9]

Methods	P_f	β	N_s	$\epsilon = \frac{\lvert\beta_e - \beta\rvert}{\beta_e}$
MCS	**0.1037**	**1.2607**	**10,000**	—
FORM	0.1091	1.2313	58	2.33%
SORM	0.1091	1.2313	60	2.33%
IS	0.1126	1.2128	1000	3.80%
DS	0.0653	1.5117	4001	19.9%
SS	0.0800	1.4051	1000	11.45%
PI-DNN	**0.0999**	**1.2821**	**0**	**1.70%**

9.4 PINN-Based Design Optimization Under Uncertainty

Reliability-based design optimization (RBDO) is proposed to optimize design that is characterized by a low probability of failure. By giving a specific level of risk and reliability, RBDO aims to determine the optimum design of the products/systems. The detailed concept of RBDO has been introduced in Chap. 7. In this section, the previously proposed PINN-RA method is applied in the operation process of RBDO to improve its performance, and then the PINN-based RBDO method will be proposed.

The mathematical formulation of the general component level of RBDO problem is introduced in Chap. 7, expressed as

$$\begin{aligned} &\text{minimize} && \mathrm{Cost}(\boldsymbol{d}) \\ &\text{subject to} && P[G_j(\boldsymbol{X})>0] \le P_{F_j}^{Tar}, j=1,\ldots,nc \\ & && \boldsymbol{d}^L \le \boldsymbol{d} \le \boldsymbol{d}^U, \boldsymbol{d} \in \mathbf{R}^{nd} \text{ and } \boldsymbol{X} \in \mathbf{R}^{nr} \end{aligned} \tag{9.38}$$

where $\boldsymbol{d}$ is the design variable, which is the mean of the nd-dimensional random variable $\boldsymbol{X}_{rv} = \{X_1, X_2, \ldots, X_{nd}\}^T$; $\boldsymbol{X} = \{\boldsymbol{X}_{rv}, \boldsymbol{X}_{rp}\}^T$, where $\boldsymbol{X}_{rp}$ represents the random parameter of the random input $\boldsymbol{X}$; P_{Fj}^{Tar} is the target probability of failure, which is achieved by reliability analysis, for the j-th constraint; nc and nd are the number of probabilistic constraints and design variables, respectively.

The traditional way of RBDO involves two loops: an outer loop that optimizes the design variables and an inner loop that evaluates the reliability constraints. Some methods use approximations surrogate methods to reduce the computation time for calculate the reliability constraints are introduced in Chap. 7. In RBDO process, surrogate models such as Kriging and support vector machine (SVM) are used to approximately estimate the reliability index or failure probability to reduce the computational cost. However, the construction of a surrogate model relies on actual system response. Acquiring response data in massive quantities incurs immeasurable computational costs in some high-dimensional nonlinear problems. To solve the abovementioned problems, a PINN model is constructed to approximate the system performance function. The response of system can be expressed by PINN model as

$$u(\boldsymbol{X}) \approx \hat{u}(\boldsymbol{X}) = N(\boldsymbol{X}, \boldsymbol{\theta}) \tag{9.39}$$

where u can be considered as the response of system, $\hat{u}$ is the predicted value of PINN model to the system response. The design variable and the random parameters can be considered as the input of the PINN model $N(*)$.

Constraint function $G_j(\boldsymbol{X})$ is in the familiar form as Eq. (9.29) proposed in the previous sections, so that the PINN model for $G_j(\boldsymbol{X})$ can be denoted as $\widehat{G}_j(\boldsymbol{X})$.

$$G_j(\boldsymbol{X}) \approx \widehat{G}_j(\boldsymbol{X}) = \widehat{u}(\boldsymbol{X}) - u_d \tag{9.40}$$

where u_d indicates the threshold value to identify the failure. Then the probabilistic constraints can be expressed using MCS as

$$P_{F_j} \equiv P[G_j(\boldsymbol{X}) > 0] \cong P\left[\widehat{G}_j(\boldsymbol{X}) > 0\right] = \frac{1}{M}\sum_{m=1}^{M} I_{\widehat{\Omega}_{F_j}}\left(x^{(m)}\right) \tag{9.41}$$

where M is the sampling size of MCS, $\boldsymbol{x}^{(m)}$ is the m-th realization of $\boldsymbol{X}$, and the failure set $\widehat{\Omega}_{F_j}$ for the PINN model is defined as $\hat{\Omega}_{F_j} \equiv \{x : \hat{G}_j(\boldsymbol{X}) > 0\}$.

For design optimization, sensitivity analysis is a crucial step as it quantifies the impact of parameter variations on the optimal objective value and the optimal point. One frequently used approach is using surrogate model to approximate the probabilistic response. However, this approach needs actual response data which comes from expensive simulations or observations, and it may provide inaccurate sensitivity information when the models cannot obtain sufficient and accurate data for training. In Chap. 7, a sampling-based stochastic sensitivity analysis method that employs a score function is introduced. Now the previous PINN model will be incorporated in current approach and then the PINN-based sensitivity analysis method will be proposed.

The definition of the probability of failure with respect to the i-th design variable μ_i yields is given in Chap. 7 as

$$\frac{\partial P_F(\psi)}{\partial \mu_i} = \frac{\partial}{\partial \mu_i}\int_{\boldsymbol{R}^{nr}} I_{\Omega_F}(\boldsymbol{x}) f_{\boldsymbol{X}}(\boldsymbol{x};\boldsymbol{\mu})dx \tag{9.42}$$

then gave the definition of first-order score function for μ_i as

$$s_{\mu_i}^{(1)}(\boldsymbol{x};\boldsymbol{\mu}) \equiv \frac{\partial \ln f_{\boldsymbol{X}}(\boldsymbol{x};\boldsymbol{\mu})}{\partial \mu_i} \tag{9.43}$$

The sensitivity of the probabilistic constraint can be transmitted to

$$\frac{\partial P_{F_j}}{\partial \mu_i} \cong \frac{1}{M}\sum_{m=1}^{M} I_{\widehat{\Omega}_{F_j}}\left(x^{(m)}\right) s_{\mu_i}^{(1)}\left(x^{(m)};\mu\right) \tag{9.44}$$

where $s_{\mu_i}^{(1)}(x^m;\mu)$ is the first-order score function, which is calculated in Sect. 7.4.3.

After the analysis of probability of failure, the design procedure can be conducted. The complete numerical procedure of the proposed RBDO based on PINN is described in four main steps:

1. Construct a PINN model that can approximate the system performance function. The PINN model is a neural network that is trained with physics-informed constraints so that actual response data from simulations or observations is not needed.
2. Use the PINN model to approximate the probabilistic constraints. The probabilistic constraints are expressed as the probability of failure, which is calculated using Monte Carlo simulation (MCS) with the PINN model.
3. Use the sensitivity analysis method to quantify the impact of parameter variations on the optimal objective value and the optimal point. The sensitivity analysis method uses a score function, which is a function that measures how sensitive the probability of failure is to the changes in the design variables. The probability of failure used in score function is also calculated using MCS with the PINN model.
4. Use an optimization algorithm that can handle both deterministic and probabilistic constraints to find the optimal design variables that minimize the objective function, subject to the probabilistic constraints and any other deterministic constraints.

Figure 9.5 is a flowchart of PINN-based RBDO.

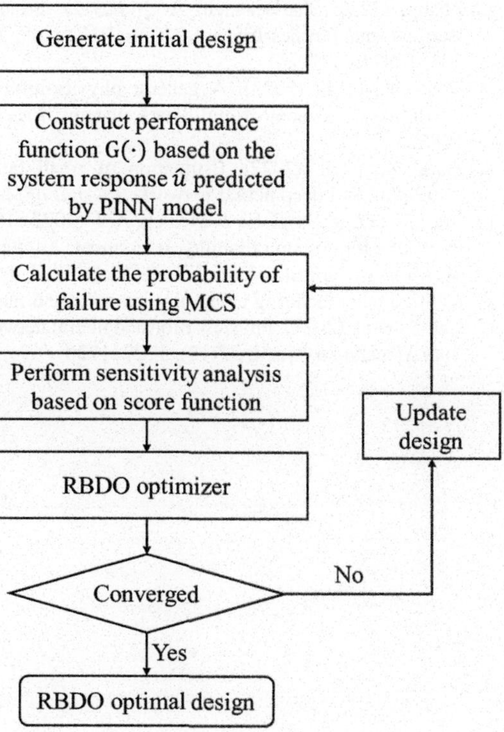

Fig. 9.5 Flowchart of PINN-based RBDO

References

1. Jiang, Z., et al. (2017). Structural reliability analysis of wind turbines: A review. *Energies, 10*, 2099.
2. Hu, W., et al. (2023). Surrogate-based time-dependent reliability analysis for a digital twin. *Journal of Mechanical Design, 145*(9), 1–36.
3. Hao, Z., et al. (2022). *Physics-informed machine learning: A survey on problems, methods and applications.*
4. Raissi, M., Perdikaris, P., & Karniadakis, G. E. (2019). Physics-informed neural networks: A deep learning framework for solving forward and inverse problems involving nonlinear partial differential equations. *Journal of Computational Physics, 378*, 686–707.
5. Cuomo, S., et al. (2022). Scientific machine learning through physics–informed neural networks: Where we are and what's next. *Journal of Scientific Computing, 92*(3), 88.
6. Nabian, M. A., et al. (2021). Efficient training of physics-informed neural networks via importance sampling. *Computer-Aided Civil and Infrastructure Engineering, 36*(8), 962–977.
7. Wu, C., et al. (2023). A comprehensive study of non-adaptive and residual-based adaptive sampling for physics-informed neural networks. *Computer Methods in Applied Mechanics and Engineering, 403*, 115671.
8. Xu, Y., et al. (2022). Physics-informed machine learning for reliability and systems safety applications: State of the art and challenges. *Reliability Engineering and System Safety, 230*, 108900.
9. Chakraborty, S. J. A. P. A. (2020). *Simulation free reliability analysis: A physics-informed deep learning based approach.*
10. Zhang, C., & Shafieezadeh, A. J. (2022). Simulation-free reliability analysis with active learning and Physics-Informed Neural Network. *Reliability Engineering and System Safety, 226*, 108716.
11. Zhou, T., et al. (2023). A generic physics-informed neural network-based framework for reliability assessment of multi-state systems. *Reliability Engineering and System Safety, 229*, 108835.
12. Peng, X., et al. (2022). Estimation of small failure probability based on adaptive subset simulation and deep neural network. *Journal of Mechanical Design, 144*(10), 101704.
13. Zhou, T., Droguett, E. L., & Mosleh, A. J. (2022). Physics-informed deep learning: A promising technique for system reliability assessment. *Applied Soft Computing Journal, 126*, 109217.
14. Wenzel, S., Slomski-Vetter, E., & Melz, T. J. M. (2022). Optimizing system reliability in additive manufacturing using physics-informed machine learning. *Machines, 10*(7), 525.
15. Cai, S., et al. (2021). Physics-informed neural networks (PINNs) for fluid mechanics: A review. *Acta Mechanica Sinica, 37*(12), 1727–1738.

Chapter 10
Engineering Applications of Design Optimization Under Uncertainty

Nomenclature

AFP	Automated fiber placement
AFROM	Auxiliary fuzzy robust optimization model
ASSA	Augmented step size adjustment
AUV	Autonomous underwater vehicle
BP	Back propagation
CFD	Computational fluid dynamics
CF	Creep-fatigue
CFEP	Carbon fiber reinforced polymer
CGF	Cumulant generating function
CLSC	Closed-loop supply chain
CS	Constant stiffness
CTS	Continuous tow shearing
EV	Electric vehicle
FEA	Finite element analysis
GA	Genetic algorithm
HHC	Home health care
HHCRSP	Home health care route scheduling problem
H-LCF	High and low creep-fatigue
HM	Horsetail matching
IGA	Isogeometric analysis
KDE	Kernel density estimation
KL	Karhunen-Loeve
LCF	Low cycle fatigue
LSF	Limit state function
MCDM	Multi-criteria decision-making
MCS	Monte Carlo simulation
MIP	Mixed integer programming

W. Hu, *Design Optimization Under Uncertainty*,
https://doi.org/10.1007/978-3-031-49208-2_10

MILP	Mixed-integer linear programming
MLE	Maximum likelihood estimation
MOMFPFRP	Multi-objective mixed fuzzy possibilistic flexible robust programming
MOPSO	Multi-objective particle swarm optimization
MRBIS	Multimodal radial-based importance sampling
MVSOSA	Mean-value second-order saddle point approximation
NSGA-II	Non-dominated sorting genetic algorithm II
OWT	Offshore wind turbine
PC	Prestressed concrete
PCE	Polynomial chaos expansion
PMA	Performance measurement approach
PS	Power system
PSs	Photovoltaic systems
PSO	Particle swarm optimization
RBDO	Reliability-based design optimization
RBF	Radial basis function
RDO	Robust design optimization
RIA	Reliability index approach
RVE	Representative volume element
SA	Saddle point
SIMP	Solid isotropic materials with penalization
SORA	Sequence optimization and reliability assessment
SRPS	Sustainable and reliable power system
TH method	Torabi and Hassini method
TS	Tabu search
TST	Trip and service times
VRB-VCS FLB	Variable-roll-blank and variable-cross-sectional shape front longitudinal beam
VS	Variable stiffness
WCA	Water cycle algorithm
YAG method	Yavari and Geraeli method

10.1 RBDO Engineering Applications

RBDO is proposed to ensure safe and reliable operation of the product/system. The use of RBDO can decrease operating and maintenance costs and increase product durability and safety. This section illustrates RBDO engineering applications in four fields: aeronautical engineering, ocean engineering, bridge engineering, and vehicle engineering.

10.1.1 Aeronautical Engineering

To ensure the reliability of aerospace structures in complex atmosphere, the design optimization of aerospace structures needs to consider fail-safe size uncertainties, whose sources can be divided into three categories [1]. The first type of uncertainty refers to the uncertainty of structural parameters, which can be described as aleatory, epistemic, or hybrid uncertainty. The second type of uncertainty is the partial collapse of the airframe when an unexpected event occurs. The last type of uncertainty is the characteristics of the debris during engine failure such as the number of impacts or the location and size of holes in the fuselage. This section reviews various engineering examples of design optimization of variable stiffness (VS) composite laminates, compressor pallets, and aircraft tail fins in aerospace structures under uncertain conditions.

10.1.1.1 VS Composites Plate

In the RBDO problem of lightweight design considering manufacture constraints of VS composite laminates, the uncertainty mainly comes from the property of heterogeneous material that consists of various composite laminates. The uncertainty distribution of material properties is related to lamination parameters and lamination thickness. Note that number of layers of composite material is discrete, so traditional RBDO method cannot solve such problems.

Hao et al. [2] investigated the RBDO problem for lightweight design of VS composite laminates with discrete random variables, and solved the problem by the idea of transforming discrete random variables into continuous random variables. In the RBDO of VS composite laminates, the weight is measured by the total thickness of laminate, and the constraint boundary is that the buckling load cannot exceed a maximum threshold. Then, a two-stage RBDO framework is built. In the first stage, gradient optimization is performed directly using lamination parameters and layer thickness to obtain an approximate optimal layer value. In the second stage, the discrete layers are transformed into continuous variables by using intermediate density variables. Then, inspired by the idea of solid isotropic materials with penalization (SIMP), the buckling analysis and sensitivity derivation are carried out by isogeometric analysis (IGA), while the reliability analysis is performed by augmented step size adjustment (ASSA) method. Finally, the optimal lightweight design is obtained through gradient optimization, which realizes the purpose of using gradient optimization in RBDO problems with discrete random variables. The optimization results show that the maximum buckling load satisfying the constraints is reduced by 18.3% and the weight is reduced by 12.5% compared to the initial design.

10.1.1.2 Compressor Disc

High-pressure compressor disc is generally subjected to mechanical stress from centrifugal load, vibration stress, and thermal stress during operation. Therefore, the failure modes of compressor disc in service include low cycle fatigue (LCF) at the center [3, 4], creep-fatigue (CF) at the rim [5], and combined high and low cycle fatigue (H-LCF) at the slot [6]. There is a strong correlation among these failure modes due to the interaction of mutual working condition and geometry structure. Meanwhile, uncertainties in geometry, applied loads, and material properties will affect the reliability of high-pressure compressors disc. To reduce the influence of these uncertainties, Liu et al. [7] proposed an RBDO method to solve the ill-posed multi-mode model of compressor disc with multiple correlated failure modes from the perspective of multi-objective.

The method they propose is performed in the framework of the double-loop strategy. First, the pair-copula function [8] is used to quantify the multivariate correlation between failure modes. Combined with the reliability allocation method [9], the reliability requirements of system-level structural targets are allocated to the reliability requirements under each failure mode at the subsystem level, thus eliminating the repeated calculation of structural reliability in the double-loop framework. Because of the complex correlation between failure modes, the lifetime of the structure cannot be fitted with the generalized distribution function. Therefore, a non-parametric kernel density estimation (KDE) method [10] is used to simulate the PDF of the structure lifetime. Finally, the improved multi-objective particle swarm optimization (MOPSO) [11, 12] algorithm integrated niche sorting strategy and dynamic inertia weight is used to solve the ill-conditioned multi-objective problem. The niche sorting strategy can reduce the fitness of similar particles to promote diversity, while the dynamic inertia weight can adjust the search speed according to the number of iterations to obtain the global optimal solution efficiently and accurately.

The optimized results show that the maximum von-Mises stress and strain of the compression discs are reduced by 1.63% and 5.93%, respectively, and the reliability design target of 99% is satisfied.

10.1.1.3 Aircraft Tail Fuselage

Design optimization of tail fuselage considering local collapse is a multi-model optimization problem, because engine failure or blade release will randomly impact the tail fuselage and cause different degrees of local collapse. Cid et al. [10] applied the concept of multi-model optimization to existing uncertainty optimization methods, and compared the advantages and disadvantages of traditional RBDO methods (performance measurement approach (PMA) and sequence optimization and reliability assessment (SORA)) and horsetail matching (HM) methods [11, 12].

In this case, the target of RBDO is to minimize the weight of aircraft while maintaining von-Mises stress under applied loads below the maximum threshold. The Young's modulus of the structural material is regarded as the uncertainty. The local collapse model considers three failure scenarios that are most harmful to the fuselage structure, including blade failure, turbine disk failure, and the unducted blade failure in open rotor engines, corresponding to eight local collapse models. To apply HM technology to multi-model optimization methods, the external penalty function method is used to weight the objective function and design constraints to transform the constrained problem into an unconstrained problem.

Using the multi-model optimization technique [13–15], PMA, SORA, and HM methods are compared for the multi-model RBDO problem of aircraft tail. The optimization results show that all methods increases the weight of the tail fuselage by 39.42%, and obtain the damage resistant structure with a reliability index of 3.719. Here the increase of weight is a compromise of assuring reliability under uncertainty. According to their conclusion, the traditional RBDO method can guarantee the reliability index applied in all applicable limit states, while HM enhances the ability of non-random dominant design with the same objective function values.

10.1.1.4 Real-World Engineering Application

In decades, the potential of fiber-reinforced composites in lightweight design has been evident in the aerospace application. With the development of composite manufacturing techniques such as automated fiber placement (AFP) and continuous tow shearing (CTS) [16], fiber paths can be arranged by pre-designed curves, as shown in Fig. 10.1.

Therefore, compared to the constant stiffness (CS) fiber path, the curved fiber path can freely distribute the rigid and flexible regions, thereby improving the

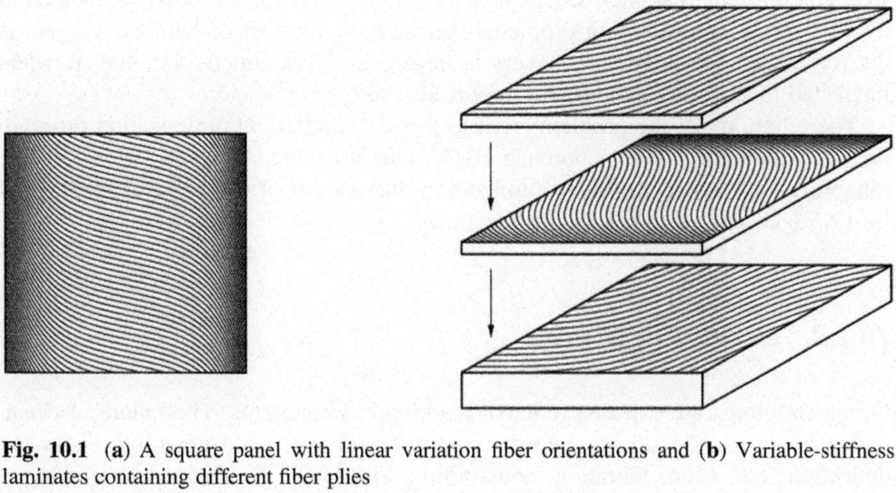

Fig. 10.1 (**a**) A square panel with linear variation fiber orientations and (**b**) Variable-stiffness laminates containing different fiber plies

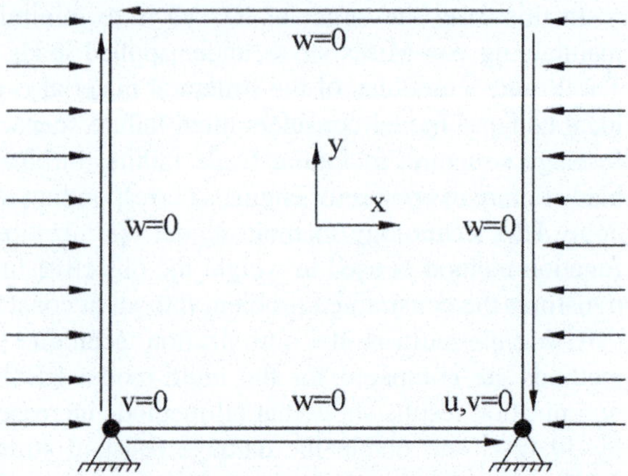

Fig. 10.2 The square panel subject to combined compression–shear load under the fully simply supported boundary condition

response of VS laminates to buckling loads. However, uncertainties of material properties and manufacturing probably cause the performance of laminate to deviate from design expectations. Hao et al. [2] studied the RBDO problem of a square panel made of carbon fiber composite T300/5208 with a length of 254 mm, and proposed a bi-stage optimization method to solve the optimal RBDO solution. The square panel subjects to combined compression and shear loads under completely simply supported boundary conditions. The square panel and load distribution are shown in Fig. 10.2.

The first stage of the bi-stage optimization method is to perform deterministic optimization with the total thickness of the CS laminate as the objective and the buckling load response within the feasible region of the panel parameters as constraints to obtain an approximate number of layers. In the second stage, RBDO is performed with the panel mass as the objective function and the buckling load response under material property uncertainty as the probabilistic constraint. It is worth mentioning that intermediate density variables are introduced in the method to avoid discrete programming problems (because the number of laminated layers is discrete), then the number of layers is treated as a continuous variable. Readers interested in specific methods can refer to [2].

The reliability of the problem is set to $\beta = 3$. The RBDO optimization program uses the reliability index approach (RIA) and uses the ASSA algorithm in the reliability analysis loop. After optimization, the weight of the laminate is reduced by 12.5% compared to the original design.

10.1.2 Ocean Engineering

Ocean structures are exposed to harsh working environments. The failure of ocean structures is usually due to fatigue, buckling, erosion, weld cracking, excessive deflection, corrosion, vibration, and fouling, etc. [17, 18]. The structural fatigue failure caused by environmental load (such as wave or sea breeze) is common. Therefore, the uncertainty of such environmental loads needs to be quantified so that reliability regarding fatigue can be well performed [19]. Existing methods cannot fully quantify the uncertainty of waves [20]. Meanwhile, the strength design process of ocean structures needs to consider the coupling effect of multiple disciplines (e.g., fluid mechanics, structural mechanics, and fracture mechanics) which may eventually affect the safety of marine structures. Three engineering examples in ocean structure optimization under uncertainties is introduced, including a lightweight optimum design of head pressure shell of an autonomous underwater vehicle (AUV), the optimization of stress distribution on a wellhead platform, and the lightweight design of support structure of an offshore wind turbine (OWT).

10.1.2.1 Autonomous Underwater Vehicle's Head Pressure Shell

The head pressure shell can provide a secure working environment for sensors and other functional systems of AUV in the unpredictable ocean wave environment. Affected by the uncertainty of ocean wave load, buckling deformation is an important failure mode of the head pressure shell.

Li et al. [21] applied the grid sandwich structure to the head pressure shell of AUV in modeling. The RBDO optimization mathematical model is established by setting the total head pressure shell weight and maximum von-Mises stress as the objective function and the strength requirement as the constraints, respectively. The back propagation (BP) neural network model is established based on result of finite element model of the head pressure shell, whose parameters are trained by particle swarm optimization (PSO) algorithm. The distribution regularity of maximum von-Mises stress is obtained by random sampling in optimum neural network. Combined with the distribution regularity of yield strength, which is assumed to obey the normal distribution, the stress-strength interference model is established. With the help of this model, the failure probability of the head pressure shell can be easily obtained by integrating. Finally, the optimization problem is solved by genetic algorithm (GA).

The proposed method finally achieves the lightweight design effect of 38.26% reduction in weight of the grid sandwich head pressure shell compared with the solid pressure shell. Meanwhile, the head pressure shell of the grid sandwich structure reaches the level of 3-σ of the random variables, which can ensure sufficient reliability of the AUV.

10.1.2.2 Wellhead Platform

The use of wellhead platform is one of the effective ways to reduce the production cost of offshore oil fields [22]. To ensure the reliability of the wellhead platform, wave load cannot be ignored. The influence of long-term wave load can cause various failure forms of wellhead platform components, such as yield, instability, fatigue, plastic failure, or large deformation. At the same time, because of the unquantifiable randomness of wave load, the reliability analysis of wellhead platform becomes a difficult task.

In this context, Meng et al. [22] investigated the RBDO problem to enhance the safety performance of wellhead platforms and proposed the RBDO-MVSOSA strategy. Firstly, Morrison formula is used to calculate the approximate wave load, which is used to establish the performance function of yield failure at each failure point of the wellhead platform. In the reliability analysis part, the mean-value second-order saddle point approximation (MVSOSA) method is used to performance function reliability analysis. As an improved method based on the saddle-point approximation (SA) method, the MVSOSA method does not require the transformation of random variables (from X-space to U-space) to search for the most probable point. The X-to-U space transformation is a nonlinear transformation, which will increase the nonlinearity of the limit state function (LSF) and finally resulting in a loss of reliability analysis accuracy. Meanwhile, the MVSOSA method has a higher accuracy of reliability analysis results compared to that using the SA method. Finally, gradient optimization algorithm is used to find the optimal solution of RBDO for wellhead platforms.

The results show that wellhead platform volume is reduced by 9% compared to the initial design, and the stress distribution is improved. However, it is worth pointing out that MVSOSA method is an SA-based method, and its reliability analysis process relies on the cumulant generating function (CGF), but only Gaussian variables are available for the analytic derivation of moment generating function (MGF), so their method is only applicable to the case where the random variables obey a Gaussian distribution.

10.1.2.3 Offshore Wind Turbine Blades

The key to RBDO of wind turbine blades is to accurately predict the fatigue life under complex load conditions such as wind load, gravity load, and centrifugal load [23]. The uncertainty of dynamic wind load is the most crucial source of uncertainty affecting the fatigue reliability of wind turbine blades. The review [24] on the structural reliability analysis of wind turbines is concluded that the method of quantification of uncertainty is a key factor in the structural reliability analysis of wind turbines.

Hu et al. [25] studied the RBDO of 5-MW wind turbine blades. The uncertainty model of dynamic wind load involves two parts: the annual wind load variation and the wind load variation in a large spatiotemporal range. The annual wind load variation is represented by a joint PDF of the 10-min average wind speed and the 10-min average turbulence intensity, while the wind load variation in a large spatiotemporal range is represented by a PDF of multiple set parameters. By using the maximum likelihood estimation (MLE) method and 249 sets of measured wind data, the 10-min fatigue damage table is obtained, and the Kriging surrogate model is used to fit these data and create a 10-min fatigue damage surrogate model. Finally, the uncertainty model of dynamic wind load is established when Riemann integral method, wind load probability table, and 10-min fatigue damage table are used to derive the 20-year fatigue damage.

Their proposed method consists of two parts. The first part is the DDO procedure. The purpose of performing DDO is to provide a good initial design for RBDO. The design variable is to control the thickness of the composite laminates and the normalized total cost of the composite materials used in the blades is set as the objective function. The constraint is chosen as the 20-year fatigue damage under the influence of dynamic wind load. Since DDO does not involve uncertainty in dynamic wind loads, wind loads are provided by means of an average wind load probability table generated using Monte Carlo simulation (MCS) methods.

The second part is the RBDO procedure. Based on the mathematical model of DDO, the uncertainty factor and failure probability are further taken into consideration. The uncertainties include the deduced uncertainties of the dynamic wind load and the uncertainty of the thickness of the composite laminates during the manufacturing process, the latter of which is assumed to follow a normal distribution. The probability constraint of RBDO is that the fatigue failure probability at the fatigue hotspot is less than the target failure probability. Due to the implicit character of 20-year fatigue damage estimation at fatigue hotspots, its sensitivity is difficult to obtain. Therefore, the sampling-based reliability method of MCS is used to calculate the fatigue failure probability. Finally, SQP algorithm is used as an optimizer.

The optimization results show that the fatigue failure probability of 50.06% in the initial design is reduced to 2.281% in the optimal design, while the cost is only increased by 3.01%.

10.1.2.4 Real-World Engineering Application

Wind energy is an important direction for the country's clean energy development. Monopile offshore wind turbines is one of the wind power equipment that converts offshore wind energy into electricity. The basic structure of a monopile offshore wind turbine is shown in Fig. 10.3 [26]. During normal operation, the monopile is subjected to four external loads: (1) Current loads; (2) Wave load; (3) Top wind turbine load; and (4) Wind pressure, as shown in Fig. 10.4. Meng et al. [27] studied the RBDO problem of monopiles under mixed uncertainty, where external dimensions are random variables and material density is an interval variable.

The optimization objective in this optimization problem is to minimize weight m of the monopile. The optimization variables are diameter and thickness of the monopile. The mathematical model of this optimization problem is shown in Eq. (10.1).

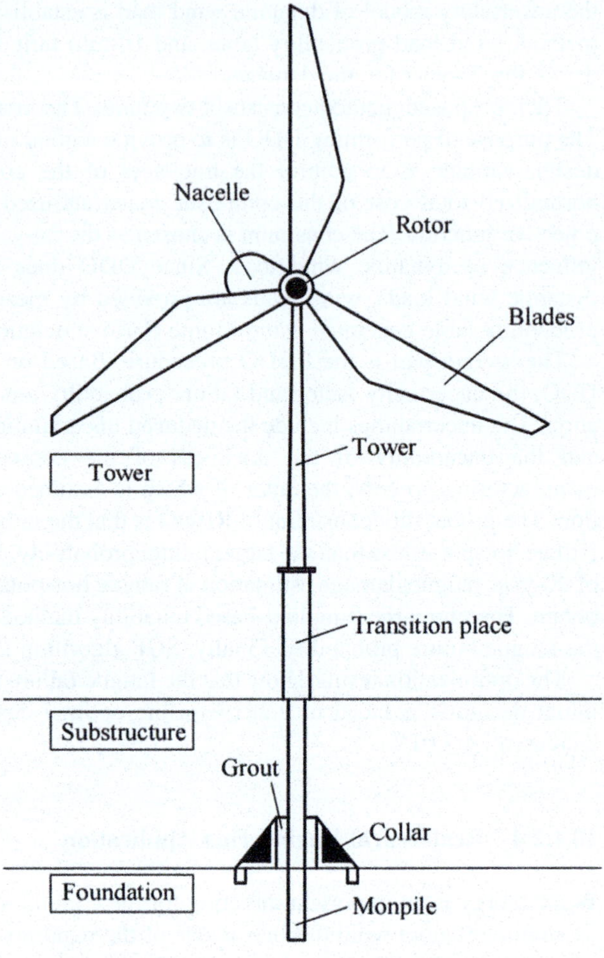

Fig. 10.3 Basic structure of monopile offshore wind turbine [26]

$$
\begin{aligned}
\min \quad & F = m(\mu_D, \mu_T) \\
s.t. \quad & P_f\{\sigma_{allow} - \sigma_{\max}[D, T, \rho] \le 0\} \le \Phi(-\beta_t) \\
& T \ge 6.36 + \frac{D}{100} \\
& 4\,\text{m} \le \mu_D \le 8\,\text{m},\ 20\,\text{mm} \le \mu_T \le 200\,\text{mm} \\
& 7840\,\text{kg/m}^3 \le \rho \le 7860\,\text{kg/m}^3
\end{aligned}
\tag{10.1}
$$

where $\sigma_{allow} = 345$ MPa represents the allowable stress of material for the monopile. $\sigma_{\max}$ represents the maximum stress on the monopile. ρ is the density of monopile material and is an interval variable. The nominal value of interval variable ρ is 7850 kg/m^3.

SORA is used to find the optimal solution, in which the water cycle algorithm (WCA) is used in the optimization cycle for global optimization. For details on the

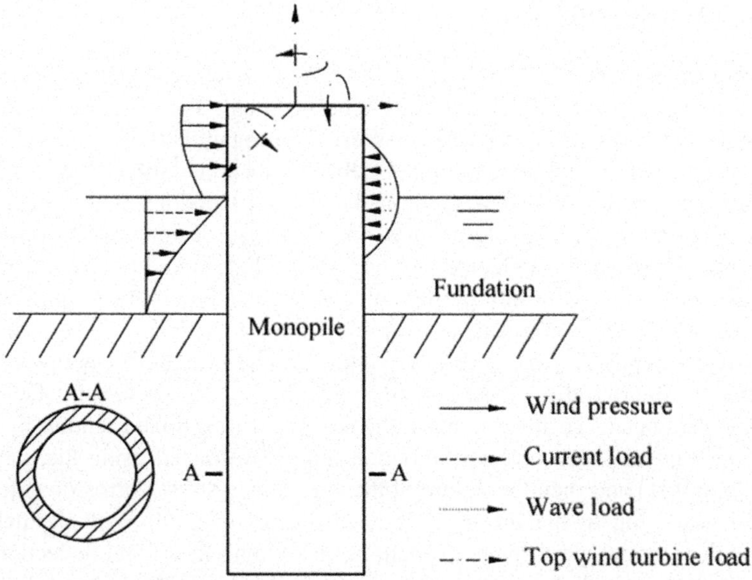

Fig. 10.4 Load distribution on offshore monopile

reliability calculation process and specific optimization methods under mixed uncertainty, refer to [27]. Compared with the original design, the weight after the optimized design is reduced by 12.78%, and the reliability reaches 99.87%.

10.1.3 Bridge Engineering

One of the most important components of bridge design is ensuring structural stability, even when low-probability events such as heavy rains, hurricanes, earthquakes, and other harsh environments occur. The service life of bridges is also affected by many factors, such as the random gravity loads exerted on the bridge by cars driving on the bridge deck, corrosion, and deformation of bridge construction materials, etc. However, it is necessary to model the uncertainty of these factors to ensure the reliability of bridge design, which is not an easy task. This section reviews engineering examples of bridge design optimization under uncertainties including corrosion, traffic loads, and flutter, respectively, and shows the way to optimize bridge designs under these uncertainties.

10.1.3.1 Bridge Corrosion

In prestressed concrete (PC) bridges, the main factor leading to severe deterioration of the bridge is corrosion of the steel reinforcement. For instance, the sudden collapse of the Ynys-y-Gwas Bridge in West Glamorgan, UK, in 1985 was caused by corrosion of the bridge reinforcement [28]. The shear and torsion bars of PC box girder bridges, especially the post-tensioned prestressed tendons, will all have pitting corrosion. Currently, few of researches of PC bridge design optimization under uncertainties have been considered. Nguyen et al. [29] studied the RBDO problem of PC bridges with uncertainty pitting damage effects. To involve the influence of shear and torsional reinforcement by pitting during the analysis, the model of maximum pit depth over the standard length of the reinforcement is invoked [30], while the maximum pitting depth at a given length can be expressed by the Gumbel statistical parameter. To involve the influence of post-tensioned tendon by pitting during the analysis, the corrosion initiation time is estimated using the method of Bentur et al. [31] and then the failure of the wire due to stress corrosion cracking is modeled according to the linear fracture mechanics method. With the help of a pitting failure model, performance functions on strength limits and deflection limits are established and reliability analysis is performed using FORM. According to the reliability analysis results, optimization is executed using the first-order Taylor series expansion and quasi-Newton optimization algorithm of MATLAB optimization toolbox. Finally, the optimal dimensional design solution of PC bridge is obtained, which considers the production cost and reliability. In the process of optimization, it is found that the target reliability index should be in the range of 3.5 ~ 5 to get the optimal solution.

10.1.3.2 Bridge Traffic Load

Traffic load is also an important factor to be considered in bridge design and evaluation. The dynamic response of a bridge under moving load is greater than that under the same horizontal static load. In addition, random fluctuation of material parameters will affect the response of bridge structure to vehicle load. Dynamic analysis of bridge structures involving material parameter uncertainties had been extensively studied in the past decades [32–34]. However, the research of RBDO method involving the interaction between bridge and vehicle in bridge design optimization is still limited. Ni et al. [35] proposed an RBDO method to carry out RBDO for a three-dimensional box-section bridge that considered the uncertainties of system material parameters (such as Young's modulus and mass density) and the influence of moving loads.

In bridge-vehicle systems, the uncertainties of material parameters are simulated as a Gaussian and/or lognormal random field, where the Gaussian random input is approximated by Karhunen-Loeve (KL) expansion, while the lognormal random input is approximated by a combination of KL expansion and polynomial chaos expansion. The moving load acting on the bridge is approximately equivalent using the Rayleigh damping model. After quantifying the related uncertainties, the maximum allowable deformation of the bridge is defined as the LSF and approximated by Kriging model. The RBDO method proposed by them is based on the framework of the double-loop strategy. The reliability analysis loop uses MCS method to calculate the failure probability, while the optimization loop uses SQP algorithm to calculate the minimum cross section area of the bridge under the preset probability constraints. Finally, the optimization results show that when the reliability of the bridge reaches 90%, the cross-section area of initial design (corresponding to the reliability of 84.496%) is increased by 20%.

10.1.3.3 Bridge Flutter

Long-span suspension bridge is a kind of flexible structure, which is easily affected by wind vibration. Therefore, flutter instability caused by wind vibration is also the important factors to consider in RBDO of long-span suspension bridges, which may lead to the collapse of bridge structures.

Kusano et al. [36] extended the study of RBDO of single box girder for suspension bridge (taking East Belt Bridge as an example) to shape optimization and size optimization under probabilistic flutter constraint [36]. In their study, aerodynamic coefficients of different deck sections are simulated through a series of computational fluid dynamics (CFD), and based on the response of CFD, a pneumatic surrogate model is constructed using Kriging model to estimate the force coefficients of different deck sections [37]. Then, the quasi-steady formula is used to define the flutter derivatives of these force coefficients in completely numerical terms. In the RBDO procedure of suspension bridge box girder shape and size, the construct of LSF considers the uncertainty of force coefficient and slope, and the ultimate wind speed at the bridge position. The RIA is adopted to solve the optimal solution in the final.

The optimization results can be obtained under different reliability indexes. When the reliability index β is reduced to 7, the beam volume decreases by 13.33% compared with the initial design where $\beta = 7.58$, and when the reliability index β is increased to 9, the beam volume increases by 10.03% compared with the initial design.

10.1.3.4 Real-World Engineering Application

Existing famous long-span suspension bridges such as Akashi Kaikyo Bridge, the Xihoumen Bridge, the Yi Sunsen Bridge, the Jiangsu Runyang Bridge, and the Tsing Ma Bridge, etc., are flexible structures. However, flexible structures are susceptible to wind vibrations, which can lead to bridge flutter instability and collapse. The

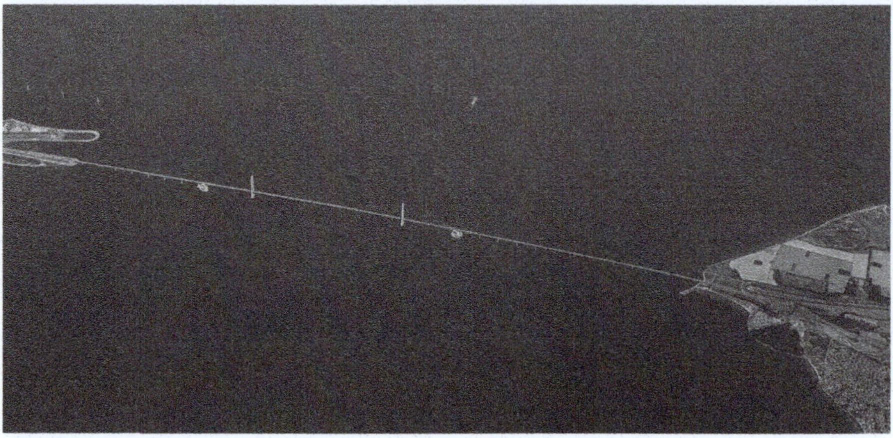

Fig. 10.5 Overall view of the Great Belt East Bridge. (Picture from Wikipedia: Great Belt Bridge – Wikipedia)

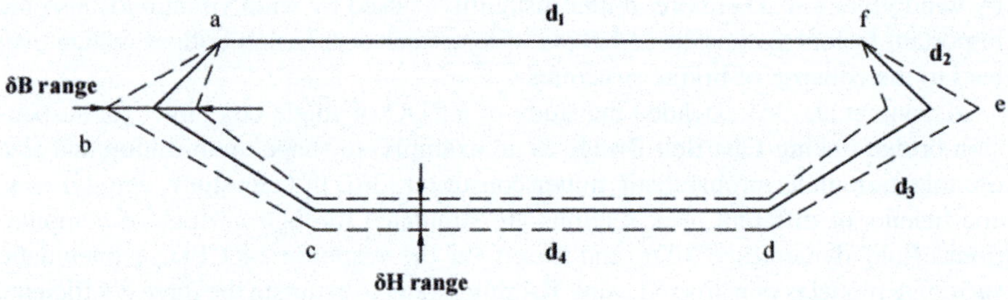

Fig. 10.6 Box girder geometry and design variables

study of Diana et al. [38] underlined the importance of bridge deck shape to the vortex shedding phenomenon while Ge and Xiang [39] studied the importance of bridge deck section on flutter velocity for cable-supported bridges. Therefore, the design optimization of the bridge deck shape has a positive effect on the resistance of the structure to wind-induced instabilities. In addition, material cost as one of the main expenses of long-span suspension bridge construction, material lightweight design will help to reduce the costs. Kusano et al. [36] took the Great Belt East Bridge in Denmark as the research object and proposed RBDO method for the design of single-box girder of suspension bridge under probabilistic flutter constraints. Figure 10.5 shows the overall view of the Great Belt East Bridge, Fig. 10.6 shows the stream-lined box girder geometry and design variables, the solid line is the initial design, and the dashed line is the shape after design optimization. The RBDO mathematical model representation of the box girder is as Eq. (10.2):

Table 10.1 RBDO results of box girder for different target reliability

β_t	Objective function value	Variation
7	2409.19	−13.33
8	2718.78	−2.19
9	3058.50	10.03
10	3628.01	30.52

$$\begin{aligned}
\min \quad & \text{Girder volume}(\delta H, \delta B, d_1, d_2, d_3, d_4) \\
\text{s.t.} \quad & g_1 : P\left[V_f(x) - x_w \le 0\right] \le P_f \\
& g_2 : -10\% \le \delta H \le 10\% \\
& g_3 : -10\% \le \delta B \le 10\% \\
& g_4 : 7\,\text{mm} \le d_i \le 25\,\text{mm}, i = 1, 2, 3, 4 \\
& g_5 : \sigma_c = 565\,\text{MPa} \\
& g_6 : \frac{z_d}{z_{\max}} - 1 \le 0 \\
\text{where} & \\
z_{\max} = & \frac{L}{500}; L = 1624\,\text{m}
\end{aligned} \tag{10.2}$$

The objective is to minimize the volume of the beam while meeting the requirements of constraint reliability. The uncertainties come from the bridge fluttering wind speed $V_f(x)$ and force coefficient x. To avoid the shape of box girder δH and δB being infeasible, the range of values for the design variables are constrained to $\pm 10\%$ of the original size. Constraint g_5 considers tensile stresses σ_c under static overload. Constraint g_6 considers the vertical displacement Z_d across the middle deck under static overload, $Z_{\max}$ is the threshold. L is the total length of the bridge. The probabilistic constraint is defined by the structural limit state function of flutter failure, where x_w is the limit wind speed at the position of the beam. The uncertainty of $V_f(x)$ is assumed to follow the Gaussian distribution, with the output of CFD simulation as the mean, and the custom parameters as standard deviation. For details, refer to [36].

The response of the limit state function is provided by the Kriging surrogate model. RIA is used to solve Eq. (10.2). The optimal solutions under different reliability index are summarized in Table 10.1.

10.1.4 Vehicle Engineering

As a complex system, the vehicles are confronted with multiple challenges throughout their lifecycle. Recent customer surveys show that in addition to cost, reliability and maintainability are key decision factors that customers seriously consider when purchasing a vehicle. Because vehicles are machines with high risk, reliability must be guaranteed during their design, manufacture, and use [40]. This section reviews 3 aspects of vehicle crashworthy design, composite battery box design, and vehicle gearbox housing design.

10.1.4.1 Variable Blank Variable Section Shape Front Longitudinal Beam

Duan et al. [40] invented a novel variable-roll-blank and variable-cross-sectional shape front longitudinal beam (VRB-VCS FLB), which has more potential than traditional FLBS in crashworthiness and lightweight design. Studies had shown that in the stamping process of VRB, transition zone center movement, sheet metal thinning [41], and rebound [42] may occur. Therefore, it is necessary to consider various uncertainties in the manufacturing process to improve the lightweight and crashworthy reliability of VRB-VCS FLBs.

They proposed a multi-objective RBDO method for VRB-VCS FLB. The objective function consists of two parts: the crashworthiness of the vehicle is measured by the peak acceleration of the collision pulse, and the lightweight of the vehicle is measured by the body weight. The performance function involves the crash pulse first step acceleration, crash pulse second step acceleration, energy absorption of VRB-VCS FLB, and dash panel intrusion. Their proposed multi-objective RBDO method is performed in the framework of the double-loop strategy. In the reliability analysis loop (the inner loop), the multimodal radial-based importance sampling (MRBIS) method [43] is used to estimate the probability of a system with multiple failure modes, which solves the problem that traditional reliability analysis methods (such as MCS and importance sampling) cannot directly evaluate the reliability of a system involving multiple failure modes. In the design optimization loop (the outer loop), the non-dominated sorting genetic algorithm II (NSGA-II) is used to generate Pareto boundaries. In the meantime, to overcome the high computational cost of a single run of crashworthiness finite element analysis (FEA), the ε-support vector regression is used as a surrogate model to reduce the computational burden.

Finally, the optimal Pareto set is obtained, while the weight and peak acceleration reduction percentage of the optimal overall performance relative to the initial state are 11.357% and 14.356%, respectively. However, additional uncertainties such as material properties, collision direction, and collision velocity have not been considered in this study, so there is still enough exploration space for RBDO studies of VRB-VCS FLBs.

10.1.4.2 Carbon Fiber Reinforced Polymer Composites

The application of carbon fiber reinforced polymer (CFRP) composites brings great challenges to the design optimization process, such as complex nonlinear material behavior, inherent uncertainties and multilayer characteristics of design variables, and multiple working conditions of structural components [44–47]. Liu et al. [48]

studied the application of CFRP composite materials to the lightweight problem of electric vehicle battery boxes, proposed an RBDO method to find the optimal combination of the microstructure parameters of battery boxes and the geometrical parameters of macrostructures.

The method consists of three parts: uncertainty quantification and propagation, reliability analysis, and optimization. In the first part, a representative volume element (RVE) model considering the damage evolution of the component is established using X-ray micro-CT scanning, while the mechanical properties of CFRP are simulated based on the homogeneity theory [49] to reveal the propagation process of uncertainty. In the second part, based on the simulation data of RVE model, a constitutive model considering tension asymmetry, anisotropy, and failure is established, which is used to analyze the stiffness and strength of the cell box structure. In the third part, to avoid the optimization falling into the local optimal solution and to give consideration to the search efficiency, OLRPSO [50] (an optimization algorithm based on improved particle swarm optimization) is selected as the optimizer to perform the optimization. The above optimization method is integrated in the framework of SORA method. The Kriging model is used to predict the mechanical response of carbon fiber cloth and evaluate its reliability.

After optimization, the weight of the new CFRP battery box cover is reduced by 22.14%, all performance indicators meet the requirements, and the reliability reaches 90%. Finally, they suggested that the stacking direction and sequence of composite laminates could be further considered in the design optimization process in future studies to better exploit the excellent mechanical properties of CFRP.

10.1.4.3 Reducer Housing

The reducer housing needs to ensure the stability of the designed structure under complex load environment. Therefore, RBDO needs to be utilized to reduce the risk of failure due to uncertainties such as design variables and materials. Xu et al. [51] investigated the RBDO problem of electric vehicle reducer housing, proposed a multi-objective reliability design optimization method considering the maximum allowable deviation range of design variables. The method uses interval theory to analyze the reliability, and describes the uncertain value by interval value. The advantage of this method is that the distribution of the exact probability of the uncertain value is not required as in probability theory, and can adapt to the reliability analysis of complex working conditions.

With minimum mass and maximum first-order natural frequency as design objectives, maximum von-Mises stress as constraint condition, and material uncertainty as the main uncertainty source, a mathematical model of RBDO for vehicle reducer housing is established. Because radial basis function (RBF) neural network has outstanding ability in nonlinear function approximation, the proposed method chooses RBF neural network as the surrogate model. Based on the finite element model response, the structural parameters of the RBF are optimized in combination with the PSO algorithm to establish an accurate alternative model for reliability analysis. In the optimization part, SQP algorithm is used to solve the lower and upper bounds of constraints, while NSGA-II is used to generate discrete Pareto solution sets. Finally, multi-criteria decision-making (MCDM) is used to sort Pareto solution sets.

Compared with the deterministic optimization results, the optimal ordering solution set shows that the minimum mass is increased by 0.64%, while the maximum first-order natural frequency and maximum von-Mises stress of the reducer housing are decreased by 4.8% and 4.1%, respectively.

10.1.4.4 Real-World Engineering Application

The application of new lightweight materials and structural design optimization are the main ways to realize lightweight vehicles. The new composite material CFRP has attracted much attention due to its low density, high specific stiffness, flexibility, and energy absorption capacity. The battery box of an electric vehicle is a typical component using composite materials, Fig. 10.7 shows the appearance of an electric vehicle battery cap. However, CFRP composite materials have obvious material and structure integration characteristics, which have complex damage evolution and destruction processes under complex external loads [52, 53]. Zhao et al. [48] redesigned the battery cap made of CFRP material based on reliability. The objective function is to minimum the mass of battery cap while the constraint functions include mode constraints, impact constraints, and drop constraints.

Based on the internal geometric structural parameters of plain woven CFRP obtained from radiographic micro-CT scanning experiments, they used Abaqus to establish an RVE model that considers component damage evolution, and combined it with the Kriging surrogate model to achieve rapid prediction of constraint response.

Fig. 10.7 Electric vehicle battery box system shape cover and related shape design variables

SORA is used to solve the optimal solution, in which the modified particle swarm optimization algorithm [50] is used in the optimization loop for global optimization. The mass of the battery cap is 4.924 kg after optimizing, which is 22.14% lighter than the original structure.

10.1.5 Summary

The following summarizes this section, as shown in Table 10.2, including application field, specific application, sources of uncertainty, optimization methods, optimization results, and references.

10.2 RDO Engineering Applications

Robust design optimization (RDO) refers to improving the adaptability to external anomalies as much as possible under the premise of ensuring the normal operation of the system, which focuses on the stability of the optimization results to the solution. The quantification of various uncertainty is currently a major challenge for RDO in engineering applications. This section reviews the relevant examples of RDO engineering applications from three aspects: energy management, logistics scheduling, and closed-loop supply chain (CLSC). Finally, summarizes them with a table.

10.2.1 Energy Management

The contradiction between power supply demand and environmental protection in power system (PS) has attracted wide attention. Overuse of traditional PS can cause serious environmental and social problems, while renewable PS are characterized by supply instability due to the intermittent nature of renewable energy supply. Considering the uncertainty of supply and demand for the RDO of PS is an effective measure to balance the contradiction. This section reviews the RDO for power systems powered by both wind and solar renewable energy sources, and for power systems powered by a mix of renewables and conventional energy sources (e.g., coal, oil, etc.).

10.2.1.1 Wind Energy

A wind turbine-driven ammonia (NH_3) synthesis plant is powered primarily by wind energy, but the supply of wind energy is volatile. Verleysen et al. [55] studied the RDO of Power-to-NH_3 process under wind uncertainty by combining uncertainty

Table 10.2 The summary of RBDO engineering applications

RBDO engineering applications					
Application field	Specific application	Sources of uncertainty	Method	Result/Contribution (↑ is increment, ↓ is decrement)	References
Aeronautical engineering	VS composite plate	1. Material properties 2. Manufacturing processes	Bi-stage RBDO framework	Weight ↓ 12.5% and maximum buckling load ↓ 18.3%.	Hao et al. [2]
	Compressor disc	1. Material properties 2. Manufacturing processes 3. Loads	Reliability allocation methods + KDE + Modify MOPSO in double-loop strategy	Maximum von-Mises ↓ 1.63% and strain ↓ 5.93%.	Liu et al. [7]
	Tail fuselage	1. Fan blade failure 2. Turbine disk failure 3. Ductless fan blade failure of open rotor engine	PMA、SORA、MH	Tail fuselage's weight ↑ 39.42% but β is equal to 3.7190.	Cid et al. [54]
Ocean engineering	AUV head pressure housing	1. Material properties 2. Manufacturing processes 3. Wave loads	BP neural network + GA	Head pressure shell's weight ↓ 38.26%, and the grid sandwich structure reaches the level of 3σ of the random variable.	Li et al. [21]
	Wellhead platform	1. Material properties 2. Manufacturing processes 3. Wave load 4. Sea state	MVSOSA + Gradient optimization in double-loop strategy	Platform volume ↑ 9% but stress distribution of the platform is improved.	Meng et al. [22]

	Offshore wind turbine blades	1. Wind load 2. Manufacturing processes	MCS + SQP in double-loop strategy	The fatigue failure probability ↓ from 50.06% to 2.281%, while the cost only increased by 3.01%.	Hu et al. [25]
Bridge engineering	PC bridge	1. Material properties 2. Corrosion	FORM + First-order Taylor series expansion and quasi-Newton optimization algorithm in double-loop strategy	Get the optimal design size with reliability index distribution between 3.5 and 5.	Nguyen et al. [29]
	3D box bridge	1. Material properties 2.Vehicle moving load	Kriging + MCS + SQP in double-loop strategy	Sectional area ↑ 20% but reliability up to 90% compared to the initial design (reliability corresponds to 84.496%).	Ni et al. [35]
	Long span bridge (Great Belt East Bridge)	1. Wind-induced vibration	Kriging + RIA	β decreased from 7.58 to 7, volume of bridge ↓ 13.33%; β increased from 7.58 to 9, volume of bridge ↑ 10.03%.	Kusano et al. [36]
Vehicle engineering	VRB-VCS FLB	1. Material properties 2. Manufacturing processes	ε-support vector regression + MRBIS + NSGA-II in double-loop strategy	The weight ↓ 11.357% and peak acceleration ↓ 14.356% relative to the initial state.	Duan et al. [40]
	CFRP composite battery case cover	1. Material properties 2. Manufacturing processes	Kriging + OLRPSO + SORA in decoupling strategy	Battery case cover's weight ↓ 22.14%, reliability up to 90%.	Liu et al. [48]
	Reducer housing	1. Material properties 2. Manufacturing processes 3. Loads	RBF + SQP + NSGA-II + MCDM	Best solution: Minimum mass of reducer housing ↓ 0.64%, maximum first natural frequency ↓ 4.8%, and maximum von-Mises stress ↓ 4.1% compare to the initial design.	Xu et al. [51]

quantization method with multi-objective genetic optimization algorithm. To maximize the average NH_3 production demand and minimize the sensitivity to NH_3 production, an optimization model is established with the current density of the electrolyzer, the H_2/N_2 ratio entering the NH_3 synthesis unit and its outlet pressure as constraints. The NH_3-based energy storage system in Aspen Plus is modeled in Python to quantify the uncertainty of wind velocity measurement error, cell temperature change, and NH_3 synthesis amount. In the process of RDO, polynomial chaos expansion (PCE) is used as surrogate model to identify the uncertainty propagation process of the system model. On this basis, MCS method is used to quantify the mean and standard deviation of the target. Finally, the robust optimization of the Power-to-NH_3 process of the system model is performed by using the NSGA-II algorithm.

The results of global sensitivity analysis on the design optimization show that the influence of temperature fluctuation of NH_3 reactor on the average NH_3 yield is 99.7%. Using the same sensitivity analysis for the most high-yield design, the effect of wind velocity measurement error and temperature change on NH_3 yield is 75.4% and 22.5%, respectively. The CoV of the optimized device is 1.46%. According to their summary, the research in the future will include analyzing the dynamic operation of the dynamic NH_3 production pathway and strengthening its levelized costs.

10.2.1.2 Solar Energy

The off-grid charging station is powered by distributed energy sources (including renewable, non-renewable, and a combination of both), which inevitably needs to meet load demand during its operation. The renewable energy is unstable, which hinders the energy supply of off-grid charging station. However, few studies have been conducted on off-grid charging stations [56–58], especially no studies on the robust model of the risk of off-grid charging stations.

Wang et al. [59] investigated the RDO problem of off-grid solar-powered charging station powered by photovoltaic systems (PVs). A novel approach is proposed which integrates robust optimization as a non-stochastic framework into mixed-integer linear programming (MILP) of deterministic models. The MILP model of the charging station is established by minimizing the annualized investment cost of the PV system, the diesel generator, and their operating cost, with the output power of the PV system and the diesel generator in the specified time period as the constraint. On the basis of the MILP model, a robust optimization model is established by considering the uncertainty of solar power generation and hydrogen production from water electrolyzer in the energy conversion process. Finally, a robust optimization method is used to search for the optimal solution.

The optimization results show that when handing the worst uncertainties, the capacity of the photovoltaic system and the diesel generator is increased by 12.78% and 33.33%, respectively. Although the total annualized cost is increased by 13.75%, the robust model of charging stations is obtained.

10.2.1.3 Power System

Sustainability and reliability are two indispensable aspects of PS design. Most PS design studies focus on economic and environmental sustainability [60–62]. In recent years, social standards related to reliability have also received increasing attention [63, 64]. Taking sustainability and reliability into consideration makes PS design more realistic. However, the design of sustainable and reliable power systems (SRPSs) is a conflicting multi-objective optimization problem that is also subject to many different types of uncertainties. Such as the inherent uncertainty of supply, demand, and energy distribution and the uncertainty affected by climate disasters [65].

Tsao et al. [66] proposed an interactive fuzzy multi-objective programming method by considering the uncertainty of power demand, intermittency of renewable energy, and production cost parameters, which combined fuzzy possibility-flexible programming, robust optimization, and Torabi and Hassini (TH) method [67]. The proposed method is called multi-objective mixed fuzzy possibilistic flexible robust programming (MOMFPFRP). The sustainability target includes the CO_2 emission cost of generator set, while the reliability target includes the average failure cost of network interruption. Then, a multi-objective MILP model is established with supply demand, assembly capacity, and system transmission power as constraints. The expected value of the objective function is quantified by the expected interval and expected value method of the triangular fuzzy number proposed by Jimenez [38], which has the advantage of not increasing the number of objective function and inequality constraints and has high computational efficiency. Then, feasibility and optimal robustness are formulated on the basis of the fuzzy sequencing method proposed by Yager [68]. Based on a set of given triangular fuzzy numbers, the multi-objective MILP model is rewritten into an auxiliary fuzzy robust optimization model (AFROM) by fuzzy possibility-flexible programming, then the AFROM model is transformed into a single objective model by TH aggregation function to solve the multi-objective conflict problem.

Finally, taking PS in southern Vietnam as an example, the optimization results show that the total installed capacity increased by about 54.42%, with an increase in the installed capacity of renewable energy (e.g., biomass from 23.35% to 39.76%) and a decrease in the installed capacity of traditional energy (e.g., coal and natural gas). Although the total cost of the power system is increased by about 4.2%, the conflict multi-objective optimization problem is solved, and the reliable and sustainable power supply of the system is achieved. According to their summary, the research can realize the intelligent planning problem of power system in the future by combining storage technology, visible control, etc.

10.2.1.4 Real-World Engineering Application

One of the innovative applications of renewable energy is to power electric vehicle charging stations [69]. An off-grid solar charging station requires energy provided by distributed energy sources. An off-grid solar charging station working system proposed in Wang's research [59] is shown in Fig. 10.8. The energy supply includes renewable and non-renewable energy sources. Renewable energy is provided by photovoltaic systems, and part of the energy generated by the photovoltaic system directly powers electric vehicles through fuel cells; while non-renewable energy is provided by diesel generators. When suffer from thunderstorms lead to insufficient solar power supply, the diesel generator supplies power to ensure the energy demand of electric vehicles. However, the unpredictability of renewable energy supply leads to probabilistic energy supply. To ensure a stable supply of energy, the investment cost of non-renewable energy power generation equipment needs to increase. Wang et al. [59] proposed a robust MILP model to relieve the negative impact of uncertainty.

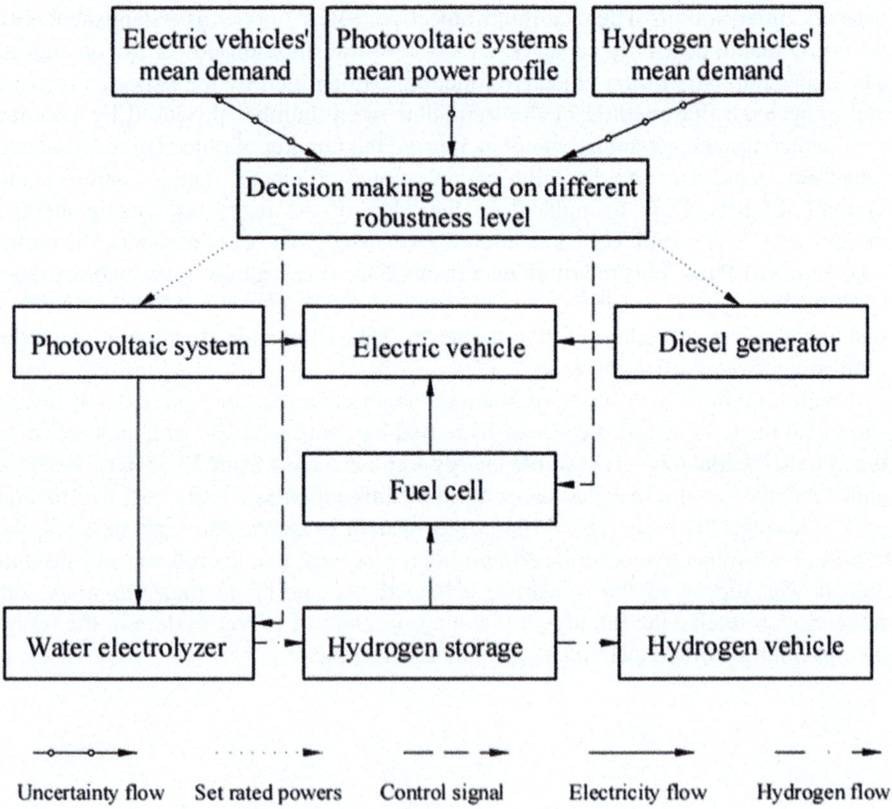

Fig. 10.8 Off-grid solar charging station working system proposed by Wang's research [59]

The objective function of the robust MILP model includes the annualized investment cost of the photovoltaic system, the diesel generator, and its operating cost. The constraints consider the daily average power generation distribution of photovoltaic systems and diesel power generation systems, hydrogen storage capacity, power demand, etc. For details, refer to [59].

The robust optimization model is solved using traditional robust optimization methods and the CPLEX solver in GAMS software. The optimization increases the total annualized cost from \$287,256 in deterministic mode to \$326,757 in robust mode, which increases 13.75%, but obtains a robust solution.

10.2.2 Logistics Scheduling

Logistics scheduling refers to the allocation, arrangement, and management of logistics resources to achieve the purpose of optimal logistics operation. However, in the actual logistics operation, there are many uncertain factors such as weather, traffic conditions, manpower, and materials, etc. These complex factors often lead to the delay of logistics transportation, cost increase, resource waste, and other problems. Therefore, the way of deal with these uncertain factors accurately and quickly becomes the key factor of logistics scheduling optimization. This section reviews the engineering application of RDO in logistics scheduling from three aspects: medical resource scheduling, emergency logistics scheduling, and logistics distribution scheduling.

10.2.2.1 Home Health Care Routing and Scheduling Problem

Considering the robustness of the optimization solution in the home health care route scheduling problem (HHCRSP) can provide a more rational decision when uncertainties interfere between operating costs and the requirement for timely service. Trip and Service Time (TST) is key factors in scheduling home health care (HHC) services. Factors such as time to diagnosis, medical skills, road conditions, weather conditions, and driving skills are among the uncertainties included in TST.

Only few studies on HHCRSP have included the above uncertainties [70–72]. Shi et al. [73] are the first study of HHCRSP with uncertainty including trip and service times (TST) from the perspective of robust optimization. The objective function is to minimize the total travel cost and the fixed cost of the caregiver, while the constraint function includes the service time window, resource scheduling, and allocation, etc. A deterministic model of mixed integer programming (MIP) is then constructed. The uncertainty set includes patient service time and business travel time. To achieve robust design, the worst case is selected in the uncertainty parameter set. The uncertainty variables are defined based on the budget uncertainty theory, while the arrival time of each caregiver is rewritten as a complex recursive function to rewrite MIP model as a robust optimization model quantitively. Finally, the robust model is solved by the Tabu Search (TS) method to avoid the case where the optimal solution search falls into local optimum in a strictly constrained problem.

Using the benchmark example proposed by Solomon et al. [74], a set of weighted total distance and total time has been minimized and the optimal solutions are obtained. The study provides guidance to HHC companies on reliable schedule development, which enhances industry competitiveness. Future research work is to combine big data with RDO to enhance the method. In addition, using the meta-heuristic optimization algorithm as the optimizer can improve the practicability of the proposed framework.

10.2.2.2 Emergency Logistics Planning and Scheduling

Currently, there is no reliable way to accurately predict the time, location, or magnitude of an earthquake. In the aftermath of an earthquake, relief activities such as tactical decisions (i.e., location of temporary facilities, mobilization level of relief supplies, and deployment of transport vehicles) and operational decisions (i.e., transport plans between temporary facilities and transport vehicles) are fraught with uncertainties. The way to deal with these uncertainties is a challenge in emergency logistics planning and scheduling [75–78].

Stochastic programming is a common method for the study of decision problems with uncertainties [77, 78]. However, the use of stochastic programming not only requires prior knowledge of the probability distribution of uncertain parameters, but also has the disadvantage of not being suitable for problems with a large number of uncertain variables.

Firstly, a deterministic MILP model of the post-disaster relief logistics system is established. The model aims to minimize the cost, which includes personnel mobilization, helicopter allocation, and transportation strategies. The constraints are composed of the maximum number of rescue personnel and key populations temporarily set up and the number of helicopters and dispatching constraints. Then, the robust optimization methods proposed Bertsimas and Sim [79] are introduced to deal with the uncertainties. By expressing the uncertain parameters of resource allocation and scheduling time as interval data, the deterministic MILP model is transformed into a stochastic model and the min-max method is used to ensure the feasibility of the solution. Readers interested in specific methods can read the references [79].

Finally, an example based on Sichuan earthquake disaster logistics is used to test the proposed model. The test results show that the model can be used to help decision-makers determine the level of mobilization of relief supplies, the initial helicopter deployment of relief supplies, and the transportation plan within the disaster area. However, the model proposed in the study mainly focuses on developing initial post-disaster relief logistics plans based on information collected immediately after a disaster. Because information pertaining to the demand level and transportation environment will be continuously updated and because relief logistics campaigns typically require multiple periods to complete [80, 81], the future work is to extend the model to multi-cycle settings, which will help provide decision-makers with the opportunity to dynamically adjust and optimize existing plans.

10.2.2.3 Electric Vehicle Logistics and Distribution Dispatch

The logistics of delivery using electric vehicles (EVs) is fraught with uncertainties. On the one hand, with the increase in the number of private vehicles, the problem of road traffic congestion has been exacerbated, which resulting in the uncertain of logistics delivery time. On the other hand, due to the limited battery life, EVs must be charged at a designated station halfway to their destination. However, the charging time is affected by customer demand scheduling, which leads to the uncertainty of the battery life of EVs. The researches on the optimization of EV logistics distribution are still in its infancy [82–87]. However, most researches on path optimization mainly focus on the deterministic conditions. Whereas, researches involving analysis of uncertain conditions mainly rely on assumptions of prior knowledge [88–90].

Ma et al. [91] studied the robust optimization of multi-distribution center distribution routes for EVs. Based on the robust discrete optimization theory of Bertsimas, taking the minimum transportation time as the goal, the uncertainty of time in the distribution section is expressed by the interval, which reduces the need for uncertain basic data. Meanwhile, a robust optimization model of EV distribution path nonlinear planning with adjustable robustness is established with the constraints of EV running path and battery capacity and consumption. Since GA can avoid the trouble of nonlinear and multimodal constraints in solving nonlinear programming problems, a three-segment hybrid coding method is designed in this study. A hybrid coding method combining distribution center, customer demand point, and distribution path is used to encode. Then the corresponding decoding method and genetic operation are designed to avoid the infeasible solution in the process of population evolution.

Finally, taking part of the road network in Xifeng District of Qingyang City as an example, a distribution scheme with high applicability and economy is obtained. Meanwhile, EV distribution schemes under different transportation conditions can be obtained by setting different robust control risk coefficients to provide decision support for the selection of EV distribution paths in multiple distribution centers.

10.2.2.4 Real-World Engineering Application

HHC can provide convenience for home treatment, elderly care, etc., alleviating the resource shortage problem caused by limited hospital beds. The most critical goal of HHC companies is to meet the needs of patients in a timely manner. However, there is a tension between the company's operating costs and the problem of delayed patient service. Delayed services lead to lower patient satisfaction and even missed

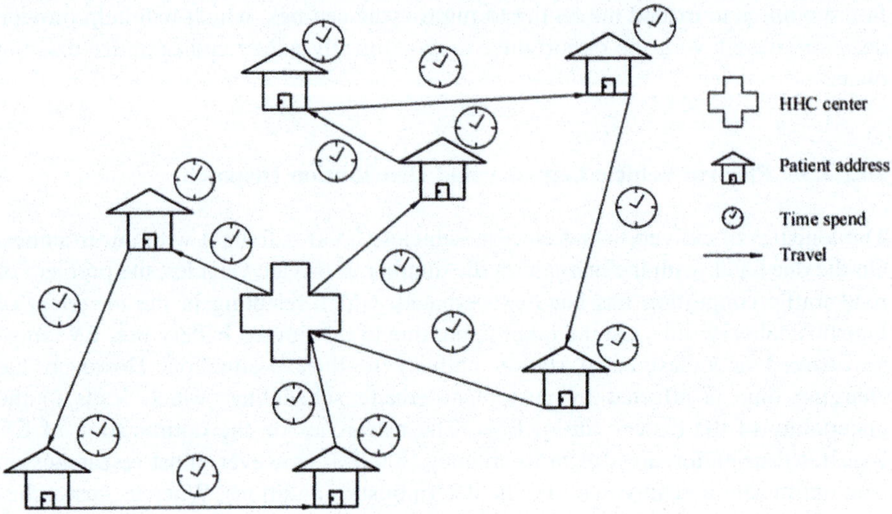

Fig. 10.9 Description of HHC route scheduling

opportunities for optimal treatment. TST is a key factor in scheduling HHC services, Fig. 10.9 describes HHC path scheduling. Fully considering the uncertainty of scheduling time, service time, etc., when formulating plans can improve service quality and reduce losses. Shi et al. [73] studied a robust optimization problem for HHC routing scheduling problem considering uncertain TST.

The objective function contains two parts. The first part is the total cost of traveling and hiring personnel. The second part is the penalty term of the maximum possible arrival time of the caregiver to the patient's home, which is derived based on budget theory and recursive functions. The constraint function includes uncertainty time windows of service time and business travel time.

The RO model is solved using the Taguchi design method combined with the tabu search algorithm. Robust optimization results compared with deterministic design optimization, the robust solution has advantages over the deterministic model in terms of delayed service percentage and average delay time, which can provide a valuable framework for HHC companies to develop robust schedules when scheduling caregivers.

10.2.3 Closed-Loop Supply Chain

Closed-loop supply chain (CLSC) network is an integrated system that includes both forward supply chain and reverse supply chain. During its design optimization process, multiple objectives sometimes contradict each other [92]. Meanwhile, CLSC's strategic and tactical decisions are heavily influenced by uncertainties such as demand, recycling rates, and recycling quality, which can create volatility in supply chain models. From the perspective of product service life, this section reviews the CLSC network robust design optimization engineering cases from three aspects: perishable goods, disposable goods, and durable goods.

10.2.3.1 Closed-Loop Supply Chain Network for Perishable Goods

Perishable goods, such as those in the dairy and pharmaceutical industries, are characterized by a limited life span and limited storage time. Therefore, the CLSC network is highly sensitive to the uncertainty of demand, which can easily cause waste and environmental pollution if improperly allocated. A specialized model for studying the cost and environmental objectives of CLSC for perishable goods under uncertainty can effectively balance the relationship between cost and pollution. Some scholars have conducted studies on the supply chain network design of perishable goods [93–95], but these studies have not considered the uncertainty of return rate and return quality.

In view of the uncertainties of demand, return rate, and quality of returned goods, Yavari et al. [96] proposed a multi-objective robust MILP optimization model for green closed-loop supply chain network of perishable goods. The double optimization objectives are to minimize the total cost of the supply chain and minimize the environmental pollution, while the constraints include 25 constraints on the multi-period, multi-product supply chain composed of suppliers, manufacturers, warehouses, retailers, and collection centers. To solve the multi-objective problem, the compromise programming is used to transform the problem into a single-objective optimization model. The robust optimization method of Bertsimas and Sim [79] and Ben-Tal et al. [97] is used to deal with these uncertainties. Due to the NP-hard nature of the problem [98], traditional algorithms can lead to inefficient consequences. They defined two linear programming models and a small-sized integer linear programming model (the distribution model). Based on solving the model, an efficient heuristic method named Yavari and Geraeli (YAG) method is developed, which proved to be effective for large complex problems in subsequent experiments.

Finally, the supply chain of an Iranian dairy company is taken as a case study. The experimental results show that the proposed method is helpful for managers to make comprehensive decisions on the forward and reverse flow of the closed-loop supply chain of perishable products. The average difference between the YAG method and the optimal solution in a reasonable time is less than 1.65%. In addition, the YAG method found the optimal solution in more than 34% instances.

10.2.3.2 Closed-Loop Supply Chain Network for Disposable Home Appliances

In the CLSC network of disposable products, erratic fluctuations in the number of returns can affect the quantity of products offered by suppliers. Effective handling of demand and cost uncertainties can facilitate the development of CLSC networks for disposable products.

Gholizadeh et al. [99] first considered the concept of grading for multi-product closed-loop supply chain by taking the closed-loop supply chain design optimization of disposable household appliances as an example. The forward chain of the model consists of four levels of suppliers, manufacturers, distributors, and customers, while the reverse chain consists of three levels of collection, recycling, and processing. In the uncertainty model, the objective function is to maximize the value of reverse products and direct supply products in the closed-loop chain network, including sales revenue of direct sales products, recovery revenue of reverse products, operating costs, and transportation costs at all points of the organization. The constraint set consists of 28 constraints, including customer demand, return quantity, raw material recovery, and procurement, etc. Uncertain parameters include demand, product return rates, shipping costs, and operating costs for each period. To reduce the difficulty of calculation, the nonlinear constraints are linearized by substituting some of the variables in the multivariable constraints with their determined boundary values. In the proposed robust optimization model, the objective function consists of three parts. The first two parts are the mean and variance of the total cost of the chain, which are used to represent the stability of the model. The third part is used to measure the equilibrium of the objective function. GA is used to solve the model. Different from general GA, the parameters of GA are adjusted with the help of orthogonal experiment and Taguchi design.

Finally, a disposable necessities company located in Amer, Iran, is taken as an example to verify the effectiveness of the method. The results show that this method can provide a solution very close to the exact solution and has good convergence, which provides the company with the optimal operation scheme of CLSC network for disposable electrical products. In addition, they mentioned that there is due to the uncertainty nature utilizing of fuzzy or stochastic programming with the approach combination and probable uncertainty.

10.2.3.3 A Closed-Loop Supply Chain Network for Durable Goods

Most of the design optimization of durable product supply chain focuses on forward supply chain. There are few researches on CLSC network.

Atabaki et al. [100] complement the design of reverse logistics engineering for durable goods, which enables the supply chain of durable products to form a closed loop. Various recycling facilities are considered in the CLSC structure of durable products. Economic costs, CO_2 emissions, and energy consumption are considered as multiple objectives, while uncertainties included stochastic uncertainties in the operating costs and demand of forward flow facilities and cognitive uncertainties in facility construction costs, reverse facility operating costs, number of returns, and recovery rates. The corresponding robust optimization model is built to make robust decisions on supplier selection, location allocation, transportation mode, assembly technology, and recycling level. The robust optimization model introduces possibility programming and scenario-based stochastic programming to deal with mixed uncertainties in the model. Scenario-based stochastic programming is used to deal with stochastic uncertainties, which is represented by probability distribution. Possibility programming is used to deal with cognitive uncertainties, which deals with the problem of cognitive uncertainty whose parameters in the data can be represented by fuzzy triangular numbers.

Finally, a numerical case of CLSC network optimization for durable products made of nickel, steel, and copper is tested. The optimization results show that the obtained optimal solution is not affected by any realization of the uncertainty and can provide a robust decision for the decision maker.

10.2.3.4 Real-World Engineering Application

Supply chain design is one of the important strategic decisions that affects a company's competitive advantage and economic growth. Figure 10.10 is a schematic diagram of the closed-loop supply chain of disposable electrical appliances. The life cycle of disposable products includes two parts: forward production and reverse recycling. The CLSC is an integrated system that integrates the forward and reverse supply chains. CLSC underlines resource recycling and extending product life cycles, which can bring positive benefits to the environment and corporate economy. Therefore, many companies turn to the design of closed-loop supply chain networks [99]. However, supply chain network design is a complex problem with multiple objectives that sometimes conflict with each other [92]. Gholizadeh et al. [99] studied a disposable electrical recycling network of a multi-layer closed-loop supply chain. By using the linearization formula [101], they proposed a simplified version of the robust optimization model, readers interested in relevant details refer to [99].

The objective function is to minimize the mean and variance of CLSR total profit and the balance of the supply chain model. The constraint function contains the parameter interaction relationships between various links in the supply chain.

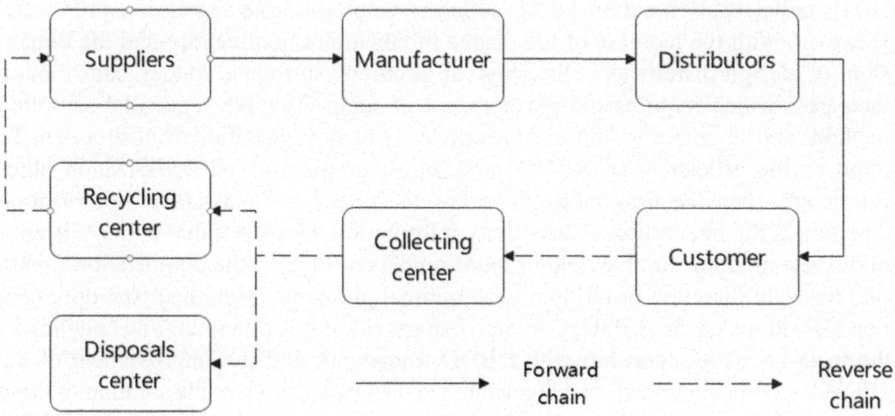

Fig. 10.10 Disposable appliances closed-loop supply chain

The optimization program is implemented in LINGO software, and Taguchi design and genetic algorithm are used to obtain robust solutions. Finally, operations research techniques are used to achieve optimal decision-making. Taking the supply chain case of a disposable household products company located in Amir, Iran, as an example, the proposed method efficiently provides a solution that is very close to the exact solution.

10.2.4 Summary

As in Sect. 10.1, we utilize a Table 10.3 to summarize the information in this section, including the application field, specific application, sources of uncertainty, optimization method, optimization results, and references.

10.3 Research Outlook of Design Optimal Under Uncertainty

10.3.1 Challenges and Prospects of RBDO

The main challenges of RBDO research in the engineering field are embodied in the expansion to the high-dimensional design space, the trouble of highly nonlinear constraints, the effective integration of sensitivity information, the processing of continuous-discrete random variable mixing, and the consideration of multi-objective optimization problems.

High-dimensional design space and high nonlinear constraints will rapidly increase the amount of computation required for reliability analysis. The surrogate model can effectively alleviate these troubles, which widely used include Kriging [103], radial basis function [104], support vector machine regression [105], etc. However, with the increase of the degree of constraint nonlinearity and the dimension of design parameters, the cost of accurate surrogate model construction increases, which requires the development of more effective sequential sampling methods and strategies to improve the efficiency of surrogate model construction. To improve the efficiency of RBDO procedures, gradient-based optimization algorithms are often the first choice. The key to correct and reasonable optimization direction is the integration of sensitivity information. However, due to the existence of discrete random variables and/or multi-modal constraints, the optimization cannot find the right direction or fall into local optimal problems. Meta-heuristic optimization algorithms (such as PSO, GA, etc.) can get rid of this dilemma. The coupling of the meta-heuristic algorithm with RBDO framework and the improvement of the efficiency of the meta-heuristic algorithm are the key to improve the solution of these problems. The phenomenon of multiple failure modes also makes the RBDO problem complicated. They can be converted into multi-objective optimization

Table 10.3 The summary of RDO engineering applications

RBDO engineering applications					
Application field	Specific application	Sources of uncertainty	Method	Result/Contribution (↑ is increment, ↓ is decrement)	Reference
Energy management	NH_3 synthesis plant	1. Wind energy 2. NH_3 reactor temperature	PCE + MCS + NSGA-II	The CoV of plant output is 1.46%.	Verleysen et al. [55]
	Photovoltaic system	1. Power distribution of photovoltaic system 2. Demand for hydrogen and electricity	Robust optimization + MILP	Total annualized cost ↑ 13.75% but charging stations are more robust.	Wang et al. [59]
	Hybrid PS	1. CO_2 emission 2. Renewable energy sustainability	MOMFPFRP	Total installed capacity ↑ 54.42%, total cost ↑ 4.2%, but the conflict multi-objective optimization problem is solved.	Tsao et al. [66]
Logistics scheduling	HHC	1. Service time 2. Travel time	Robust design optimization + TS	It is the first to study the HHCRSP problem with uncertain travel and service time from a robust optimization, and the study provides guidance to HHC companies on reliable schedule development, which enhances industry competitiveness.	Shi et al. [73]
	Post-earthquake relief activities	1. Demand level 2. Transit time	Bertsimas and Sim's method [79]	Using interval theory to quantify the uncertainty. The test results show that the model can be used to help decision-makers determine the level of mobilization of relief supplies, the initial helicopter deployment of relief supplies, and the transportation plan within the disaster area.	Liu et al. [102]

(continued)

Table 10.3 (continued)

RBDO engineering applications					
Application field	Specific application	Sources of uncertainty	Method	Result/Contribution ($\uparrow$ is increment, $\downarrow$ is decrement)	Reference
	EV path planning	1. Delivery time 2. Delivery distance	Bertsimas's robust discrete optimization theory + modify GA	EV distribution schemes under different transportation conditions can be obtained by setting different robust control risk coefficients to provide decision support for the selection of EV distribution paths in multiple distribution centers.	Ma et al. [91]
CLSC	Perishable goods	1. Demand for goods 2. Return rate 3. Quality of return	Bertsimas and Sim's method [79] + Ben-Tal's method [97] + YAG	The average difference with the optimal solution is less than 1.65%. The YAG method found an optimal solution more than 34% of the time. Improve the efficiency of large-scale problem solving.	Yavari et al. [96]
	Disposable household appliances	1. Product demand 2. Transportation costs 3. Operating cost	GA + orthogonal experiment + Taguchi design	First proposed the hierarchical concept of multi-product CLSC. The results show that this method can provide a solution very close to the exact solution and has good convergence, which provides the company with the optimal operation scheme of CLSC network for disposable electrical products.	Gholizadeh et al. [99]
	Durable product	1. Random uncertainty (Demand level; Operating cost) 2. Cognitive uncertainty (Plant establishment costs; Reverse plant operating costs; The number of returned goods; Recovery rate)	Possibility planning + scenario-based stochastic planning	The reverse logistics engineering of durable goods is designed and CLSC of durable goods is formed. The optimization results show that the obtained optimal solution is not affected by any realization of the uncertainty and can provide a robust decision for the decision maker.	Atabaki et al. [100]

problems, then need to develop appropriate decision criteria to sort the Pareto solution set.

In terms of optimization strategy selection, the double-loop strategy is usually used as the optimization method for RBDO in new engineering applications because of its advantages of good generality and easy implementation in any general optimization software. Due to the nested relationship between reliability analysis loop and optimization loop, the calculation of reliability analysis is repeated. Using single-loop strategy (such as SLSV, etc.) or decoupling strategy (such as SORA, etc.) can effectively avoid unnecessary calculation, but the key is to develop advanced methods for reasonable equivalent approximation of constraints.

In the future, efforts regarding RBDO should focus on providing and executing efficient and reliable numerical programs using sound/improved theoretical algorithms and appropriate tools. The development of new simulation schemes for reliability and sensitivity evaluation provides a further choice for the development of new RBDO methods for engineering practice, and the advanced meta-heuristic algorithm is a further expansion of RBDO application field. Different methods provide different advantages and difficulties for the optimization process, and selecting the right advanced theory and method for the characteristics of the problem is the key to improving the efficiency of RBDO. Overcoming these challenges can lead to significant progress in the field and ultimately help in complex decision-making processes in real-world situations.

10.3.2 Challenges and Prospects of RDO

RDO models usually contain multiple objective functions. With the gradual expansion of the scale of design, the problem of design optimization is also becoming more and more complicated. In the future, RDO will consider constraints more comprehensively. Different noise types, incomplete and missing data, multimodal problems, and even the integration of multidisciplinary constraints make RDO implementation more difficult. This requires the design team to have efficient computational and simulation tools that can address the design optimization needs of large-scale scenarios.

The meta-heuristic optimization algorithm is helpful to solve the problem of large amount of computation and to escape the local optimal trap of multi-modal problems. Advanced uncertainty quantization technology can quantify various uncertainty variables. In the future, RDO can be combined with cloud computing and big data to mine the rules and patterns in the design process, and visualization technology and collaborative design platform can be used to realize the simulation, visualization, and collaboration of robust design optimization, promote the information exchange and integration among multiple disciplines, so as to build a complete and practical optimization model and improve the parallel processing and computing capabilities of RDO.

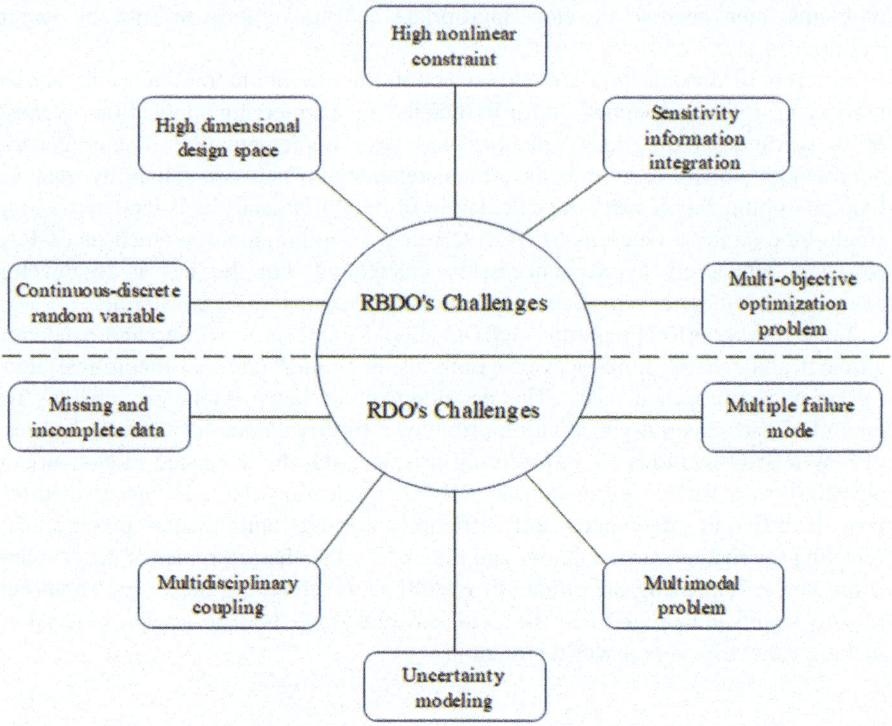

Fig. 10.11 Challenges of RBDO and RDO

As the usage of RDO becomes more widespread, the future research and exploration of RDO can be developed in the aspects of uncertainty modeling, non-convex optimization methods, and robustness indicators. By introducing uncertainty modeling, the uncertain factors are predicted and controlled, which improves the robustness of the model and the adaptability of the model to unknown situations. Exploring non-convex optimization methods enables RDO to adapt to more complex models and problems. Finally, the development of more generalized robustness indicators helps ensure that RDO is not limited in a particular application domain or problem (Fig. 10.11).

References

1. Cid Bengoa, C. (2021). *Probabilistic fail-safe size optimization of aerospace structures under several sources of uncertainty.*
2. Hao, P., et al. (2021). Efficient reliability-based design optimization of composite structures via isogeometric analysis. *Reliability Engineering System Safety, 209*, 107465.

3. Zhu, S.-P., et al. (2013). Bayesian framework for probabilistic low cycle fatigue life prediction and uncertainty modeling of aircraft turbine disk alloys. *Probabilistic Engineering Mechanics, 34*, 114–122.
4. Hu, D., et al. (2019). Effect of inclusions on low cycle fatigue lifetime in a powder metallurgy nickel-based superalloy FGH96. *International Journal of Fatigue, 118*, 237–248.
5. Hu, D., et al. (2016). Creep-fatigue behavior of turbine disc of superalloy GH720Li at 650 C and probabilistic creep-fatigue modeling. *Materials Science Engineering A, 670*, 17–25.
6. Zhu, S.-P., et al. (2017). A combined high and low cycle fatigue model for life prediction of turbine blades. *Materials, 10*(7), 698.
7. Liu, X., et al. (2021). Reliability-based design optimization approach for compressor disc with multiple correlated failure modes. *Aerospace Science and Technology, 110*, 106493.
8. Aas, K., et al. (2009). Pair-copula constructions of multiple dependence. *Insurance: Mathematics Economics, 44*(2), 182–198.
9. Nowak, A. S., & Collins, K. R. (2012). *Reliability of structures*. CRC Press.
10. Cid Bengoa, C., et al. (2020). *Multi-model optimization approach of aircraft structures under uncertainty using horsetail matching and RBDO methods*. In AIAA Scitech 2020 Forum. Orlando, FL.
11. Chengwei, F., et al. (2021). Whole-process design and experimental validation of landing gear lower drag stay with global/local linked driven optimization strategy. *Chinese Journal of Aeronautics, 34*(2), 318–328.
12. Meng, D., et al. (2019). Structural reliability analysis and uncertainties-based collaborative design and optimization of turbine blades using surrogate model. *Fatigue Fracture of Engineering Materials Structures, 42*(6), 1219–1227.
13. Baldomir, A., et al. (2012). *Size optimization of shell structures considering several incomplete configurations*. In 53rd AIAA/ASME/ASCE/AHS/ASC structures, structural dynamics and materials conference 20th AIAA/ASME/AHS adaptive structures conference 14th AIAA.
14. Cid, C., Baldomir, A., & Hernandez, S. (2016). *Reliability based design optimization of structures considering several incomplete configurations*. In 17th AIAA/ISSMO multidisciplinary analysis and optimization conference.
15. Cid Bengoa, C., et al. (2018). Multi-model reliability-based design optimization of structures considering the intact configuration and several partial collapses. *Structural Multidisciplinary Optimization, 57*, 977–994.
16. Kim, B. C., Weaver, P. M., & Potter, K. (2014). Manufacturing characteristics of the continuous tow shearing method for manufacturing of variable angle tow composites. *Composites Part A: Applied Science and Manufacturing, 61*, 141–151.
17. Scheu, M. N., et al. (2019). A systematic failure mode effects and criticality analysis for offshore wind turbine systems towards integrated condition based maintenance strategies. *Ocean Engineering, 176*, 118–133.
18. Martinez-Luengo, M., & Shafiee, M. (2019). Guidelines and cost-benefit analysis of the structural health monitoring implementation in offshore wind turbine support structures. *Energies, 12*(6), 1176.
19. Yeter, B., Garbatov, Y., & Soares, C. G. (2015). Fatigue reliability of an offshore wind turbine supporting structure accounting for inspection and repair. In *Analysis and design of marine structures V* (pp. 751–762). CRC Press.
20. Ćorak, M., et al. (2022). Uncertainties of wave data collected from different sources in the Adriatic Sea and consequences on the design of marine structures. *Ocean Engineering, 266*, 112738.
21. Li, N., et al. (2021). Optimal design and strength reliability analysis of pressure shell with grid sandwich structure. *Ocean Engineering, 223*, 108657.
22. Meng, D., et al. (2020). *Reliability-based optimisation for offshore structures using saddlepoint approximation*. In Proceedings of the Institution of Civil Engineers-Maritime Engineering. Thomas Telford.

23. Hu, W., et al. (2016). Integrating variable wind load, aerodynamic, and structural analyses towards accurate fatigue life prediction in composite wind turbine blades. *Structural Multidisciplinary Optimization, 53*, 375–394.
24. Jiang, Z., et al. (2017). Structural reliability analysis of wind turbines: A review. *Energies, 10*(12), 2099.
25. Hu, W., Choi, K., & Cho, H. (2016). Reliability-based design optimization of wind turbine blades for fatigue life under dynamic wind load uncertainty. *Structural Multidisciplinary Optimization, 54*, 953–970.
26. Larsenon, K. (2020). *Corrosion risks and mitigation strategies for offshore wind turbine foundations*.
27. Meng, D., et al. (2023). A novel hybrid adaptive kriging and water cycle algorithm for reliability-based design and optimization strategy: Application in offshore wind turbine monopile. *Computer Methods in Applied Mechanics and Engineering, 412*, 116083.
28. Forde, M. C. (1998). Bridge research in Europe. *Construction Building Materials, 12*(2–3), 85–91.
29. Nguyen, V.-S., et al. (2013). Reliability-based optimisation design of post-tensioned concrete box girder bridges considering pitting corrosion attack. *Structure Infrastructure Engineering, 9*(1), 78–96.
30. Stewart, M. G. (2009). Mechanical behaviour of pitting corrosion of flexural and shear reinforcement and its effect on structural reliability of corroding RC beams. *Structural Safety, 31*(1), 19–30.
31. Bentur, A., Berke, N., & Diamond, S. (1997). *Steel corrosion in concrete: fundamentals and civil engineering practice*. CRC Press.
32. Wu, S., & Law, S. (2011). Dynamic analysis of bridge with non-Gaussian uncertainties under a moving vehicle. *Probabilistic Engineering Mechanics, 26*(2), 281–293.
33. Mao, J., et al. (2016). Random dynamic analysis of a train-bridge coupled system involving random system parameters based on probability density evolution method. *Probabilistic Engineering Mechanics, 46*, 48–61.
34. Ni, P., et al. (2019). Using polynomial chaos expansion for uncertainty and sensitivity analysis of bridge structures. *Mechanical Systems Signal Processing, 119*, 293–311.
35. Ni, P., et al. (2021). Reliability based design optimization of bridges considering bridge-vehicle interaction by kriging surrogate model. *Engineering Structures, 246*, 112989.
36. Kusano, I., et al. (2020). Reliability based design optimization for bridge girder shape and plate thicknesses of long-span suspension bridges considering aeroelastic constraint. *Journal of Wind Engineering and Industrial Aerodynamics, 202*, 104176.
37. Montoya, M. C., et al. (2018). CFD-based aeroelastic characterization of streamlined bridge deck cross-sections subject to shape modifications using surrogate models. *Journal of Wind Engineering Industrial Aerodynamics, 177*, 405–428.
38. Diana, G., et al. (2013). Wind tunnel tests and numerical approach for long span bridges: The Messina bridge. *Journal of Wind Engineering and Industrial Aerodynamics, 122*, 38–49.
39. Ge, Y., & Xiang, H. (2008). Recent development of bridge aerodynamics in China. *Journal of Wind Engineering and Industrial Aerodynamics, 96*(6–7), 736–768.
40. Duan, L., et al. (2019). Multi-objective reliability-based design optimization for the VRB-VCS FLB under front-impact collision. *Structural Multidisciplinary Optimization, 59*, 1835–1851.
41. Zhang, H., & Zhang, X. (2016). Crashworthiness performance of conical tubes with nonlinear thickness distribution. *Thin-Walled Structures, 99*, 35–44.
42. Lu, R., et al. (2017). Simulation of springback variation in the U-bending of tailor rolled blanks. *Journal of the Brazilian Society of Mechanical Sciences Engineering, 39*, 4633–4647.
43. Duan, L., et al. (2017). Multi-objective system reliability-based optimization method for design of a fully parametric concept car body. *Engineering Optimization, 49*(7), 1247–1263.
44. Liu, Q., et al. (2013). Lightweight design of carbon twill weave fabric composite body structure for electric vehicle. *Composite Structures, 97*, 231–238.

45. Belingardi, G., Beyene, A. T., & Koricho, E. G. (2013). Geometrical optimization of bumper beam profile made of pultruded composite by numerical simulation. *Composite Structures, 102*, 217–225.
46. Hesse, S., Lukaszewicz, D.-J., & Duddeck, F. (2015). A method to reduce design complexity of automotive composite structures with respect to crashworthiness. *Composite Structures, 129*, 236–249.
47. Kalantari, M., Dong, C., & Davies, I. J. (2016). Multi-objective robust optimisation of unidirectional carbon/glass fibre reinforced hybrid composites under flexural loading. *Composite Structures, 138*, 264–275.
48. Liu, Z., et al. (2018). Reliability-based design optimization of composite battery box based on modified particle swarm optimization algorithm. *Composite Structures, 204*, 239–255.
49. Charalambakis, N. (2010). Homogenization techniques and micromechanics. A survey and perspectives. *Applied Mechanics Reviews, 63*(3).
50. Jin, R., Chen, W., & Sudjianto, A.. (2003). *An efficient algorithm for constructing optimal design of computer experiments*. In International design engineering technical conferences and computers and information in engineering conference.
51. Xu, X., et al. (2022). Multi-objective reliability-based design optimization for the reducer housing of electric vehicles. *Engineering Optimization, 54*(8), 1324–1340.
52. Li, X., et al. (2016). Effect of strain rate on the mechanical properties of carbon/epoxy composites under quasi-static and dynamic loadings. *Polymer Testing, 52*, 254–264.
53. Wu, Y., et al. (2017). Dynamic crash responses of bio-inspired aluminum honeycomb sandwich structures with CFRP panels. *Composites Part B: Engineering, 121*, 122–133.
54. Cid Bengoa, C., et al. (2020). *Multi-model optimization approach of aircraft structures under uncertainty using Horsetail Matching and RBDO methods*. In AIAA Scitech 2020 Forum.
55. Verleysen, K., et al. (2020). How can power-to-ammonia be robust? Optimization of an ammonia synthesis plant powered by a wind turbine considering operational uncertainties. *Fuel, 266*, 117049.
56. Ugirumurera, J., & Haas, Z. J. (2017). Optimal capacity sizing for completely green charging systems for electric vehicles. *IEEE Transactions on Transportation Electrification, 3*(3), 565–577.
57. Xie, R., et al. (2018). Planning fully renewable powered charging stations on highways: A data-driven robust optimization approach. *IEEE Transactions on Transportation Electrification, 4*(3), 817–830.
58. Mehrjerdi, H. (2019). Off-grid solar powered charging station for electric and hydrogen vehicles including fuel cell and hydrogen storage. *International Journal of Hydrogen Energy, 44*(23), 11574–11583.
59. Wang, Y., et al. (2020). Robust design of off-grid solar-powered charging station for hydrogen and electric vehicles via robust optimization approach. *International Journal of Hydrogen Energy, 45*(38), 18995–19006.
60. Sharafi, M., & ElMekkawy, T. Y. (2015). Stochastic optimization of hybrid renewable energy systems using sampling average method. *Renewable Sustainable Energy Reviews, 52*, 1668–1679.
61. Li, S., Coit, D. W., & Felder, F. (2016). Stochastic optimization for electric power generation expansion planning with discrete climate change scenarios. *Electric Power Systems Research, 140*, 401–412.
62. Muela, E., Schweickardt, G., & Garces, F. (2007). Fuzzy possibilistic model for medium-term power generation planning with environmental criteria. *Energy Policy, 35*(11), 5643–5655.
63. Santoyo-Castelazo, E., & Azapagic, A. (2014). Sustainability assessment of energy systems: integrating environmental, economic and social aspects. *Journal of Cleaner Production, 80*, 119–138.
64. Iddrisu, I., & Bhattacharyya, S. C. (2015). Sustainable Energy Development Index: A multidimensional indicator for measuring sustainable energy development. *Renewable Sustainable Energy Reviews, 50*, 513–530.

65. You, A. D., et al. (2011). A study of electrical security risk assessment system based on electricity regulation. *Energy Policy, 39*(4), 2062–2074.
66. Tsao, Y.-C., & Thanh, V.-V. (2020). A multi-objective fuzzy robust optimization approach for designing sustainable and reliable power systems under uncertainty. *Applied Soft Computing, 92*, 106317.
67. Torabi, S. A., & Hassini, E. (2008). An interactive possibilistic programming approach for multiple objective supply chain master planning. *Fuzzy Sets Systems, 159*(2), 193–214.
68. Yager, R. R. (1981). A procedure for ordering fuzzy subsets of the unit interval. *Information Sciences, 24*(2), 143–161.
69. Fathabadi, H. (2017). Novel wind powered electric vehicle charging station with vehicle-to-grid (V2G) connection capability. *Energy Conversion and Management, 2020*, 229–239.
70. Yuan, B., Liu, R., & Jiang, Z. (2015). A branch-and-price algorithm for the home health care scheduling and routing problem with stochastic service times and skill requirements. *International Journal of Production Research, 53*(24), 7450–7464.
71. Liu, R., Yuan, B., & Jiang, Z. (2019). A branch-and-price algorithm for the home-caregiver scheduling and routing problem with stochastic travel and service times. *Flexible Services Manufacturing Journal, 31*, 989–1011.
72. Lanzarone, E., & Matta, A. (2014). Robust nurse-to-patient assignment in home care services to minimize overtimes under continuity of care. *Operations Research for Health Care, 3*(2), 48–58.
73. Shi, Y., Boudouh, T., & Grunder, O. (2019). A robust optimization for a home health care routing and scheduling problem with consideration of uncertain travel and service times. *Transportation Research Part E: Logistics Transportation Review, 128*, 52–95.
74. Solomon, M. M. (1987). Algorithms for the vehicle routing and scheduling problems with time window constraints. *Operations Research, 35*(2), 254–265.
75. Caunhye, A. M., Nie, X., & Pokharel, S. (2012). Optimization models in emergency logistics: A literature review. *Socio-Economic Planning Sciences, 46*(1), 4–13.
76. Özdamar, L., & Ertem, M. A. (2015). Models, solutions and enabling technologies in humanitarian logistics. *European Journal of Operational Research, 244*(1), 55–65.
77. Liberatore, F., et al. (2013). Uncertainty in humanitarian logistics for disaster management. A review. In *Decision aid models for disaster management emergencies* (pp. 45–74). Springer.
78. Hoyos, M. C., Morales, R. S., & Akhavan-Tabatabaei, R. (2015). OR models with stochastic components in disaster operations management: A literature survey. *Computers Industrial Engineering, 82*, 183–197.
79. Bertsimas, D., & Sim, M. (2004). The price of robustness. *Operations Research, 52*(1), 35–53.
80. Özdamar, L., Ekinci, E., & Küçükyazici, B. (2004). Emergency logistics planning in natural disasters. *Annals of Operations Research, 129*, 217–245.
81. Bozorgi-Amiri, A., & Khorsi, M. (2016). A dynamic multi-objective location–Routing model for relief logistic planning under uncertainty on demand, travel time, and cost parameters. *The International Journal of Advanced Manufacturing Technology, 85*, 1633–1648.
82. Artmeier, A., et al. (2010). *The optimal routing problem in the context of battery-powered electric vehicles*. In CPAIOR Workshop on Constraint Reasoning and Optimization for Computational Sustainability (CROCS).
83. Worley, O., Klabjan, D., & Sweda, T. M.. (2012). *Simultaneous vehicle routing and charging station siting for commercial electric vehicles*. In 2012 IEEE international electric vehicle conference. IEEE.
84. Yi, T., et al. (2020). Joint optimization of charging station and energy storage economic capacity based on the effect of alternative energy storage of electric vehicle. *Energy, 208*, 118357.
85. Schneider, M., Stenger, A., & Goeke, D. (2014). The electric vehicle-routing problem with time windows and recharging stations. *Transportation Science, 48*(4), 500–520.
86. Afroditi, A., et al. (2014). Electric vehicle routing problem with industry constraints: trends and insights for future research. *Transportation Research Procedia, 3*, 452–459.

87. Strehler, M., Merting, S., & Schwan, C. (2017). Energy-efficient shortest routes for electric and hybrid vehicles. *Transportation Research Part B: Methodological, 103*, 111–135.
88. Caplice, C., & Mahmassani, H. S. (1992). Aspects of commuting behavior: Preferred arrival time, use of information and switching propensity. *Transportation Research Part A: Policy Practice, 26*(5), 409–418.
89. Wang, Y., et al. (2012). Location optimization of multiple distribution centers under fuzzy environment. *Journal of Zhejiang University Science A, 13*(10), 782.
90. Wang, Y., et al. (2015). Vehicle routing problem based on a fuzzy customer clustering approach for logistics network optimization. *Journal of Intelligent Fuzzy Systems, 29*(4), 1427–1442.
91. Ma, C., et al. (2018). Distribution path robust optimization of electric vehicle with multiple distribution centers. *PLoS One, 13*(3), e0193789.
92. Srivastava, S. K. (2008). Network design for reverse logistics. *Omega, 36*(4), 535–548.
93. Hasani, A., Zegordi, S. H., & Nikbakhsh, E. (2012). Robust closed-loop supply chain network design for perishable goods in agile manufacturing under uncertainty. *International Journal of Production Research, 50*(16), 4649–4669.
94. Govindan, K., et al. (2014). Two-echelon multiple-vehicle location–routing problem with time windows for optimization of sustainable supply chain network of perishable food. *International Journal of Production Economics, 152*, 9–28.
95. Keyvanshokooh, E., Ryan, S. M., & Kabir, E. (2016). Hybrid robust and stochastic optimization for closed-loop supply chain network design using accelerated Benders decomposition. *European Journal of Operational Research, 249*(1), 76–92.
96. Yavari, M., & Geraeli, M. (2019). Heuristic method for robust optimization model for green closed-loop supply chain network design of perishable goods. *Journal of Cleaner Production, 226*, 282–305.
97. Ben-Tal, A., El Ghaoui, L., & Nemirovski, A. (2009). *Robust optimization* (Vol. Vol. 28). Princeton University Press.
98. Ibaraki, T., & Katoh, N. (1988). *Resource allocation problems: Algorithmic approaches*. MIT Press.
99. Gholizadeh, H., Tajdin, A., & Javadian, N. (2020). A closed-loop supply chain robust optimization for disposable appliances. *Neural Computing Applications, 32*, 3967–3985.
100. Atabaki, M. S., Mohammadi, M., & Naderi, B. (2020). New robust optimization models for closed-loop supply chain of durable products: Towards a circular economy. *Computers Industrial Engineering, 146*, 106520.
101. Glover, F., & Woolsey, E. (1974). Converting the 0-1 polynomial programming problem to a 0-1 linear program. *Operations Research, 22*(1), 180–182.
102. Liu, Y., et al. (2018). Robust optimization for relief logistics planning under uncertainties in demand and transportation time. *Applied Mathematical Modelling, 55*, 262–280.
103. Sacks, J., et al. (1989). Design and analysis of computer experiments. *Statistical Science, 4*(4), 409–423.
104. Musavi, M. T., et al. (1992). On the training of radial basis function classifiers. *Neural Networks, 5*(4), 595–603.
105. Smola, A. J., & Schölkopf, B. (2004). A tutorial on support vector regression. *Statistics Computing, 14*, 199–222.

Index

W. Hu, *Design Optimization Under Uncertainty*,
https://doi.org/10.1007/978-3-031-49208-2

The manufacturer's authorised representative in the EU is Springer Nature Customer Service Centre GmbH, Europaplatz 3, 69115 Heidelberg, Germany. If you have any concerns regarding our products, please contact ProductSafety@springernature.com

Printed and bound by CPI Group (UK) Ltd, Croydon, CR0 4YY
28/09/2026
02224085-0004